Lecture Notes in Physics

Edited by H. Araki, Kyoto, J. Ehlers, München, K. Hepp, Zürich
R. Kippenhahn, München, D. Ruelle, Bures-sur-Yvette
H. A. Weidenmüller, Heidelberg, J. Wess, Karlsruhe and J. Zittartz, Köln
Managing Editor: W. Beiglböck

346

D. Lüst
S. Theisen

Lectures on String Theory

Springer-Verlag Berlin Heidelberg GmbH

Authors

Dieter Lüst
Stefan Theisen
CERN, CH-1211 Geneva 23, Switzerland

ISBN 978-3-662-13743-7   ISBN 978-3-540-46857-8 (eBook)
DOI 10.1007/978-3-540-46857-8

Originally published by Springer-Verlag Berlin Heidelberg New York in 1989

2153/3140-543210 – Printed on acid-free paper

Preface

These Lectures on String Theory are an extended version of
lectures we gave at the Max-Planck-Institut für Physik und Astro-
physik in Munich in fall and winter 1987/88. They were meant to
be an introduction to the subject and this is also the intention of
these notes.

We have not attempted to give a complete list of references. Instead, at the end of each chapter we give a short list of the original
papers and some reviews we found helpful and were familiar with.
An indispensable general reference is the book by Green, Schwarz
and Witten: *Superstring Theory,* 2 Vols., Cambridge University
Press, 1987. It also contains a more complete set of references,
in particular for our chapters 2, 3, 5 and 7 – 9. Early reviews are
collected in *Dual Resonance Theory*, ed. M. Jacob, Physics Reports
Reprints Vol. I, North Holland, 1974.

We would like to thank our collaborators S. Ferrara, I.-G. Koh,
J. Lauer, W. Lerche, B. Schellekens and G. Zoupanos, from whom
we have learned much of the subject presented here. We also ac-
knowledge helpful discussions with many other people and thank
in particular L. Alvarez-Gaumé, W. Buchmüller, L. Castellani, P.
Forgács, F. Gieres, G. Horowitz, P. Howe, L. Ibañez, T. Jacobsen, J.
Kubo, H. Nicolai, H.-P. Nilles, R. Rohm, M. Schmidt, J. Schnittger,
J. Schwarz, A. Shapere, M. Srednicki, R. Stora, N. Warner, P. West,
A. Wipf and R. Woodard. We thank P. Breitenlohner for supply-
ing his powerful TEX macro package and J. Paxon for her help in
preparing the manuscript.

CERN, Geneva, 1989
Dieter Lüst
Stefan Theisen

Table of Contents

Chapter 1

Introduction

String theory is currently one of the main activities among high energy theorists. It began at the end of the 1960's as an attempt to explain the spectrum of hadrons and their interactions. It was however discarded as a theory of strong interactions, a development which was supported by the rapid success of quantum chromodynamics. One problem was the existence of a critical dimension, which is 26 for the bosonic string and 10 for the fermionic string. Another obstacle in the interpretation of string theory as the theory of strong interactions was the existence of a massless spin two particle which is not present in the hadronic world.

In 1974 Scherk and Schwarz suggested to turn the existence of the massless spin two particle into an advantage for string theory by interpreting this particle as the graviton, the field quantum of gravitation. This implies that the string tension has to be related to the characteristic mass scale of gravity, namely the Planck mass $M_P = \sqrt{\hbar c/G} \simeq 10^{19} \mathrm{GeV}$. They also recognized that at low energies this graviton interacts according to the covariance laws of general relativity. In this way string theory could, at least in principle, achieve a unification of gravitation with all the other interactions in a quantized theory.

At that time it was only known how to incorporate non-abelian gauge symmetries in open string theories. Moreover, any open string theory with local interactions which consist of splitting and joining of strings automatically also contains closed strings with the massless spin two state in its spectrum. These theories are however plagued by gravitational and gauge anomalies which were believed to be fatal.

The renewed interest in string theory started again in 1984 when Green and Schwarz showed that the open superstring is anomaly free if and only if the gauge group is $SO(32)$. In addition they found that the ten-dimensional supersymmetric Einstein-Yang-Mills field theory is anomaly free for the gauge group $SO(32)$ and also for the phenomenologically more interesting group $E_8 \times E_8$, which is however excluded in the open string theory. This puzzle was resolved soon after by Gross, Harvey, Martinec and Rohm with the formulation of the heterotic string. It is a theory of closed strings only and represents the most economical way of incorporating both gravitational and gauge interactions. The allowed gauge symmetries $E_8 \times E_8$ or $SO(32)$ arise in a way different from the open string due to the incorporation of the so-called Kac-Moody algebras, which are infinite-dimensional extensions of ordinary Lie algebras.

On the other hand, the heterotic string is formulated as a ten-dimensional theory and obviously fails to reproduce an important experimentally established fact, namely that we live in four-dimensional (almost) flat Minkowski space-time.

The first approach to obtain four-dimensional string theories was along the old ideas of Kaluza and Klein, namely to consider compactifications of the ten-dimensional heterotic string theory. In this case, four string coordinates are uncompactified, whereas the remaining six are curled up and describe a tiny compact space whose size is of the order of the Planck length. The internal space can however not be arbitrarily chosen; the requirement of preserving conformal invariance puts severe constraints on it. Analyzing these constraints it turns out that the internal six-dimensional space must have vanishing Ricci-curvature. Examples are tori or the so-called Calabi-Yau manifolds.

More recently, it was discovered how one can construct (heterotic) string theories directly in four dimensions without ever referring to any compactification scheme. This opened a wide range of possibilities to obtain consistent string theories in four dimensions. In the most general four-dimensional string theories the part which refers to the extra dimensions (above four), which is needed because of conformal invariance, is replaced by a general conformal field theory. This internal conformal field theory has to obey some additional consistency requirements (like modular invariance, as we will discuss in some detail), but does however not need to admit an interpretation as a compact six-dimensional space. Unfortunately there exists a huge number of consistent internal conformal field theories, destroying the once celebrated uniqueness of string theory. In addition there exists so far no compelling principle which determines the number of space-time dimensions to be four. All dimensions below ten seem to be on an equal footing. However, the uniqueness in string theory could still be true in the sense that all different models are just different ground states, i.e. different classical solutions of an unique second quantized string theory. Then one specific string vacuum with a specific (hopefully correct) choice of gauge group and number (hopefully four) of flat space-time dimensions could be singled out by an underlying dynamical principle. At the moment, this is however wishful thinking and all ideas in this direction must be considered as pure speculations. Nevertheless, many of the four-dimensional heterotic string models exhibit promising aspects for phenomenology. However, within the context of string theory the word phenomenology should not be taken too seriously. At present one should only expect an explanation of generic features of the observed world, such as the presence of chiral fermions, the number of generations etc.

These lecture notes are intended to provide some of the tools which are necessary for the construction of four-dimensional (heterotic) string theories. Our main emphasis is on the relation to conformal field theory. One of the constructions of four-dimensional heterotic strings, namely the covariant lattice construction, will be discussed in detail. For an outline of the topics covered we refer to the table of contents. The selection we made was dictated by limitations of space and time and by our preferences.

Chapter 2

The Classical Bosonic String

Even though we will eventually be interested in understanding string theory at the quantum level and want to be able to discuss the interaction of several strings, it will turn out to be useful to start two steps back and treat the free classical string. Doing this we will set up the Lagrangian formalism which is essential for the path-integral quantization treated in Chapter 3 and solve the classical equations of motion of a single free string. These solutions will be used for the quantization in terms of operators, which we will discuss in detail in the next chapter.

2.1 The relativistic particle

Before treating the relativistic string, we will, as a warm up exercise, first study the free relativistic particle of mass m moving in a d-dimensional Minkowski space. Its action is simply the length of its world-line[1]

$$S = -m \int_{s_0}^{s_1} \mathrm{d}s = -m \int_{\tau_0}^{\tau_1} \mathrm{d}\tau \Big[-\frac{\mathrm{d}x^\mu}{\mathrm{d}\tau} \frac{\mathrm{d}x^\nu}{\mathrm{d}\tau} \eta_{\mu\nu} \Big]^{1/2}, \qquad (2.1)$$

where τ is an arbitrary parametrization of the world-line, whose embedding in d-dimensional Minkowski space is described by real functions $x^\mu(\tau)$, $\mu = 1, \ldots, d$ (we use the metric $\eta_{\mu\nu} = \mathrm{diag}(- + \cdots +))$. The action eq.(2.1) is invariant under τ-reparametrizations. Under infinitesimal reparametrizations x^μ transforms like

---

[1] It is easy to generalize the action to the case of a particle moving in a curved background by simply replacing the Minkowski metric $\eta_{\mu\nu}$ by a general metric $g_{\mu\nu}$.

$$\delta x^{\mu}(\tau) = \xi(\tau)\partial_{\tau}x^{\mu}(\tau)\,. \tag{2.2}$$

Invariance holds as long as $\xi(\tau_0) = \xi(\tau_1) = 0$. The momentum conjugate to $x^{\mu}(\tau)$ is

$$p^{\mu} = \frac{\partial L}{\partial \dot{x}^{\mu}} = m\frac{\dot{x}^{\mu}}{\sqrt{-\dot{x}^2}}\,, \tag{2.3}$$

where the dot denotes derivative with respect to τ and $\dot{x}^2 = \eta_{\mu\nu}\dot{x}^{\mu}\dot{x}^{\nu}$. Eq.(2.3) immediately leads to the following constraint equation

$$\phi \equiv p^2 + m^2 = 0. \tag{2.4}$$

Constraints which, as the one above, follow from the definition of the conjugate momenta without the use of the equations of motion, are called primary constraints. Their number is equal to the number of zero eigenvalues of the matrix $\frac{\partial p_{\mu}}{\partial \dot{x}^{\nu}} = \frac{\partial^2 L}{\partial \dot{x}^{\nu}\partial \dot{x}^{\mu}}$, which, in the case of the free relativistic particle, is just one, the corresponding eigenvector being $\dot{x}^{\mu}$. We note that the absence of zero eigenvalues is necessary (via the inverse function theorem) to express the "velocities" $\dot{x}^{\mu}$ uniquely in terms of the "momenta" and "coordinates", p^{μ} and x^{μ}. Systems where the rank of $\frac{\partial^2 L}{\partial \dot{x}^{\mu}\partial \dot{x}^{\nu}}$ is not maximal, thus implying the existence of primary constraints, are called singular. For these systems the τ-evolution is governed by the Hamiltonian $H = H_{\mathrm{can}} + \sum c_n \phi_n$, where H_{can} is the canonical Hamiltonian, the ϕ_n an irreducible set of primary constraints and the c_n constants in the coordinates and momenta. This is so since the canonical Hamiltonian is well defined only on the submanifold of phase space defined by the primary constraints and can be arbitrarily extended off that manifold. For the free relativistic particle we find that $H_{\mathrm{can}} = \frac{\partial L}{\partial \dot{x}^{\mu}}\dot{x}^{\mu} - L$ vanishes identically and the dynamics is completely determined by the constraint, eq.(2.4). The condition $H_{\mathrm{can}} = 0$ implies the existence of a zero eigenvalue of $\frac{\partial^2 L}{\partial \dot{x}^{\mu}\partial \dot{x}^{\nu}} = \frac{\partial}{\partial \dot{x}^{\nu}}H_{\mathrm{can}}$. This is always the case for systems with "time"-reparametrization invariance and follows from the fact that the "time" evolution of an arbitrary function $f(x,p)$, given by $\frac{df}{d\tau} = \frac{\partial f}{\partial \tau} + \{f, H\}_{P.B.}$, should also be valid for $\tilde{\tau} = \tilde{\tau}(\tau)$ on the constrained phase-space; here $\{\,,\,\}_{P.B.}$ is the usual Poisson bracket defined by $\{f,g\}_{P.B.} = \left(\frac{\partial f}{\partial x}\frac{\partial g}{\partial p} - \frac{\partial f}{\partial p}\frac{\partial g}{\partial x}\right)$. From this we also see that a particular choice of the constants c_n corresponds to a particular gauge choice which, for the relativistic particle, means a choice for the "time" variable τ. We write

$$H = \frac{N}{2m}(p^2 + m^2) \tag{2.5}$$

and find that

$$\frac{\mathrm{d}x^\mu}{\mathrm{d}\tau} = \{x^\mu, H\}_{P.B.} = \frac{N}{m}p^\mu = \frac{N\dot{x}^\mu}{\sqrt{-\dot{x}^2}}, \tag{2.6}$$

from which follows that $\dot{x}^2 = -N^2$. Thus, the choice $N = 1$ corresponds to choosing as the parameter τ the proper time.

At this point it is appropriate to introduce the concept of first and second class constraints. If $\{\phi_k\}$ is the collection of all constraints and if $\{\phi_a, \phi_k\}_{P.B.} = 0$, $\forall k$ upon application of the constraints, we say that ϕ_a is first class. Otherwise it is called second class. First class constraints are associated with gauge conditions.

For the relativistic particle the constraint given in eq.(2.4) is trivially first class and reflects τ reparametrization invariance.

Classically we can describe the free relativistic particle by an alternative action which has two advantages over eq.(2.1): (i) it does not contain a square root and (ii) it allows the generalization to the massless case. This is achieved by introducing an auxiliary variable $e(\tau)$, which can be viewed as an einbein on the world-line. The action then becomes

$$S = \frac{1}{2}\int_{\tau_0}^{\tau_1}\mathrm{d}\tau\, e\left(e^{-2}\left(\frac{\mathrm{d}x^\mu}{\mathrm{d}\tau}\right)^2 - m^2\right). \tag{2.7}$$

We derive the equations of motion

$$\frac{\delta S}{\delta e} = 0 \;\Rightarrow\; \dot{x}^2 + e^2 m^2 = 0\,,$$

$$\frac{\delta S}{\delta x^\mu} = 0 \;\Rightarrow\; \frac{\mathrm{d}}{\mathrm{d}\tau}(e^{-1}\dot{x}^\mu) = 0\,. \tag{2.8}$$

Since the equation of motion for e is purely algebraic, we can solve it for e and substitute it back into the action eq.(2.7) to obtain the action eq.(2.1), thus showing their classical equivalence.[2] We note that since $\frac{\partial^2 L}{\partial \dot{x}^\mu \partial \dot{x}^\nu} = e^{-1}\eta^{\mu\nu}$ has maximal rank we now do not have primary constraints. The constraint equation $p^2 + m^2 = 0$ does not follow from the definition of the conjugate momenta alone; in addition one has to use the equations of motion. Constraints of this kind are called secondary constraints. But since it is first class it implies a symmetry. Indeed, the action eq.(2.7) is invariant under τ-reparametrizations under which

$$\delta x^\mu = \xi \partial_\tau x^\mu$$
$$\delta e = \delta_\tau(\xi e)$$

(2.9)

and we can use τ reparametrization invariance to go to the gauge $e = 1/m$. If we then naively used the gauge fixed action to find the equations of motion we would find $\ddot{x}^\mu = 0$, whose solutions are all straight lines in Minkowski space, which we know to be incorrect. This simply means that we cannot use the reparametrization freedom to fix e and then forget about it. We rather have to use the gauge fixed equation of motion for e, which reads $T \equiv \dot{x}^2 + 1 = 0$, as a constraint. This excludes all time-like and light-like lines and identifies the parameter τ in this particular gauge as the proper time. (In the massless case we set $e = 1$ and have to supplement the equation $\ddot{x}^\mu = 0$ by the constraint $T \equiv \dot{x}^2 = 0$, which leaves only the light-like world-lines.) Note that the equation of motion, $\ddot{x}^\mu = 0$, does not imply $T = 0$, but it implies that $\frac{dT}{d\tau} = 0$, i.e. $T = 0$ is a constraint on the initial data and is conserved.

---

[2] It is important to point out that classical equivalence does not necessarily imply quantum equivalence.

2.2 The Nambu-Goto action

Let us now turn to the string. The generalization of eq.(2.1) to a one-dimensional object is to take as its action the area of the world-sheet swept out by the string, i.e.

$$
\begin{aligned}
S_{NG} &= -T \int \mathrm{d}A \\
&= -T \int \mathrm{d}^2\sigma \left[-\det \frac{\partial X^\mu}{\partial \sigma^\alpha} \frac{\partial X^\nu}{\partial \sigma^\beta} \eta_{\mu\nu} \right]^{1/2} \\
&= -T \int \mathrm{d}^2\sigma \left[(\dot{X} \cdot X')^2 - \dot{X}^2 X'^2 \right]^{1/2} \\
&\equiv -T \int \mathrm{d}^2\sigma \sqrt{-\Gamma},
\end{aligned}
\tag{2.10}
$$

where $\sigma^\alpha = (\sigma, \tau)$ are the two parameters describing the world-sheet; the dot denotes derivative with respect to τ and the prime derivative with respect to σ. $X^\mu(\sigma, \tau)$, $\mu = 1, \ldots, d$, are maps of the world-sheet into d-dimensional Minkowski space and T a constant of mass dimension two, the string tension. $\Gamma_{\alpha\beta} = \frac{\partial X^\mu}{\partial \sigma^\alpha} \frac{\partial X^\nu}{\partial \sigma^\beta} \eta_{\mu\nu}$ is the metric on the world-sheet inherited from the underlying d-dimensional Minkowski space the string is moving in and $\Gamma < 0$ is its determinant. The requirement that Γ be negative means that at each point the world-sheet has one time-like or light-like and one space-like tangent vector. This is necessary for causal propagation of the string. Requiring $\dot{X}^\mu + \lambda X'^\mu$ to be time-like and space-like when λ is varied gives $\Gamma < 0$. The action eq.(2.10) was first considered by Nambu [1] and Goto [2], hence the subscript NG. Being the area of the world-sheet, the action is invariant under reparametrizations:

$$
\delta x^\mu(\sigma, \tau) = \xi^\alpha \partial_\alpha x^\mu(\sigma, \tau)
\tag{2.11}
$$

as long as $\xi^\alpha = 0$ on the boundary. X^μ transforms as a scalar under reparametrizations of the world-sheet.[3]

---

[3] A general tensor density of rank, say $(1,1)$, and weight w transforms under

We will choose the parameter σ such that $0 \leq \sigma \leq \bar{\sigma}$ where $\bar{\sigma} = \pi$ for open strings and $\bar{\sigma} = 2\pi$ for closed strings. To derive the equations of motion for the string we vary its trajectory, keeping initial and final positions fixed; i.e. $\delta X^{\mu}(\tau_0) = 0 = \delta X^{\mu}(\tau_1)$ but at the ends of the open string $\delta X^{\mu}(\sigma, \tau)$ is arbitrary. We then get

$$\frac{\partial}{\partial \tau} \frac{\partial L}{\partial \dot{X}^{\mu}} + \frac{\partial}{\partial \sigma} \frac{\partial L}{\partial X'^{\mu}} = 0 \qquad (2.12)$$

together with the edge condition for the open string

$$\frac{\partial L}{\partial X'^{\mu}} = 0 \qquad \qquad \text{at } \sigma = 0, \pi \qquad (2.13)$$

and the periodicity condition for the closed string

$$X^{\mu}(\sigma + 2\pi) = X^{\mu}(\sigma). \qquad (2.14)$$

Physically the edge condition means that no momentum flows off the ends of an open string. This will become clear below. Due to the square root in the action the equations of motion are rather complicated; the canonical momentum is

$$\Pi^{\mu} = \frac{\partial L}{\partial \dot{X}_{\mu}} = -T \frac{(\dot{X} \cdot X') X'^{\mu} - (X')^2 \dot{X}^{\mu}}{\left[(X' \cdot \dot{X})^2 - (\dot{X})^2 (X')^2 \right]^{1/2}}. \qquad (2.15)$$

The matrix $\dfrac{\partial^2 L}{\partial \dot{X}^{\nu} \partial \dot{X}^{\mu}} = \dfrac{\partial}{\partial \dot{X}^{\nu}} \Pi_{\mu}$ has (for each value of σ) two zero eigenvalues with the corresponding eigenvectors $\dot{X}^{\mu}$ and X'^{μ}. The resulting primary constraints are

---

reparametrizations $\sigma^{\alpha} \to \bar{\sigma}^{\alpha}(\sigma, \tau)$ of the world-sheet as

$$t_{\alpha}^{\beta}(\sigma, \tau) \to \tilde{t}_{\alpha}^{\beta}(\sigma, \tau) = \left| \frac{\partial(\sigma, \tau)}{\partial(\bar{\sigma}, \bar{\tau})} \right|^{w} \frac{\partial \bar{\sigma}^{\gamma}}{\partial \sigma^{\alpha}} \frac{\partial \sigma^{\beta}}{\partial \bar{\sigma}^{\delta}} t_{\gamma}^{\delta}(\bar{\sigma}, \bar{\tau}),$$

where the first factor is the Jacobian of the transformation. For infinitesimal transformations $\bar{\sigma}^{\alpha}(\sigma, \tau) = \sigma^{\alpha} + \xi^{\alpha}(\sigma, \tau)$, this gives

$$\delta t_{\alpha}^{\beta}(\sigma, \tau) = \tilde{t}_{\alpha}^{\beta}(\sigma, \tau) - t_{\alpha}^{\beta}(\sigma, \tau) = (\xi^{\gamma} \partial_{\gamma} - w \partial_{\gamma} \xi^{\gamma}) t_{\alpha}^{\beta} + t_{\gamma}^{\beta} \partial_{\alpha} \xi^{\gamma} - t_{\alpha}^{\delta} \partial_{\delta} \xi^{\beta}.$$

The generalization to tensors of arbitrary rank is obvious.

$$\Pi^\mu X'_\mu = 0 \tag{2.16}$$

and

$$\Pi^2 + T^2 X'^2 = 0. \tag{2.17}$$

They are called Virasoro constraints and play an important role in string theory, as we will see later. The canonical Hamiltonian, $H = \int_0^{\bar\sigma} \mathrm{d}\sigma(\dot{X} \cdot \Pi - L)$, is easily seen to vanish identically and hence the dynamics is completely determined by the constraints.

2.3 The Polyakov action and its symmetries

In the same way as we could express the action for the relativistic particle by introducing a metric on the world-line, we can introduce a metric $h_{\alpha\beta}(\sigma,\tau)$ on the world-sheet and write

$$
\begin{aligned}
S_P &= -\frac{T}{2} \int \mathrm{d}^2\sigma \sqrt{h}\, h^{\alpha\beta} \partial_\alpha X^\mu \partial_\beta X^\nu \eta_{\mu\nu} \\
&= -\frac{T}{2} \int \mathrm{d}^2\sigma \sqrt{h}\, h^{\alpha\beta} \Gamma_{\alpha\beta}
\end{aligned}
\tag{2.18}
$$

where $h = -\det h_{\alpha\beta}$. This form of the string action is the starting point of the path-integral quantization of Polyakov [3], hence the subscript P. The action is easy to generalize to a string moving in a curved background by replacing the Minkowski metric $\eta_{\mu\nu}$ by a general metric $g_{\mu\nu}(X)$. In this general form the action constitutes a non-trivial quantum field theory, a non-linear sigma model. We will always choose $g_{\mu\nu} = \eta_{\mu\nu}$ which can be considered as the first term of a perturbative expansion around a flat background. This is of course a severe limitation and a complete theory would determine its own background in which the string is propagating, much in the same way as in general relativity where the metric of space-time is determined by the matter content according to Einstein's equations.

We now define the two-dimensional energy-momentum tensor in the usual way as the response of the system to changes in the metric under which $\delta S = -T \int d^2\sigma \sqrt{h} T_{\alpha\beta} \delta h^{\alpha\beta}$; i.e.

$$T_{\alpha\beta} = -\frac{1}{T} \frac{1}{\sqrt{h}} \frac{\delta S}{\delta h^{\alpha\beta}}. \qquad (2.19)$$

Using $\delta h = -h_{\alpha\beta}(\delta h^{\alpha\beta})h$ we find

$$T_{\alpha\beta} = \frac{1}{2}\partial_\alpha X^\mu \partial_\beta X_\mu - \frac{1}{4}h_{\alpha\beta}h^{\gamma\delta}\partial_\gamma X^\mu \partial_\delta X_\mu \qquad (2.20)$$

and the equations of motion are

$$T_{\alpha\beta} = 0 \qquad (2.21a)$$

$$\frac{1}{\sqrt{h}}\partial_\alpha(\sqrt{h}h^{\alpha\beta}\partial_\beta X^\mu) = 0 \qquad (2.21b)$$

with the appropriate boundary or periodicity conditions. Energy-momentum conservation, $\nabla^\alpha T_{\alpha\beta} = 0$, is also easily verified with the help of the equation of motion for X^μ. ∇_α is a covariant derivative with the usual Christoffel connection $\Gamma^\gamma_{\alpha\beta} = \frac{1}{2}h^{\gamma\delta}(\partial_\alpha h_{\beta\delta} + \partial_\beta h_{\alpha\delta} - \partial_\delta h_{\alpha\beta})$. From the vanishing of the energy-momentum tensor we derive $\det(\partial_\alpha X^\mu \partial_\beta X_\mu) = \frac{1}{4}h(h^{\gamma\delta}\partial_\gamma X_\mu \partial_\delta X^\mu)^2$ which, when inserted into S_P, shows the classical equivalence of the Polyakov and the Nambu-Goto action.

We can now check that the constraints, eqs.(2.16) and (2.17), which were primary in the Nambu-Goto formulation, follow here only if we use the equation of motion $T_{\alpha\beta} = 0$, i.e. they are secondary. This is the same situation we encountered in the case of the relativistic particle.

Before discussing the symmetries of the Polyakov action we want to point out that the two metrics on the world-sheet, namely the one inherited from the embedding space, $\Gamma_{\alpha\beta} = \partial_\alpha X^\mu \partial_\beta X^\nu \eta_{\mu\nu}$, entering the Nambu-Goto action and the intrinsic metric $h_{\alpha\beta}$, entering the Polyakov action, are a priori unrelated. The Polyakov action is *not* the area of the world-sheet measured with the intrinsic metric, which would simply be $\int d^2\sigma \sqrt{h}$ and could be added to S_P as a cosmological term (cf. below). But, for any real symmetric 2×2 matrix A we have the inequality $(\mathrm{tr}A)^2 \geq 4\det A$ with equality for $A \propto 1$. Choosing $A^\alpha{}_\beta = h^{\alpha\gamma}\Gamma_{\gamma\beta}$ it follows that $S_P \geq S_{NG}$. Equality holds if and only if $h_{\alpha\beta} \propto \Gamma_{\alpha\beta}$, i.e. if the two metrics are conformally related. This is the case if the equation of motion for $h_{\alpha\beta}$ is satisfied.

We can now ask whether there are other terms one could add to S_P. The only possibilities compatible with d-dimensional Poincaré invariance and power counting renormalizability of the two-dimensional theory are[4]

$$S_1 = \lambda_1 \int \mathrm{d}^2\sigma \sqrt{h} \qquad (2.22)$$

which is the cosmological term mentioned above, and

$$S_2 = \frac{\lambda_2}{4\pi} \int \mathrm{d}^2\sigma \sqrt{h}R \qquad (2.23)$$

where R is the curvature scalar for the metric $h_{\alpha\beta}$. S_2 is the two-dimensional Gauss-Bonnet term, i.e. the integrand is a total derivative and consequently does not contribute to the classical equations of motion. Inclusion of the cosmological term would lead to the equation of motion $T_{\alpha\beta} = -\frac{\lambda_1}{2T}h_{\alpha\beta}$, from which we conclude that $\lambda_1 h^{\alpha\beta}h_{\alpha\beta} = 0$. This is unacceptable unless $\lambda_1 = 0$. We will thus consider the action S_P, eq.(2.18), which is the action of a collection of d massless real scalar fields (X^μ) coupled to gravity ($h_{\alpha\beta}$) in two dimensions.

Let us now discuss the symmetries of the Polyakov action.

(i) global symmetries:

- Poincaré invariance:

---

[4] For the open string there are in fact further possible terms besides S_1 and S_2, which are defined on the boundary of the world-sheet. It turns out that they can also be discarded and we will not discuss them here.

$$\delta X^\mu = a^\mu{}_\nu X^\nu + b^\mu \quad (a_{\mu\nu} = -a_{\nu\mu})$$
$$\delta h_{\alpha\beta} = 0 \tag{2.24}$$

(ii) local symmetries:

- reparametrization invariance

$$\delta X^\mu = \xi^\alpha \partial_\alpha X^\mu$$

$$\delta h_{\alpha\beta} = \xi^\gamma \partial_\gamma h_{\alpha\beta} + \partial_\alpha \xi^\gamma h_{\gamma\beta} + \partial_\beta \xi^\gamma h_{\alpha\gamma}$$
$$= \nabla_\alpha \xi_\beta + \nabla_\beta \xi_\alpha \tag{2.25}$$

$$\delta \sqrt{h} = \partial_\alpha (\xi^\alpha \sqrt{h})$$

- Weyl rescaling

$$\delta h_{\alpha\beta} = 2\Lambda h_{\alpha\beta}$$
$$\delta X^\mu = 0 \tag{2.26}$$

Here ξ^α and Λ are arbitrary (infinitesimal) functions of (σ, τ) and $a_{\mu\nu}$ and b_μ are constants. From eq.(2.24) we see that X^μ is a Minkowski space vector whereas $h_{\alpha\beta}$ is a scalar. Under reparametrizations of the worldsheet, eq.(2.25), the X^μ are world-sheet scalars, $h_{\alpha\beta}$ a world-sheet tensor and $\sqrt{h}$ a scalar density of weight -1. The scale transformation of the world-sheet metric, eq.(2.26), is the infinitesimal version of $h_{\alpha\beta}(\sigma, \tau) \to \Omega^2(\sigma, \tau) h_{\alpha\beta}(\sigma, \tau)$ for $\Omega^2(\sigma, \tau) = e^{2\Lambda(\sigma, \tau)} \sim 1 + 2\Lambda(\sigma, \tau)$.

One immediate important consequence of Weyl invariance of the action is the tracelessness of the energy-momentum tensor:

$$h^{\alpha\beta} T_{\alpha\beta} = 0 \tag{2.27}$$

which is satisfied by the expression eq.(2.20) without invoking the equations of motion. It is not difficult to see that this has to be so. Consider an action which depends on a metric and a collection of fields ϕ_i which transform under Weyl rescaling as $h_{\alpha\beta} \to e^{2\Lambda} h_{\alpha\beta}$ and $\phi_i \to e^{d_i \Lambda} \phi_i$. If the action is scale invariant, i.e. if $S[e^{2\Lambda} h_{\alpha\beta}, e^{d_i \Lambda} \phi] = S[h_{\alpha\beta}, \phi]$, then

$$0 = \delta S = \int \mathrm{d}^2\sigma \left\{ -2 \frac{\delta S}{\delta h^{\alpha\beta}} h^{\alpha\beta} e^{-2\Lambda} + \sum_i d_i \frac{\delta S}{\delta \phi_i} \phi_i e^{d_i \Lambda} \right\} \delta\Lambda. \qquad (2.28)$$

If we now use the equations of motion for ϕ_i, tracelessness of the energy-momentum tensor is immediate. We note that it follows without the use of the equations of motion if and only if $d_i = 0$, $\forall i$. This is for instance the case for the Polyakov action of the bosonic string but will not be satisfied for the fermionic string action in Chapter 7.

The local invariances allow for a convenient gauge choice for the world-sheet metric $h_{\alpha\beta}$, called conformal (also orthonormal) gauge. Reparametrization invariance is used to choose coordinates such that locally $h_{\alpha\beta} = \Omega^2(\sigma, \tau)\eta_{\alpha\beta}$, with $\eta_{\alpha\beta}$ being the two-dimensional Minkowski metric defined by $\mathrm{d}s^2 = -\mathrm{d}\tau^2 + \mathrm{d}\sigma^2$. It is not hard to show that this can always be done. Indeed, for any two-dimensional Lorentzian metric $h_{\alpha\beta}$, consider two null vectors at each point. In this way we get two vector fields and their integral curves which we label by σ^+ and σ^-. Then, $\mathrm{d}s^2 = -\Omega^2 \mathrm{d}\sigma^+ \mathrm{d}\sigma^-$; $h_{++} = h_{--} = 0$ since the curves are null. Now let $\sigma^\pm = \tau \pm \sigma$ from which it follows that $\mathrm{d}s^2 = \Omega^2(-\mathrm{d}\tau^2 + \mathrm{d}\sigma^2)$. A choice of coordinate system in which the two dimensional Lorentzian metric is conformally flat, i.e. in which

$$\mathrm{d}s^2 = \Omega^2(-\mathrm{d}\tau^2 + \mathrm{d}\sigma^2) = -\Omega \mathrm{d}\sigma^+ \mathrm{d}\sigma^- \qquad (2.29)$$

is called a conformal gauge. The world-sheet coordinates $\sigma^\pm$ introduced above are called light-cone, isothermal or conformal coordinates. We can now use Weyl invariance to set $h_{\alpha\beta} = \eta_{\alpha\beta}$. The components of the metric are then $\eta_{+-} = \eta_{-+} = -\frac{1}{2}$, $\eta^{+-} = \eta^{-+} = -2$, $\eta_{++} = \eta_{--} = \eta^{++} = \eta^{--} = 0$. Also, $\partial_\pm = \frac{1}{2}(\partial_\tau \pm \partial_\sigma)$ and indices are raised and lowered according to $v^+ = -2v_-$ and $v^- = -2v_+$.

It is now important to note that reparametrizations satisfying $\nabla_\alpha \xi_\beta + \nabla_\beta \xi_\alpha \propto h_{\alpha\beta}$ can be compensated by a Weyl rescaling. Expressed in light-cone coordinates the conformal gauge preserving diffeomorphisms are those which satisfy $\partial_+ \xi^- = \partial_- \xi^+ = 0$, i.e. $\xi^\pm = \xi^\pm(\sigma^\pm)$.[5] (Here we have used that $\nabla_+ \xi_+ = h_{+-} \nabla_+ \xi^- = h_{+-} \partial_+ \xi^-$ since the only non-vanishing Christoffel symbols in conformal gauge are $\Gamma^+_{++} = 2\partial_+ \Lambda$ and $\Gamma^-_{--} = 2\partial_- \Lambda$.) Indeed, instead of $\sigma^\pm$ we could as well have chosen $\tilde{\sigma}^\pm = \tilde{\sigma}^\pm(\sigma^\pm)$, or, in infinitesimal form, $\tilde{\sigma}^\pm = \sigma^\pm + \xi^\pm(\sigma^\pm)$. Note that the transformation $\sigma^\pm \to \tilde{\sigma}^\pm(\sigma^\pm)$ corresponds to $\binom{\tau}{\sigma} \to \binom{\tilde{\tau}}{\tilde{\sigma}} = \frac{1}{2}[\tilde{\sigma}^+(\tau+\sigma) \pm \tilde{\sigma}^-(\tau-\sigma)]$; i.e. any $\tilde{\tau}$ satisfying the two-dimensional wave equation will do the job. (This will allow us to go to the so-called light cone gauge (cf. Chapter 3 below).)

It is easy to see that the conformal gauge is unique to two dimensions. In $d > 0$ dimensions a metric $h_{\alpha\beta}$, being symmetric, has $\frac{1}{2}d(d+1)$ independent components. Reparametrization invariance allows to fix d of them, leaving $\frac{1}{2}d(d-1)$ components. In two dimensions this is enough to go to conformal gauge. We then still have an extra local symmetry, namely Weyl transformations, which allows us to eliminate the remaining metric component. This also shows that gravity in two dimensions is trivial in the sense that the graviton can be gauged away completely. For $d > 2$, Weyl invariance, even if present as for instance in conformal gravity, won't suffice.[6]

The argument given above that conformal gauge is always possible was a local statement. We will now set up a global criterion and consider the general case with gauge condition

$$h_{\alpha\beta} = e^{2\phi} \hat{h}_{\alpha\beta}. \tag{2.30}$$

In conformal gauge $\hat{h}_{\alpha\beta} = \eta_{\alpha\beta}$. Under reparametrizations and Weyl rescaling the metric changes as

---

[5] When we go to Euclidean coordinates on the world-sheet this corresponds to the conformal transformations. More about this later.

[6] Note that the action for the relativistic particle was not Weyl invariant; there reparametrization invariance was sufficient to eliminate the one metric degree of freedom.

ing the metric changes as

$$\delta h_{\alpha\beta} = \nabla_\alpha \xi_\beta + \nabla_\beta \xi_\alpha + 2\Lambda h_{\alpha\beta}$$
$$= (P\xi)_{\alpha\beta} + 2\tilde\Lambda h_{\alpha\beta} \qquad (2.31)$$

where the operator P maps vectors into symmetric traceless tensors according to

$$(P\xi)_{\alpha\beta} = \nabla_\alpha \xi_\beta + \nabla_\beta \xi_\alpha - (\nabla_\gamma \xi^\gamma) h_{\alpha\beta}, \qquad (2.32)$$

and we have defined $2\tilde\Lambda = 2\Lambda + \nabla_\gamma \xi^\gamma$. The decomposition into symmetric traceless and trace part is orthogonal with respect to the inner product $(\delta h^{(1)}|\delta h^{(2)}) = \int d^2\sigma \sqrt{h} h^{\alpha\gamma} h^{\beta\delta} \delta h^{(1)}_{\alpha\beta} \delta h^{(2)}_{\gamma\delta}$. The trace part of $\delta h_{\alpha\beta}$ can always be cancelled by a suitable choice of Λ. It then follows that for the gauge eq.(2.30) to be possible globally, there must exist a globally defined vector field ξ^α such that

$$(P\xi)_{\alpha\beta} = t_{\alpha\beta} \qquad (2.33)$$

for arbitrary symmetric traceless $t_{\alpha\beta}$. If the operator P has zero modes, i.e. if there exist vector fields ξ_0 such that $P\xi_0 = 0$, then for any solution ξ we also have the solution $\xi + \xi_0$. In this case the gauge fixing is not complete and those reparametrizations which can be absorbed by a Weyl rescaling are still allowed, as we have already seen above. The equation

$$(P\xi)_{\alpha\beta} = \nabla_\alpha \xi_\beta + \nabla_\beta \xi_\alpha - h_{\alpha\beta} \nabla_\gamma \xi^\gamma = 0$$

is the conformal Killing equation and its solutions are called conformal Killing vectors. The adjoint of P, $P^\dagger$, maps traceless symmetric tensors to vectors via

$$(P^\dagger t)_\alpha = -2\nabla^\beta t_{\alpha\beta}. \qquad (2.34)$$

Now, zero modes of $P^\dagger$ correspond to symmetric traceless tensors which cannot be written as $(P\xi)_{\alpha\beta}$ for any vector field ξ. Indeed, if $(P^\dagger t_0)_\alpha = 0$, then for all ξ^α, $(\xi, P^\dagger t_0) = -(P\xi, t_0) = 0$. This means that zero modes of $P^\dagger$ correspond to metric deformations which cannot be absorbed by reparametrizations and Weyl rescaling. If they do not exist, the gauge is possible globally.

This applies in particular to the conformal gauge; here the condition is that the equations $\partial_- t_{++} = 0$ and $\partial_+ t_{--} = 0$ have no globally defined solutions. We will further discuss the solutions of these equations in Chapter 6.

In conformal gauge the Polyakov action simplifies to

$$
\begin{aligned}
S_P &= -\frac{T}{2} \int d^2\sigma \, \eta^{\alpha\beta} \partial_\alpha X^\mu \partial_\beta X_\mu \\
&= \frac{T}{2} \int d^2\sigma \, (\dot{X}^2 - X'^2) \\
&= 2T \int d^2\sigma \, \partial_+ X \partial_- X \ .
\end{aligned}
\tag{2.35}
$$

Varying with respect to X^μ such that $\delta X^\mu(\tau_0) = 0 = \delta X^\mu(\tau_1)$, $\delta X^\mu(\sigma = 0, \bar{\sigma})$ arbitrary (open string) and $\delta X^\mu(\sigma + 2\pi) = \delta X^\mu(\sigma)$ (closed string), we obtain

$$
\delta S_P = T \int d^2\sigma \, \delta X^\mu (\partial_\sigma^2 - \partial_\tau^2) X_\mu - T \int_{\tau_0}^{\tau_1} d\tau \, X'_\mu \delta X^\mu \Big|_{\sigma=0}^{\sigma=\bar{\sigma}}
\tag{2.36}
$$

where the surface term is absent for the closed string. We then get the following equations of motion:

$$
(\partial_\tau^2 - \partial_\sigma^2) X^\mu = 4 \partial_+ \partial_- X^\mu = 0
\tag{2.37a}
$$

with

$$
\begin{aligned}
X^\mu(\sigma + 2\pi) &= X^\mu(\sigma), &\quad \text{(closed string)} \\
X'_\mu \big|_{\sigma=0,\pi} &= 0 \ . &\quad \text{(open string)}
\end{aligned}
\tag{2.37b}
$$

In both cases the equation of motion is the two-dimensional massless wave equation with the general solution

$$
X^\mu(\sigma, \tau) = X_R^\mu(\sigma^-) + X_L^\mu(\sigma^+)
\tag{2.38}
$$

where $X_{R,L}^\mu$ are arbitrary functions of their respective arguments, subject only to periodicity or boundary conditions. They describe the "right"- and "left"-moving modes of the string respectively. In the case of the closed string the left- and right-moving components are completely independent

18

for the unconstrained system, an observation which is crucial for the formulation of the heterotic string. This is however not the case for the open string where the boundary condition mixes left- with right-movers through reflection at the ends of the string.

We still have to impose on the solutions of the equations of motion the constraints resulting from the gauge fixed equations of motion for the metric: we have to require that the energy momentum tensor vanishes; i.e.

$$T_{01} = T_{10} = \frac{1}{2}(\dot{X} \cdot X') = 0 \tag{2.39a}$$

$$T_{00} = T_{11} = \frac{1}{4}(\dot{X}^2 + X'^2) = 0 \tag{2.39b}$$

which can be alternatively expressed as

$$\frac{1}{2}(\dot{X} \pm X')^2 = 0. \tag{2.40}$$

In light-cone coordinates the constraints become

$$T_{++} = \frac{1}{2}\partial_+ X \cdot \partial_+ X = 0 \tag{2.41a}$$

$$T_{--} = \frac{1}{2}\partial_- X \cdot \partial_- X = 0 \tag{2.41b}$$

$$T_{+-} = T_{-+} = 0 \tag{2.41c}$$

where $T_{++} = \frac{1}{2}(T_{00} + T_{01})$, $T_{--} = \frac{1}{2}(T_{00} - T_{01})$; eq.(2.41c) expresses the tracelessness of the energy momentum tensor. In terms of the left- and right-movers the constraints eq.(2.41a,b) become $\dot{X}_R^2 = \dot{X}_L^2 = 0$. Energy momentum conservation, i.e. $\nabla^\alpha T_{\alpha\beta} = 0$, becomes

$$\partial_- T_{++} + \partial_+ T_{-+} = 0$$
$$\partial_+ T_{--} + \partial_- T_{+-} = 0 \tag{2.42}$$

which, using eq.(2.41c) simply states that

$$\partial_- T_{++} = 0$$
$$\partial_+ T_{--} = 0 \tag{2.43}$$

i.e.

$$T_{++} = T_{++}(\sigma^+) \qquad \text{and} \qquad T_{--} = T_{--}(\sigma^-). \qquad (2.44)$$

From eq.(2.39b), together with the condition that $\partial_\sigma X^\mu = 0$ at the end of an open string, we learn that the ends of an open string move at the speed of light.

The conservation equations (2.43) imply the existence of an infinite number of conserved charges. In fact, for any function $f(\sigma^+)$ we have $\partial_-(f(\sigma^+)T_{++}) = 0$ and the corresponding conserved charges are

$$L_f = 2T \int_0^{\bar\sigma} d\sigma \; f(\sigma^+)T_{++}(\sigma^+) \qquad (2.45)$$

and likewise for the right-movers.

The Hamiltonian for the string in conformal gauge is

$$\begin{aligned}
H &= \int_0^{\bar\sigma} d\sigma(\dot X \cdot \Pi - L) \\
&= \frac{T}{2} \int_0^{\bar\sigma} d\sigma(\dot X^2 + X'^2) \\
&= T \int_0^{\bar\sigma} d\sigma\big((\partial_+ X)^2 + (\partial_- X)^2\big)
\end{aligned} \qquad (2.46)$$

where, as before, the canonical momentum is $\Pi^\mu = \partial L/\partial \dot X_\mu = T\dot X^\mu$. We note that the Hamiltonian is just one of the constraints. This was to be expected from our discussion of constrained systems in the context of the relativistic particle. Indeed, we saw that the canonical Hamiltonian derived from the Nambu-Goto action vanishes identically and the τ-evolution is completely governed by the constraints, i.e.

$$H = \int_0^{\bar\sigma} d\sigma\big\{N_1(\sigma,\tau)\Pi \cdot X' + N_2(\sigma,\tau)(\Pi^2 + T^2 X'^2)\big\} \qquad (2.47)$$

where N_1 and N_2 are arbitrary functions of σ and τ. Using the basic equal τ Poisson brackets

$$\{X^\mu(\sigma,\tau), X^\nu(\sigma',\tau)\}_{P.B.} = \{\Pi^\mu(\sigma,\tau), \Pi^\nu(\sigma',\tau)\}_{P.B.} = 0$$

$$\{X^\mu(\sigma,\tau), \Pi^\nu(\sigma',\tau)\}_{P.B.} = \eta^{\mu\nu}\delta(\sigma - \sigma') \qquad (2.48)$$

we find

$$\dot{X}^\mu = N_1 X'^\mu + 2N_2 \Pi^\mu \tag{2.49}$$

and

$$\dot{\Pi}^\mu = \partial_\sigma(N_1 \Pi^\mu + 2T^2 N_2 X'^\mu). \tag{2.50}$$

If we chose $N_1 = 0$ and $N_2 = \frac{1}{2T}$, eqs.(2.49) and (2.50) lead to the equation of motion $(\partial_\sigma^2 - \partial_\tau^2)X^\mu = 0$ which we have obtained previously from the action in conformal gauge. This means that choosing $N_1 = 0$ and $N_2 = \frac{1}{2T}$ corresponds to fixing the conformal gauge. With this choice for the functions N_1 and N_2 we also get the Hamiltonian eq.(2.46).

In conformal gauge the Poisson brackets are

$$\{X^\mu(\sigma,\tau), X^\nu(\sigma',\tau)\}_{P.B.} = \{\dot{X}^\mu(\sigma,\tau), \dot{X}^\nu(\sigma',\tau)\}_{P.B.} = 0$$

$$\{X^\mu(\sigma,\tau), \dot{X}^\nu(\sigma',\tau)\}_{P.B.} = \frac{1}{T}\eta^{\mu\nu}\delta(\sigma - \sigma'). \tag{2.51}$$

Using them and the explicit expression for T_{++} we can readily show that the charges L_f of eq.(2.45) generate transformations $\sigma^+ \to \sigma^+ + f(\sigma^+)$, i.e. those reparametrizations which do not lead out of conformal gauge. Indeed,

$$\{L_f, X(\sigma)\}_{P.B.} = -f(\sigma^+)\partial_+ X(\sigma). \tag{2.52}$$

So far we have only discussed issues connected with world-sheet symmetries. However, invariance under d-dimensional global Poincaré transformations, eq.(2.24), leads, via the Noether theorem, to two conserved currents; invariance under translations gives the energy momentum current

$$P_\mu^\alpha = -T\sqrt{h}h^{\alpha\beta}\partial_\beta X_\mu, \tag{2.53}$$

whereas invariance under Lorentz rotations gives the angular momentum current

$$J_{\mu\nu}^\alpha = -T\sqrt{h}h^{\alpha\beta}(X_\mu\partial_\beta X_\nu - X_\nu\partial_\beta X_\mu) = X_\mu P_\nu^\alpha - X_\nu P_\mu^\alpha. \tag{2.54}$$

Using the equations of motion it is easy to check the conservation of P_μ^α and $J_{\mu\nu}^\alpha$. The total conserved charges (momentum and angular momentum) are obtained by integrating the currents over a space-like section of the world-sheet, say $\tau = 0$. Then the total momentum in conformal gauge is

$$P_\mu = \int_0^{\bar{\sigma}} d\sigma \; P_\mu^\tau = T \int_0^{\bar{\sigma}} d\sigma \; \partial_\tau X_\mu(\sigma) \tag{2.55}$$

and the total angular momentum is

$$J_{\mu\nu} = \int_0^{\bar{\sigma}} d\sigma \, J_{\mu\nu}^\tau = T \int_0^{\bar{\sigma}} d\sigma (X_\mu \partial_\tau X_\nu - X_\nu \partial_\tau X_\mu). \tag{2.56}$$

It is easy to see that P_μ and $J_{\mu\nu}$ are conserved. Indeed, $\frac{\partial P_\mu}{\partial \tau} = \int_0^{\bar{\sigma}} d\sigma \, \partial_\tau^2 X_\mu = \int_0^{\bar{\sigma}} d\sigma \, \partial_\sigma^2 X_\mu = \partial_\sigma X_\mu(\sigma = \bar{\sigma}) - \partial_\sigma X_\mu(\sigma = 0)$ which vanishes for the closed string by periodicity and for the open string because of the boundary condition. Hence our earlier statement that the open string boundary conditions have the physical interpretation that no momentum flows off the ends of the string. Conservation of the total angular momentum is also easy to check.

With the help of the Poisson brackets eq.(2.51) it is straightforward to verify that P^μ and $J^{\mu\nu}$ generate the Poincaré algebra:

$$
\begin{aligned}
\{P^\mu, P^\nu\}_{P.B.} &= 0\,, \\
\{P^\mu, J^{\rho\sigma}\}_{P.B.} &= \eta^{\mu\sigma} P^\rho - \eta^{\mu\rho} P^\sigma\,, \\
\{J^{\mu\nu}, J^{\rho\sigma}\}_{P.B.} &= \eta^{\mu\rho} J^{\nu\sigma} + \eta^{\nu\sigma} J^{\mu\rho} - \eta^{\nu\rho} J^{\mu\sigma} - \eta^{\mu\sigma} J^{\nu\rho}\,.
\end{aligned}
\tag{2.57}
$$

2.4 Oscillator expansions

Let us now solve the equations of motion, taking into account the boundary conditions. We will do this for the unconstrained system. The constraints then have to be imposed on the solutions. We have to distinguish between the closed and the open string and will treat them in turn.

(i) <u>closed string</u>

The general solution of the two-dimensional wave equation compatible with the periodicity condition $X^\mu(\sigma, \tau) = X^\mu(\sigma + 2\pi, \tau)$ is (cf. eq.(2.38))

$$X^\mu(\sigma, \tau) = X_R^\mu(\tau - \sigma) + X_L^\mu(\tau + \sigma) \tag{2.58}$$

where

$$X_R^\mu(\tau - \sigma) = \frac{1}{2}x^\mu + \frac{1}{4\pi T}p^\mu(\tau - \sigma) + \frac{i}{\sqrt{4\pi T}}\sum_{n \neq 0}\frac{1}{n}\alpha_n^\mu e^{-in(\tau - \sigma)} \tag{2.58a}$$

$$X_L^\mu(\tau + \sigma) = \frac{1}{2}x^\mu + \frac{1}{4\pi T}p^\mu(\tau + \sigma) + \frac{i}{\sqrt{4\pi T}}\sum_{n \neq 0}\frac{1}{n}\bar\alpha_n^\mu e^{-in(\tau + \sigma)} \tag{2.58b}$$

with arbitrary Fourier modes α_n^μ and $\bar\alpha_n^\mu$. Our notation is such that the α_n^μ are positive frequency modes for $n < 0$ and negative frequency modes for $n > 0$. The normalizations have been chosen for later convenience. The requirement that $X^\mu(\sigma, \tau)$ be a real function implies that x^μ and p^μ are real and that

$$\alpha_{-n}^\mu = \left(\alpha_n^\mu\right)^\dagger \qquad \text{and} \qquad \bar\alpha_{-n}^\mu = \left(\bar\alpha_n^\mu\right)^\dagger. \tag{2.59}$$

If we define $\alpha_0^\mu = \bar\alpha_0^\mu = \frac{1}{\sqrt{4\pi T}}p^\mu$, we can write

$$\partial_- X_R^\mu = \dot X_R^\mu = \frac{1}{\sqrt{4\pi T}}\sum_{n=-\infty}^{+\infty}\alpha_n^\mu e^{-in(\tau - \sigma)} \tag{2.60a}$$

$$\partial_+ X_L^\mu = \dot X_L^\mu = \frac{1}{\sqrt{4\pi T}}\sum_{n=-\infty}^{+\infty}\bar\alpha_n^\mu e^{-in(\tau + \sigma)} \tag{2.60b}$$

from which we find that

$$P^\mu = T \int_0^{2\pi} d\sigma\, \dot X^\mu(\sigma) = p^\mu \tag{2.61}$$

i.e. p^μ is the center of mass momentum of the string. From

$$\frac{1}{2\pi}\int_0^{2\pi} d\sigma\, X^\mu(\sigma, \tau = 0) = x^\mu \tag{2.62}$$

we learn that x^μ is the center of mass position of the string at $\tau = 0$. Also, using the expression for the total angular momentum, we find

$$J^{\mu\nu} = T \int_0^{2\pi} d\sigma (X^\mu \dot{X}^\nu - X^\nu \dot{X}^\mu) = l^{\mu\nu} + E^{\mu\nu} + \bar{E}^{\mu\nu} \tag{2.63}$$

with

$$l^{\mu\nu} = x^\mu p^\nu - x^\nu p^\mu \tag{2.63a}$$

and

$$E^{\mu\nu} = -i \sum_{n=1}^{\infty} \frac{1}{n}(\alpha^\mu_{-n}\alpha^\nu_n - \alpha^\nu_{-n}\alpha^\mu_n) \tag{2.63b}$$

with a similar expression for $\bar{E}^{\mu\nu}$. From the Poisson brackets eq.(2.51) we easily derive the brackets for the α^μ_n, $\bar{\alpha}^\mu_n$, x^μ and p^μ:

$$\{\alpha^\mu_m, \alpha^\nu_n\}_{P.B.} = \{\bar{\alpha}^\mu_m, \bar{\alpha}^\nu_n\}_{P.B.} = -im\delta_{m+n}\eta^{\mu\nu} \tag{2.64a}$$

$$\{\bar{\alpha}^\mu_m, \alpha^\nu_n\}_{P.B.} = 0 \tag{2.64b}$$

$$\{x^\mu, p^\nu\}_{P.B.} = \eta^{\mu\nu} \tag{2.64c}$$

where we have introduced the notation $\delta_m = \delta_{m,0}$. x^μ and p^μ, the center of mass position and momentum, are canonically conjugate. The Hamiltonian, expressed in terms of oscillators, is

$$H = \frac{1}{2} \sum_{n=-\infty}^{+\infty} (\alpha_{-n} \cdot \alpha_n + \bar{\alpha}_{-n} \cdot \bar{\alpha}_n). \tag{2.65}$$

We have seen above that the Virasoro constraints in conformal gauge are simply $T_{++} = 0$ and $T_{--} = 0$ and that the conservation of energy-momentum gives rise to an infinite number of conserved charges eq.(2.45) with a similar expression for the right-movers. We now choose for the functions $f(\sigma^\pm)$ a complete set satisfying the periodicity condition appropriate for the closed string: $f_m(\sigma^\pm) = \exp(im\sigma^\pm)$ for all integers m. We then define the Virasoro operators as the corresponding charges at $\tau = 0$[7]:

---

[7] Since the constraints are first class and $H_{\text{can}} = 0$, it is clear that they are constant in τ (up to the constraints); this is indeed easily verified.

$$L_m = 2T \int_0^{2\pi} d\sigma \; e^{-im\sigma} T_{--}$$

$$= T \int_0^{2\pi} d\sigma \; e^{-im\sigma} (\partial_- X)^2$$

$$= \frac{1}{2} \sum_n \alpha_{m-n} \cdot \alpha_n , \qquad\qquad (2.66a)$$

$$\bar{L}_m = 2T \int_0^{2\pi} d\sigma \; e^{+im\sigma} T_{++}$$

$$= T \int_0^{2\pi} d\sigma \; e^{+im\sigma} (\partial_+ X)^2$$

$$= \frac{1}{2} \sum_n \bar{\alpha}_{m-n} \cdot \bar{\alpha}_n. \qquad\qquad (2.66b)$$

They satisfy the reality condition

$$L_n = L^\dagger_{-n} \quad , \qquad \bar{L}_n = \bar{L}^\dagger_{-n}. \qquad\qquad (2.67)$$

Comparing with eq.(2.65) we find that the Hamiltonian is simply

$$H = L_0 + \bar{L}_0. \qquad\qquad (2.68)$$

The general τ evolution operator would have been $H = \sum_n (c_n L_n + \bar{c}_n \bar{L}_n)$; the choice implied by eq.(2.68), $c_n = \bar{c}_n = \delta_n$ corresponds to the conformal gauge. Using the basic Poisson brackets it is easy to show that the constraint $T \int_0^{2\pi} d\sigma \dot{X} \cdot X' = (\bar{L}_0 - L_0)$ generates rigid σ-translations. (This can already be seen from eqs.(2.47), (2.49) and (2.50).) Since on a closed string no point is special, we need to require that $L_0 - \bar{L}_0 = 0$. It is through this condition that the left-movers know about the right-movers. The Virasoro operators satisfy an algebra called the Virasoro algebra:

$$\{L_m, L_n\}_{P.B.} = -i(m-n)L_{m+n},$$

$$\{\bar{L}_m, \bar{L}_n\}_{P.B.} = -i(m-n)\bar{L}_{m+n}, \qquad\qquad (2.69)$$

$$\{\bar{L}_m, L_n\}_{P.B.} = 0.$$

It is straightforward to verify. This algebra is nothing but the Fourier decomposition of the (equal τ) algebra of the Virasoro constraints:

$$\{T_{--}(\sigma), T_{--}(\sigma')\}_{P.B.} = -\frac{1}{2T}\{T_{--}(\sigma) + T_{--}(\sigma')\}\partial_\sigma \delta(\sigma - \sigma')\,,$$

$$\{T_{++}(\sigma), T_{++}(\sigma')\}_{P.B.} = \frac{1}{2T}\{T_{++}(\sigma) + T_{++}(\sigma')\}\partial_\sigma \delta(\sigma - \sigma')\,, \qquad (2.70)$$

$$\{T_{++}(\sigma), T_{--}(\sigma')\}_{P.B.} = 0\,.$$

It is useful to recognize that if we replace the Poisson brackets by Lie brackets, a realization of the Virasoro algebra is given by the vector fields $\bar{L}_n = e^{in\sigma^+}\partial_+$ and $L_n = e^{in\sigma^-}\partial_-$. They are the generators of the reparametrizations $\sigma^\pm \to \sigma^\pm + f_n(\sigma^\pm)$. It we define the variable $z = e^{i\sigma^-} \in S^1$, we get $L_n = iz^{n+1}\partial_z$, which are reparametrizations of the circle S^1.

(ii) <u>open string</u>

Here we have to impose the boundary condition $X'^\mu = 0$ at the ends of the string, i.e. at $\sigma = 0$ and $\sigma = \pi$. The general solution of the wave equation, subject to these boundary conditions, is

$$X^\mu(\sigma,\tau) = x^\mu + \frac{1}{\pi T}p^\mu \tau + \frac{i}{\sqrt{\pi T}}\sum_{n\neq 0}\frac{1}{n}\alpha_n^\mu e^{-in\tau}\cos n\sigma \qquad (2.71)$$

from which we get

$$\partial_\pm X^\mu = \frac{1}{2}(\dot{X}^\mu \pm X'^\mu) = \frac{1}{2\sqrt{\pi T}}\sum_{n=-\infty}^{+\infty}\alpha_n^\mu e^{-in(\tau\pm\sigma)} \qquad (2.72)$$

where we have defined $\alpha_0^\mu = \frac{1}{\sqrt{\pi T}}p^\mu$. As in the case of the closed string we easily show that x^μ and p^μ are the center of mass position and momentum of the open string. The total angular momentum is given by

$$J^{\mu\nu} = l^{\mu\nu} + E^{\mu\nu} \qquad (2.73)$$

with $l^{\mu\nu}$ and $E^{\mu\nu}$ as in eqs.(2.63a) and (2.63b). We again find

$$\{\alpha_m^\mu, \alpha_n^\mu\}_{P.B.} = -im\delta_{m+n}\eta^{\mu\nu}\,,$$

$$\{x^\mu, p^\nu\}_{P.B.} = \eta^{\mu\nu}\,. \qquad (2.74)$$

In terms of the oscillators the Hamiltonian for the open string is

$$H = \frac{1}{2} \sum_{n=-\infty}^{+\infty} \alpha_{-n} \cdot \alpha_n. \tag{2.75}$$

The open string boundary conditions mix left- with right-movers and consequently T_{++} with T_{--}. We define the Virasoro operators for the open string as

$$
\begin{aligned}
L_m &= 2T \int_0^\pi d\sigma \left(e^{im\sigma} T_{++} + e^{-im\sigma} T_{--} \right) \\
&= \frac{T}{4} \int_0^\pi d\sigma \left(e^{im\sigma} (\dot{X}^\mu + X'^\mu)^2 + e^{-im\sigma} (\dot{X}^\mu - X'^\mu)^2 \right) \\
&= \frac{T}{4} \int_{-\pi}^\pi d\sigma e^{im\sigma} (\dot{X}^\mu + X'^\mu)^2 \\
&= \frac{1}{2} \sum_{-\infty}^{+\infty} \alpha_{m-n} \cdot \alpha_n.
\end{aligned}
\tag{2.76}
$$

Note that in the third line we have extended the integration region from $0 \le \sigma \le \pi$ to the interval $-\pi \le \sigma \le +\pi$ on which the functions $e^{im\sigma}$ are periodic. This is possible since $X'^\mu(\sigma) = -X'^\mu(-\sigma)$. The L_m are a complete set of conserved charges respecting the open string boundary conditions. Comparison with eq.(2.75) gives

$$H = L_0, \tag{2.77}$$

which, as in the closed string case, reflects the fact that we are in conformal gauge. The L_m satisfy the Virasoro algebra

$$\{L_m, L_n\}_{P.B.} = -i(m - n)L_{m+n}. \tag{2.78}$$

2.5 Examples of classical string solutions

At the end of this chapter let us try to get some understanding for the solutions of the classical string equations of motion subject to the constraints.

Since in conformal gauge the coordinate functions X^μ satisfy the wave equation, we can use the remaining gauge freedom to set $X^0 = t = \kappa\tau$ for some constant κ. The X^i, $i = 1, \ldots, d-1$ then satisfy

$$(\partial_\sigma^2 - \partial_\tau^2)X^i = 0 \tag{2.79}$$

with solution

$$X^i(\sigma,\tau) = \tfrac{1}{2}a^i(\sigma + \tau) + \tfrac{1}{2}b^i(\sigma - \tau). \tag{2.80}$$

The constraint $\dot{X} \cdot X' = -\dot{X}^0 X'^0 + \dot{X}^i X'^i = 0$ leads to $a'^2 = b'^2$ and $\dot{X}^2 + X'^2 = 0$ to $\frac{1}{2}(a'^2 + b'^2) = \kappa^2$. Combined this gives

$$a'^2 = b'^2 = \kappa^2. \tag{2.81}$$

The simplest example of an open string is given by ($0 \le \sigma \le \pi$)

$$
\begin{aligned}
X^1 &= L\cos\sigma\cos\tau &\qquad X^0 &= t = L\tau \\
X^2 &= L\cos\sigma\sin\tau \\
X^i &= 0, &\qquad i &= 3,\ldots,d-1
\end{aligned}
\tag{2.82}
$$

which clearly satisfies the constraints and the edge condition. It describes a straight string of length $2L$ rotating around its midpoint in the (X^1, X^2)-plane. Its total (spatial) momentum vanishes and its energy is $E = P^0 = L\pi T$ from which we derive the mass $M^2 = -P^\mu P_\mu = (L\pi T)^2$. The angular momentum is $J = J_{12} = \frac{1}{2}L^2\pi T$ and we find that $J = \frac{1}{2\pi T}M^2 \equiv \alpha' M^2$. This is a straight line in the (M^2, J) plane with slope $\alpha' = (2\pi T)^{-1}$, called a Regge trajectory. It can actually be shown that for any classical open string solution $J \le \alpha' M^2$. (In the gauge chosen here and in the center of mass frame $J^2 = \frac{1}{2}J_{ij}J^{ij}$, $i,j = 1,\ldots,d-1$.)

For the closed string ($0 \le \sigma \le 2\pi$) the periodicity requirement leads to $a(\sigma + 2\pi) = a(\sigma)$ and $b(\sigma + 2\pi) = b(\sigma)$. From $\boldsymbol{X}(\sigma + \pi, \tau + \pi) = \frac{1}{2}a(\sigma + \tau + 2\pi) + \frac{1}{2}b(\sigma - \tau) = \frac{1}{2}a(\sigma + \tau) + \frac{1}{2}b(\sigma - \tau)$ we find that the period of a closed string is π. For an initially static closed string configuration, i.e. one that satisfies $\dot{\boldsymbol{X}}(\sigma, \tau = 0) = 0$, we find $\boldsymbol{X}(\sigma,\tau) = \frac{1}{2}(a(\sigma + \tau) + a(\sigma - \tau))$.

After half a period, i.e. at $\tau = \frac{\pi}{2}$, $\boldsymbol{X}(\sigma, \frac{\pi}{2}) = \boldsymbol{X}(\sigma + \pi, \frac{\pi}{2})$, the loop doubles up and goes around itself twice: $\boldsymbol{X}(\sigma) = \boldsymbol{X}(\sigma + \pi)$. A simple closed string configuration is

$$X^1 = \tfrac{1}{2}R\big(\cos(\sigma + \tau) + \cos(\sigma - \tau)\big) = R\cos\sigma\cos\tau \,,$$

$$X^2 = \tfrac{1}{2}R\big(\sin(\sigma + \tau) + \sin(\sigma - \tau)\big) = R\sin\sigma\cos\tau \,, \qquad (2.83)$$

$$t = R\tau \,.$$

At $t = 0$ it represents a circular string of radius R in the (X^1, X^2)-plane, centered around the origin. Its energy is $E = 2\pi RT$. Linear and angular momentum vanish. At $\tau = \frac{\pi}{2}$ it has collapsed to a point and at $t = \pi$ it has expanded again to its original size. Similar to the open string case, one can show that a general classical closed string configuration satisfies $J \leq \tfrac{1}{2}\alpha' M^2$.

The slope parameter α' or, equivalently, the string tension T are the only dimensionful parameters in the theory. To simplify notation we will choose a system of units where $\alpha' = \frac{1}{2}$ or, equivalently, $T = \frac{1}{\pi}$ for the open string and $\alpha' = 2$ or $T = \frac{1}{4\pi}$ for the closed string. We can then, if necessary, reintroduce unambiguously powers of α' to get dimensionally correct expressions.

REFERENCES

[1] Y. Nambu, Lectures at the Copenhagen symposium, 1970.

[2] T. Goto, *"Relativistic quantum mechanics of one-dimensional mechanical continuum and subsidiary condition of dual resonance model"*, Prog. Theor. Phys. **46** (1971) 1560.

[3] A.M. Polyakov, *"Quantum geometry of bosonic strings"*, Phys. Lett. **103B** (1981) 207.

Chapter 3

The Quantized Bosonic String

The quantization of the bosonic string, which is the subject of this chapter, will lead us to the notion of the critical dimension ($d = 26$). Its discovery was of great importance for the further development of string theory. The first indication that $d = 26$ plays a special role appeared in a paper by Lovelace [1]. Goddard, Goldstone, Rebbi and Thorn [2] quantized the string in light-cone gauge and showed that the quantum theory is Lorentz invariant only for $d = 26$. The decoupling of negative norm states (ghosts) in the critical dimension was shown in two different proofs of the no-ghost theorem by Brower [3] and Goddard and Thorn [4]. The modern path integral quantization started with the paper by Polyakov [5].

3.1 Canonical quantization of the bosonic string

In this section we will discuss the first quantization of the bosonic string in terms of operators, i.e. we will consider the functions $X^\mu(\sigma, \tau)$ as quantum mechanical operators. This is equivalent to the transition from classical mechanics to quantum mechanics in first quantization via canonical commutation relations for the coordinates and their canonically conjugate momenta. We replace Poisson brackets by commutators according to

$$\{\ ,\ \}_{P.B.} \rightarrow \frac{1}{i} [\ ,\ ] \tag{3.1}$$

In this way we obtain[1]

---

[1] Our notation is the same for quantum as for classical quantities. Only when confu-

$$[X^\mu(\sigma,\tau), \dot{X}^\nu(\sigma',\tau)] = \frac{i}{T}\eta^{\mu\nu}\delta(\sigma-\sigma'),$$

$$[X^\mu(\sigma,\tau), X^\nu(\sigma',\tau)] = [\dot{X}^\mu(\sigma,\tau), \dot{X}^\nu(\sigma',\tau)] = 0. \tag{3.2}$$

The Fourier expansion coefficients in eqs.(2.58) and (2.71) are now operators for which the following commutation relations hold:

$$[x^\mu, p^\nu] = i\eta^{\mu\nu}$$
$$[\alpha_m^\mu, \alpha_n^\nu] = [\bar{\alpha}_m^\mu, \bar{\alpha}_n^\nu] = m\delta_{m+n}\eta^{\mu\nu} \tag{3.3}$$
$$[\bar{\alpha}_m^\mu, \alpha_n^\nu] = 0.$$

For the open string the $\bar{\alpha}_m^\mu$ are of course absent. The hermiticity condition (2.59) is still valid, now following from the hermiticity of the operators $X^\mu(\sigma,\tau)$. If we rescale the α_m^μ's and define $a_m^\mu = \frac{1}{\sqrt{m}}\alpha_m^\mu$, $a_m^{\dagger\mu} = \frac{1}{\sqrt{m}}\alpha_{-m}^\mu$ $(m > 0)$, then the a_m^μ satisfy the usual harmonic oscillator commutation relations $[a_m^\mu, a_n^{\nu\dagger}] = \delta_{m,n}\eta^{\mu\nu}$. From $(\alpha_{-m}\alpha_m)\alpha_{\pm m} = \alpha_{\pm m}(\alpha_{-m}\alpha_m \mp m)$ we find that the negative frequency modes α_m, $m > 0$ are lowering operators and the positive frequency modes α_m, $m < 0$ are raising operators. The corresponding number operator for the m'th mode $(m > 0)$ is $N_m =: \alpha_m\alpha_{-m} := \alpha_{-m}\alpha_m$ where the normal ordering symbol means that we put negative frequency modes to the right of positive frequency modes. One now defines the oscillator ground state as the state which is annihilated by all the lowering operators. This does not, however, completely specify the state; we can choose it to be an eigenstate of the center of mass momentum operator with eigenvalue p^μ. If we denote this state by $|0;\, p^\mu\rangle$, we have

$$\alpha_m^\mu|0, p^\mu\rangle = 0 \qquad \text{for } m > 0$$
$$\hat{p}^\mu|0, p^\mu\rangle = p^\mu|0, p^\mu\rangle. \tag{3.4}$$

But we do have a problem now. Since the Minkowski metric $\eta^{\mu\nu}$ has $\eta^{00} = -1$, we get $[\alpha_m^0, \alpha_{-m}^0] = [\alpha_m^0, \alpha_m^{0\dagger}] = -m$ and states of the form

---

sions are possible do we denote operators by hatted symbols.

$\alpha^0_{-m}|0\rangle$ with $m > 0$ satisfy $\langle 0|\alpha^0_m \, \alpha^0_{-m}|0\rangle = -m\langle 0|0\rangle < 0$; i.e. these states have negative norm. They are called ghosts[2]. Negative norm states are bad news since they are in conflict with the probabilistic interpretation of quantum mechanics. However, just as we had to impose the constraints on the solutions of the classical equations of motion, we have to impose them, now as operators, as subsidiary conditions on the states. We then hope that the ghosts decouple from the physical Hilbert space. Indeed, one can prove a no-ghost theorem which states that the ghosts decouple in 26 dimensions (i.e. $d = 26$) if the normal ordering constant to be discussed below is -1. We will not prove this theorem here, but instead arrive at the same consistency conditions by different means (c.f. below and Chapter 5).

Let us now determine the propagators for the fields $X^\mu(\sigma, \tau)$. As usual, we define them as

$$\langle X^\mu(\sigma, \tau) X^\nu(\sigma', \tau')\rangle = T[X^\mu(\sigma, \tau) X^\nu(\sigma', \tau')] - N[X^\mu(\sigma, \tau) X^\nu(\sigma', \tau')]$$

$$(3.5)$$

where T denotes time-ordering and N normal ordering. Zero modes need special care. We define $: p^\nu x^\mu := x^\mu p^\nu$. This corresponds to the choice of a translationally invariant in-vacuum $p^\mu|0\rangle = 0$. If we define the variables $(z, \bar{z}) = (e^{i(\tau-\sigma)}, e^{i(\tau+\sigma)}) \in S^1 \times S^1$ we find for the closed string $(\tau > \tau')$[3]

$$\langle X^\mu_L(\sigma, \tau) X^\nu_L(\sigma', \tau')\rangle = \frac{1}{4}\alpha'\eta^{\mu\nu}\ln\bar{z} - \frac{1}{2}\alpha'\eta^{\mu\nu}\ln(\bar{z} - \bar{z}') \qquad (3.6a)$$

$$\langle X^\mu_R(\sigma, \tau) X^\nu_R(\sigma', \tau')\rangle = \frac{1}{4}\alpha'\eta^{\mu\nu}\ln z - \frac{1}{2}\alpha'\eta^{\mu\nu}\ln(z - z') \qquad (3.6b)$$

$$\langle X^\mu_R(\sigma, \tau) X^\nu_L(\sigma', \tau')\rangle = -\frac{1}{4}\alpha'\eta^{\mu\nu}\ln z \qquad (3.6c)$$

---

[2] These ghosts are not to be confused with the Faddeev-Popov ghosts of Section 3.4.

[3] In Chapter 4 we will make a Wick rotation to a Euclidean world-sheet and z and $\bar{z}$ will be complex conjugates of each other.

$$\langle X_L^\mu(\sigma,\tau) X_R^\nu(\sigma',\tau')\rangle = -\frac{1}{4}\alpha'\eta^{\mu\nu}\ln\bar{z} \tag{3.6d}$$

and

$$\langle X^\mu(\sigma,\tau) X^\nu(\sigma',\tau')\rangle = -\frac{1}{2}\alpha'\eta^{\mu\nu}\big(\ln(z-z') + \ln(\bar{z}-\bar{z}')\big). \tag{3.7}$$

The non-vanishing of eqs.(3.6.c,d) is due to the fact that X_L and X_R have common zero mode operators. If we define fields

$$X_R^\mu(z) = x_R^\mu - ip_R^\mu \ln z \; + \; \text{oscillators}$$
$$X_L^\mu(\bar{z}) = x_L^\mu - ip_L^\mu \ln\bar{z} \; + \; \text{oscillators} \tag{3.8}$$

where

$$[x_L^\mu, p_L^\nu] = [x_R^\mu, x_R^\nu] = i\eta^{\mu\nu} \tag{3.9a}$$

but

$$[x_L^\mu, p_R^\nu] = [x_R^\mu, p_L^\nu] = 0 \tag{3.9b}$$

we find

$$\langle X_L^\mu(\sigma,\tau) X_L^\nu(\sigma',\tau')\rangle = -\eta^{\mu\nu}\ln(\bar{z}-\bar{z}')$$
$$\langle X_R^\mu(\sigma,\tau) X_R^\nu(\sigma',\tau')\rangle = -\eta^{\mu\nu}\ln(z-z') \tag{3.10}$$

and vanishing cross terms. Here we have set $\alpha' = 2$. We note that the propagators for $X_L + X_R$ are the same in both cases. Treating left-and right-movers as completely independent fields will be the key ingredient for the construction of the heterotic string, which we will discuss in Chapter 10. Finally, for the open string propagator we find

$$\langle X^\mu(\sigma,\tau) X^\nu(\sigma',\tau')\rangle = -\frac{1}{2}\alpha'\big\{\ln(z-z')(\bar{z}-\bar{z}') + \ln(z-\bar{z}')(\bar{z}-z')\big\}. \tag{3.11}$$

Let us now turn to the constraints. In the classical theory they were shown to correspond to $T_{++} = T_{--} = 0$ or, expressed through the Fourier components, $L_m = \bar{L}_m = 0$ (in the open string case the $\bar{L}_m$'s are absent). However, in the quantum theory any expression containing non-commuting operators is ill-defined without specifying an operator ordering prescription.

This applies in particular to L_0. (The other L_m's are safe.) Classically it was given by $L_0 = \frac{1}{2} \sum_{n=-\infty}^{+\infty} \alpha_{-n}\alpha_n$. In the quantum theory we define the L_m's by their normal ordered expressions, i.e.

$$L_m = \frac{1}{2} \sum_{n=-\infty}^{+\infty} : \alpha_{m-n} \cdot \alpha_n : \tag{3.12}$$

and, in particular

$$L_0 = \frac{1}{2}\alpha_0^2 + \sum_{n=1}^{\infty} \alpha_{-n} \cdot \alpha_n \ . \tag{3.13}$$

We then include an as yet undetermined normal ordering constant a in all formulas containing L_0, i.e. we replace L_0 by $(L_0 - a)$. We now have to determine the algebra of the L_n's. Due to normal ordering the calculation has to be done with great care; the details can be found in the appendix to this chapter. We find the Virasoro algebra

$$[L_m, L_n] = (m - n)L_{m+n} + \frac{c}{12}m(m^2 - 1)\delta_{m+n}. \tag{3.14}$$

c is called the central charge. The term proportional to c arises as a quantum effect. Here, $c = \eta^\mu{}_\mu = d$ is the dimension of the embedding space, i.e. the number of free scalar fields (on the world sheet). This means that each free scalar field contributes one unit to the central charge. In later chapters we will derive the contribution of other fields, such as free world-sheet fermions and Faddeev-Popov ghosts, to the central charge.

Note that the term in the anomaly linear in m can be changed by redefining $L_m \rightarrow L_m - \alpha\delta_m$ which leads to an anomaly $\left(\frac{c}{12}m^3 + (2\alpha - \frac{c}{12})m\right)\delta_{m+n}$. This shift also changes the normal ordering constant to $a \rightarrow a - \alpha$. Only the relation between the normal ordering constant and the linear term in the anomaly has an invariant meaning. The quantum version of the Virasoro algebra in the form of eq.(2.70) is (for $\alpha' = 2$)

$$[T_{++}(\sigma), T_{++}(\sigma')] = 2\pi i\{T_{++}(\sigma) + T_{++}(\sigma')\}\partial_\sigma\delta(\sigma - \sigma') - \frac{i\pi c}{6}\partial_\sigma^3\delta(\sigma - \sigma')$$

$$[T_{--}(\sigma), T_{--}(\sigma')] = -2\pi i\{T_{--}(\sigma) + T_{--}(\sigma')\}\partial_\sigma\delta(\sigma - \sigma') + \frac{i\pi c}{6}\partial_\sigma^3\delta(\sigma - \sigma')$$

$$[T_{++}(\sigma), T_{--}(\sigma')] = 0$$

$$(3.15)$$

which corresponds to the choice $\alpha = \frac{c}{24}$. It is now easy to see that even though in the classical theory the constraints are $L_m = 0$, $\forall\, m$, this cannot be implemented on quantum mechanical states $|\phi\rangle$ since

$$\langle\phi|[L_m,\ L_{-m}]|\phi\rangle = \langle\phi|2mL_0|\phi\rangle + \frac{d}{12}m(m^2 - 1)\langle\phi|\phi\rangle$$

i.e. we cannot require $L_m|\phi\rangle = 0$, $\forall m$. The most we can do is to demand that on physical states

$$L_m|\text{phys}\rangle = 0 \qquad m > 0 \qquad\qquad (3.16a)$$

$$(L_0 - a)|\text{phys}\rangle = 0 \qquad\qquad\qquad (3.16b)$$

i.e. the positive frequency components annihilate physical states. This is consistent since the L_m for $m > 0$ form a closed subalgebra, and the requirement $L_m|\text{phys}\rangle = 0$, for $m > 0$ only, effectively incorporates all constraints since with $L_m = L_{-m}^\dagger$ we find that[4]

$$\langle\text{phys}'|L_n|\text{phys}\rangle = 0 \qquad \forall\, n \neq 0. \qquad\qquad (3.17)$$

For the closed string we have in addition the $\bar{L}_m$'s. They also satisfy a Virasoro algebra and commute with the L_m's. We impose the conditions eq.(3.16) also for the $\bar{L}_m$'s and in addition

$$(L_0 - \bar{L}_0)|\text{phys}\rangle = 0. \qquad\qquad (3.18)$$

---

[4]Note that the situation is very similar to the one in the quantization of electromagnetism. There we can only impose the positive frequency part of the gauge condition $\partial \cdot A = 0$ on physical states which suffices to get $\langle\text{phys}'|\partial \cdot A|\text{phys}\rangle = 0$. In this restricted Hilbert space longitudinal and scalar photons decouple.

The reason for this constraint is that the unitary operator

$$U_\delta = e^{i\delta(L_0 - \bar{L}_0)} \tag{3.19}$$

satisfies

$$U_\delta^\dagger X^\mu(\sigma, \tau) U_\delta = X^\mu(\sigma + \delta, \tau) \tag{3.20}$$

as is straightforward to show; i.e. it generates rigid σ translations. This already follows from our discussion in Chapter 2 of the motions generated by the constraints. Since no point on a closed string should be distinct, we have to impose eq.(3.18). (This also follows from eq.(3.16b) and the equivalent condition for $\bar{L}_0$ if $a = \bar{a}$.)

For the open string $L_0 = \sum_{n=1}^\infty \alpha_{-n}^\mu \, \alpha_n^\mu + \alpha' p^\mu p_\mu$ and $p^\mu p_\mu = -m^2$. Then the condition $(L_0 - a)|\mathrm{phys}\rangle = 0$ implies that $(\sum_{n=1}^\infty \alpha_{-n} \cdot \alpha_n - a)|\mathrm{phys}\rangle = \alpha' m^2 |\mathrm{phys}\rangle$; hence condition (3.16b) is called the mass-shell condition and the mass operator for the open string is

$$\alpha' m^2 = (N - a) \tag{3.21}$$

where we have defined the level number N as

$$N = \sum_{m>0} N_m = \sum_{m>0} \alpha_{-m} \cdot \alpha_m. \tag{3.22}$$

For closed strings we find from $L_0 = \sum_{n=1}^\infty \alpha_{-n}^\mu \alpha_n^\mu + \dfrac{\alpha'}{4} p^2$ and the corresponding expression for $\bar{L}_0$

$$m^2 = -p^\mu p_\mu = m_L^2 + m_R^2 \tag{3.23}$$

where

$$\begin{aligned} \alpha' m_L^2 &= 2(\bar{N} - a) \\ \alpha' m_R^2 &= 2(N - a) \end{aligned} \tag{3.24}$$

and

$$m_L^2 = m_R^2$$

as a consequence of $L_0 - \bar{L}_0 = 0$. We see that in both cases the mass of the ground state ($N = \bar{N}$) is determined by the normal ordering constant.

We want to note that the normal ordering constants in the expressions for the angular momentum operators drop out and one easily verifies the Poincaré algebra (2.57) as an operator algebra, after replacing the Poisson brackets by commutators.

3.2 Light cone quantization of the bosonic string

It is now possible to choose a gauge, called the light cone gauge, in which the Virasoro constraint equations can be solved explicitly and the theory can be described in terms of physical degrees of freedom only. However, light cone gauge is a non-covariant gauge. But, since the formulation in light cone gauge is obtained from a manifestly Lorentz invariant theory via gauge fixing, one might expect that d-dimensional Lorentz invariance is automatic (though not manifest). However, as we will see shortly, this is true in the quantum theory only if $d = 26$ and $a = 1$.

In going to light cone gauge we use the residual gauge freedom that was left after fixing the conformal gauge by choosing $\tau \propto X^+$. The constant of proportionality will be determined shortly. The light cone coordinates are defined to be $X^\pm = \frac{1}{\sqrt{2}}(X^0 \pm X^{d-1})$ and X^i, $i = 1, \ldots, d - 2$. The X^i are the transverse coordinates. The scalar product in terms of light cone components is $V \cdot W = V^i W^i - V^+ W^- - V^- W^+$ and indices are raised and lowered according to $V^+ = -V_-$, $V^- = -V_+$ and $V^i = V_i$. To go to light cone gauge is possible since the X^μ satisfy the wave equation. (Cf. the discussion in Section 2.3.) Let us now determine the constant of proportionality. From eq.(2.55) we find

$$X^+ = \begin{cases} \alpha' p^+ \tau & \text{(closed string)}, \\ 2\alpha' p^+ \tau & \text{(open string)}. \end{cases} \tag{3.25}$$

This means that $\alpha_n^+ = \bar{\alpha}_n^+ = \sqrt{\frac{\alpha'}{2}}p^+\delta_n$ for the closed string and $\alpha_n^+ = \sqrt{2\alpha'}p^+\delta_n$ for the open string. With this identification the constraint equations $(\dot{X}^\mu \pm X'^\mu)^2 = 0$ become

$$\partial_\pm X^- = \begin{cases} \frac{1}{\alpha'p^+}(\partial_\pm X^i)^2 & \text{(closed string)}, \\[2mm] \frac{1}{2\alpha'p^+}(\partial_\pm X^i)^2 & \text{(open string)}; \end{cases} \tag{3.26}$$

i.e. we can solve the X^- in terms of the X^i so that in light cone gauge both X^+ and X^- are eliminated, leaving only the transverse components X^i as independent variables. We can now express α_n^- in terms of the α_n^i. Expanding eq.(3.26) in Fourier modes we find

$$\alpha_n^- = \frac{1}{\sqrt{2\alpha'}p^+}\Big(\sum_{m=-\infty}^{+\infty} : \alpha_{n-m}^i \alpha_m^i : -2a\delta_n\Big)$$

$$\bar{\alpha}_n^- = \frac{1}{\sqrt{2\alpha'}p^+}\Big(\sum_{m=-\infty}^{+\infty} : \bar{\alpha}_{n-m}^i \bar{\alpha}_m^i : -2a\delta_n\Big) \tag{3.27}$$

for the closed string and

$$\alpha_n^- = \frac{1}{2\sqrt{2\alpha'}p^+}\Big(\sum_{m=-\infty}^{+\infty} : \alpha_{n-m}^i \alpha_m^i : -2a\delta_n\Big) \tag{3.28}$$

for the open string. Here we have again introduced a normal ordering constant. The mass operator is now obtained from $m^2 = (2p^+p^- - p^ip^i)$ by use of the relation between p^μ and α_0^μ and the expression for α_0^- in terms of the transverse oscillators. We obtain

$$m^2 = \frac{2}{\alpha'}\Big\{ \sum_{n>0}(\alpha_{-n}^i\alpha_n^i + \bar{\alpha}_{-n}^i\bar{\alpha}_n^i - 2a\Big\} \tag{3.29}$$

for the closed string, and

$$m^2 = \frac{1}{\alpha'}\Big\{ \sum_{n>0}\alpha_{-n}^i\alpha_n^i - a\Big\} \tag{3.30}$$

for the open string. This also follows directly from the covariant expressions since in light cone gauge $\alpha_n^+ = \bar{\alpha}_n^+ = 0$ for $n \neq 0$. The normal ordering constants in eqs.(3.27) and (3.28) are the same as the ones in eqs.(3.21) and (3.24).

The action in light cone gauge is simply the restriction of the covariant action eq.(2.35) to the independent physical (transverse) degrees of freedom:

$$S_{\text{l.c.}} = \frac{T}{2} \int d^2\sigma \left((\dot{X}^i)^2 - (X'^i)^2 \right). \tag{3.31}$$

The canonical Hamiltonian following from the light cone action is

$$\begin{aligned}
H &= \frac{1}{2} \sum_{n=-\infty}^{+\infty} : \alpha_{-n}^i \alpha_n^i : -a \\
&= \frac{1}{2}(p^i)^2 + N - a
\end{aligned} \tag{3.32}$$

for the open string, and

$$\begin{aligned}
H &= \frac{1}{2} \sum_{n=-\infty}^{+\infty} : (\alpha_{-n}^i \alpha_n^i + \bar{\alpha}_{-n}^i \bar{\alpha}_n^i) : -2a \\
&= (p^i)^2 + N + \bar{N} - 2a
\end{aligned} \tag{3.33}$$

for the closed string. Note that these expressions do not follow by substituting (3.27) and (3.28) into eqs.(2.65) and (2.75). This would give vanishing results since the covariant Hamiltonian is just one of the constraints and eqs.(3.27) and (3.28) solutions of them.

3.3 Spectrum of the bosonic string

Next let us look at the spectrum of the theory. The states are generated by acting with the transverse oscillators on the oscillator ground state. We have to distinguish between open and closed strings and will discuss them in turn.

open string spectrum

The ground state $|0, p^i\rangle$ is unique up to Lorentz boosts. Its mass is given by its eigenvalue of the mass operator $\alpha' m^2 |0, p^i\rangle = -a|0, p^i\rangle$. The first excited state is $\alpha^i_{-1}|0, p^j\rangle$; it is a $d-2$ dimensional vector of the transverse rotation group $SO(d-2)$. Lorentz invariance requires that physical states fall into representations of the little group of the Lorentz group $SO(d-1,1)$, which is $SO(d-1)$ for massive particles and $SO(d-2)$ for massless particles. Above state is a vector of $SO(d-2)$, i.e. it must be massless. Acting on it with the mass operator we get

$$0 = \alpha' m^2 \big(\alpha^i_{-1}|0, p^j\rangle\big) = (1-a)\alpha^i_{-1}|0, p^j\rangle \tag{3.34}$$

i.e. space-time Lorentz invariance requires that the normal ordering constant be $a = 1$. Recall that it arose from normal ordering the expression $\sum_{n\neq 0} \alpha^i_{-n}\alpha^i_n$ appearing in α^-_0. We write

$$\begin{aligned}
\sum_{n\neq 0} \alpha^i_{-n}\alpha^i_n &= \sum_{n\neq 0} :\alpha^i_{-n}\alpha^i_n: +(d-2)\sum_{n=1}^{\infty} n \\
&= 2\Big\{ \sum_{n=1}^{\infty} \alpha^i_{-n}\alpha^i_n + \frac{d-2}{2}\sum_{n=1}^{\infty} n \Big\}.
\end{aligned} \tag{3.35}$$

The last sum in this expression is however undefined and must be regularized. We do this using ζ-function regularization. It can be shown that this regularization respects modular invariance, an important consistency condition to be introduced in Chapter 6. Consider the sum $\sum_{n=1}^{\infty} n^{-s} = \zeta(s)$, where $\zeta(s)$ is Riemann's zeta function. It converges for $s > 1$ and has a unique analytic continuation at $s = -1$ where it has the value $\zeta(-1) = -1/12$, i.e. $a = -\frac{d-2}{24}$. From Lorentz invariance we have found above that $a = 1$, which tells us that $d = 26$. To summarize, Lorentz invariance of the quantized bosonic string theory requires $a = 1$ and $d = 26$. A more rigorous argument, which also relies on Lorentz invariance, is to check the closure

Table 3.1: The five lowest mass levels of the oriented open bosonic string

level	$\alpha'(\text{mass})^2$	states and their $SO(24)$ representation contents	little group	representation contents with respect to the little group
0	-1	$\|0\rangle$ $\bullet$ (1)	$SO(25)$	$\bullet$ (1)
1	0	$\alpha^i_{-1}\|0\rangle$ $\Box$ (24)	$SO(24)$	$\Box$ (24)
2	$+1$	$\alpha^i_{-2}\|0\rangle$ $\alpha^i_{-1}\alpha^j_{-1}\|0\rangle$ $\Box$ $\Box\Box+\bullet$ (24) (299) + (1)	$SO(25)$	$\Box\Box$ (324)
3	$+2$	$\alpha^i_{-3}\|0\rangle$ $\alpha^i_{-2}\alpha^j_{-1}\|0\rangle$ $\alpha^i_{-1}\alpha^j_{-1}\alpha^k_{-1}\|0\rangle$ $\Box$ $\begin{smallmatrix}\Box\\\Box\end{smallmatrix}+\Box\Box+\bullet$ $\Box\Box\Box+\Box$ (24) (276) + (299) + (1) (2576) + (24)	$SO(25)$	$\Box\Box\Box+\begin{smallmatrix}\Box\\\Box\end{smallmatrix}$ (2900) + (300)
4	$+3$	$\alpha^i_{-4}\|0\rangle$ $\alpha^i_{-3}\alpha^j_{-1}\|0\rangle$ $\alpha^i_{-2}\alpha^j_{-2}\|0\rangle$ $\Box$ $\Box\Box+\begin{smallmatrix}\Box\\\Box\end{smallmatrix}+\bullet$ $\Box\Box+\bullet$ (24) (299) + (276) + (1) (299) + (1) $\alpha^i_{-2}\alpha^j_{-1}\alpha^k_{-1}\|0\rangle$ $\alpha^i_{-1}\alpha^j_{-1}\alpha^k_{-1}\alpha^l_{-1}\|0\rangle$ $2\times\Box+\Box\Box\Box+\begin{smallmatrix}\Box\Box\\\Box\end{smallmatrix}$ $\Box\Box\Box\Box+\Box\Box+\bullet$ 2 × (24) + (2576) + (4576) (17250) + (299) + (1)	$SO(25)$	$\Box\Box\Box\Box+\begin{smallmatrix}\Box\Box\\\Box\end{smallmatrix}$ (20150) + (5175) $+\Box\Box+\bullet$ + (324) + (1)

of the Lorentz algebra; the commutator $[M^{i-}, M^{j-}]$ is critical. In contrast to the Lorentz generators in covariant gauge, M^{i-} contains normal ordered expressions due to the appearance of α^-_n. The actual calculation of the commutator is quite tedious and will not be presented here. A simple consequence of $a = 1$ is that the ground state satisfies $\alpha' m^2 = -1$, i.e. it is a tachyon. The presence of a tachyon is not necessarily a fatal problem for the theory. It means that the ground state is unstable and some other, stable ground state might exist. Another way to get rid of the tachyon is to introduce anticommuting degrees of freedom whose normal ordering constant cancels that of the commuting degrees of freedom of the bosonic string. This is indeed what is done in the superstring theory. In table 3.1 we have collected the light cone states of the open bosonic string up to the fourth level. It is demonstrated how, for the massive states, the light cone states, which are tensors of $SO(24)$, combine uniquely into representations of $SO(25)$. It can be shown that this occurs at all mass levels and to depend crucially on the choice $a = 1$ and $d = 26$. Since at level n with mass $\alpha' m^2 = (n - 1)$ we always have a state described by a symmetric tensor of rank n we find that the maximal spin at each level is $j_{\max} = n = \alpha' m^2 + 1$. In general the states will satisfy $j \leq \alpha' m^2 + 1$ and, since j and m^2 are quantized, all states lie on Regge trajectories, with the tachyon lying on the leading trajectory.

<u>closed string spectrum</u>

Since in the case of the closed string we can excite both left- and right-moving degrees of freedom, its states are simply tensor products of the open string states, subject to the constraint $L_0 - \bar{L}_0 = 0$.

This simply means that the excitation level in both sectors has to be the same. The ground state is again a scalar tachyon $|0\rangle$ with mass $\alpha' m^2 = -4a$. The first excited state is $\alpha^i_{-1} \bar{\alpha}^j_{-1} |0\rangle$ with mass $\alpha' m^2 = 4(1 - a)$. We can decompose this state into irreducible representations of the transverse rotation group $SO(d-2)$, the little group for massless states in d dimensions, as follows:

$$\alpha^i_{-1} \bar{\alpha}^j_{-1} |0\rangle = \alpha^{[i}_{-1} \bar{\alpha}^{j]}_{-1} |0\rangle + [\alpha^{(i}_{-1} \bar{\alpha}^{j)}_{-1} - \tfrac{1}{d-2} \delta^{ij} \alpha^k_{-1} \bar{\alpha}^k_{-1}] |0\rangle + \tfrac{1}{d-2} \delta^{ij} \alpha^k_{-1} \bar{\alpha}^k_{-1} |0\rangle \tag{3.36}$$

Table 3.2: The three lowest mass levels of the oriented closed bosonic string

level	$\alpha'(\text{mass})^2$	states and their $SO(24)$ representation contents	little group	representation contents with respect to the little group
0	-4	$\lvert 0\rangle$ $\bullet$ (1)	$SO(25)$	$\bullet$ (1)
1	0	$\alpha^i_{-1}\,\bar\alpha^j_{-1}\lvert 0\rangle$ $\square \times \square$ $(24)\ \ (24)$	$SO(24)$	$\square\square$ (299) $\times$ $\begin{smallmatrix}\square\\\square\end{smallmatrix}$ (276) $\times$ $\bullet$ (1)
2	$+4$	$\alpha^i_{-2}\,\bar\alpha^j_{-2}\lvert 0\rangle$ $\alpha^i_{-1}\alpha^j_{-1}\bar\alpha^k_{-1}\bar\alpha^l_{-1}\lvert 0\rangle$ $\square \times \square$ $(\square\square + \bullet) \times (\square\square + \bullet)$ $(24)\ \ (24)$ $(299)+(1)\ \ (299)+(1)$ $\alpha^i_{-2}\,\bar\alpha^j_{-1}\bar\alpha^k_{-1}\lvert 0\rangle$ $\alpha^i_{-1}\alpha^j_{-1}\bar\alpha^k_{-2}\lvert 0\rangle$ $\square \times (\square\square + \bullet)$ $(\square\square + \bullet) \times \square$ $(24)\ \ (299)+(1)$ $(299)+(1)\ \ (24)$	$SO(25)$	$\square\square$ (324) $\times$ $\square\square$ (324) $=$ $\square\square\square\square$ (20150) $+$ $\boxplus$ (32175) $+$ (hook: $\square\square\square\square\square$ with one $\square$ below first) (52026) $+$ $\square\square$ (324) $+$ $\begin{smallmatrix}\square\\\square\end{smallmatrix}$ (300) $+$ $\bullet$ (1)

where indices in parentheses and brackets are symmetrized and anti-symmetrized, respectively. As for the open string we conclude that $a = 1$ and $d = 26$. Then these states describe a massless spin two particle, an antisymmetric tensor and a massless scalar. The spectrum for the first three levels is displayed in table 3.2. Again, the massive states combine into representations of the little group $SO(25)$. The relation between the maximal spin and the mass is now $j_{\max} = \frac{1}{2}\alpha' m^2 + 2$.

One can now distinguish between orientable and unorientable strings. The concept of orientability can be made precise by defining an unitary operator T which reverses the orientation of a string, i.e.

$$T^\dagger X^\mu(\sigma,\tau)T = X^\mu(\bar\sigma - \sigma,\tau). \tag{3.37}$$

(Recall that the parameter σ was defined to be in the range $0 \leq \sigma \leq \bar\sigma$.)

Expressed in terms of oscillators this means

$$T^\dagger \alpha_n^\mu T = (-1)^n \alpha_n^\mu \qquad \text{(open string)}$$
$$T^\dagger \alpha_n^\mu T = \qquad \bar\alpha_n^\mu \qquad \text{(closed string)}.$$

$$(3.38)$$

States of unoriented strings must be invariant under T; this means that the spectrum of an unoriented open string consists of states with even mode number only, and the spectrum of unoriented closed strings of states symmetric under the interchange of left- and right-moving oscillators. This means for the open string that the odd levels are absent and especially that the unoriented open string has no massless states. For the closed string only the symmetric and singlet piece of the massless states survive. For historical reason the closed oriented bosonic string theory is referred to as the extended Shapiro-Virasoro model, whereas the unoriented theory is called the restricted Shapiro-Virasoro model. From here on we will only consider oriented strings.

Let us now try to interpret the mass spectra. As mentioned in the introduction, any interacting string theory with local interactions has to contain closed strings. Looking at the closed string spectrum we see that a massless spin two particle will always be present. It is very suggestive to identify it with the graviton, i.e. the gauge particle of the ubiquitous gravitational interaction. If we do this we have to relate the string scale set by the slope parameter to the Planck scale, i.e. $\alpha' \sim G$ where $G = M_{\mathrm{P}}^{-2}$ is Newtons constant and M_{P} the Planck mass. It is of course one of the attractive and encouraging features of string theory that it necessarily contains gravity.[5] This however also means that the massive states, since their mass is now an integer multiple of the Planck mass, cannot be identified with known particles or hadronic resonances as was the original motivation for string

---

[5] The presence of a massless spin two particle is a priori not sufficient to have gravity. We will however show in the last chapter that at low energies it couples to matter and to itself like the graviton of general relativity.

theory. But this is all right since the higher mass and spin resonances of hadronic physics have by now found an adequate description by QCD and the prospect of having a consistent quantum theory including gravity has lead to a shift in the interpretation of string theory from the hadronic scale (100 MeV) to the Planck scale ($10^{19} GeV$). The other massless states of the closed string, the singlet and the antisymmetric tensor piece of eq.(3.36) can be interpreted as a dilaton and an antisymmetric tensor particle, both well know from e.g. Kaluza-Klein theory. What about the states of the open string? The massless vector could be interpreted as a gauge boson if we can associate a non-abelian charge in the adjoint representation of the gauge group to the open string. This can indeed be done if we attach to one end of the string the charge of the fundamental representation and to the other end the charge of its complex conjugate representation. This is the method of Chan and Paton. However, consistency of the interacting theory restricts the possible gauge groups to only one, namely $SO(32)$. We will not go into details of the Chan-Paton method and why $SO(32)$ is singled out. In Chapter 10 we will learn how to get non-abelian gauge symmetries in a theory of only closed strings.

At the end of this discussion of a possible contact of string theory with known physics we have to say a word of caution. The suggested interpretation of the massless particles can of course only hold if they have the interactions appropriate to gravitons, gauge bosons etc. This means that the theory has to be gauge invariant and especially generally coordinate invariant as a 26 dimensional theory. This is not a priori obvious and will be demonstrated in Chapter 15. Also, we still have to find a way to go from 26 to 4 dimensions. We will address this important question in Chapter 14.

3.4 Covariant path integral quantization

Path integral quantization has proven useful for theories with local symmetries, e.g. gauge theories. As such it is also applicable in string theory and an alternative to the non-covariant light-cone gauge quantization. The starting point is the Polyakov action, eq.(2.18). As discussed in Chapter 2, the induced metric $\Gamma_{\alpha\beta} = \partial_\alpha X^\mu \partial_\beta X_\mu$ and the intrinsic world-sheet metric $h_{\alpha\beta}$ are related only through the classical equations of motion, $T_{\alpha\beta} = 0$. Quantum mechanically this does not need to be so. In fact, just as we integrate in the Feynman path integral approach to quantum mechanics over all paths, not just the classical ones, we have to integrate over $h_{\alpha\beta}$ and the embeddings X^μ. One must however find a measure for the functional integrations which respects the symmetries of the classical theory, which are reparametrizations and Weyl rescaling. If this cannot be achieved, the quantum theory will have extra degrees of freedom. One then fixes the gauge using the Faddeev-Popov procedure. Indeed, the measures for the integrations over the metrics and the embeddings are both not conformally invariant. However, Polyakov has shown that the conformal anomalies cancel in 26 dimensions. So it is necessary to go to the critical dimension for the scale factor to decouple. We will not present Polyakov's analysis here but will derive the critical dimension by the requirement that the central term in the Virasoro algebra cancels when the contributions from the Faddeev-Popov ghosts are included. Cancellation of the central charge is equivalent to conformal invariance. We will assume in the following discussion that we are in the critical dimension in which the combined integration measure is conformally invariant.

Consider the vacuum to vacuum amplitude or partition function[6]

---

[6]Later we will calculate scattering amplitudes as correlation functions with this partition function.

$$Z = \int \mathcal{D}h(\sigma,\tau)\mathcal{D}X^\mu(\sigma,\tau)e^{iS_P[h,X]}. \tag{3.39}$$

The integration measures in eq.(3.39) are defined by means of the norms

$$\begin{aligned}
||\delta h|| &= \int \mathrm{d}^2\sigma\,\sqrt{h}h^{\alpha\beta}h^{\gamma\delta}\delta h_{\alpha\gamma}\delta h_{\beta\delta} \\
||\delta X|| &= \int \mathrm{d}^2\sigma\,\sqrt{h}\delta X^\mu\delta X_\mu
\end{aligned} \tag{3.40}$$

in the same way as for finite dimensional spaces the metric $\mathrm{d}s^2 = g_{ij}\mathrm{d}x^i\mathrm{d}x^j$ leads to the volume element $\sqrt{g}\,\mathrm{d}^n x$. We see that neither measure is invariant under rescaling of $h_{\alpha\beta}$. The measure in eq.(3.39) is not complete. Factors involving the volume of the symmetry group will be discussed below and in Chapter 6.

As described in Chapter 2, we can use conformal reparametrizations to go to a gauge in which the metric is equivalent to a fixed reference metric $\hat{h}_{\alpha\beta}$; i.e. our gauge condition is

$$h_{\alpha\beta} = e^{2\phi}\hat{h}_{\alpha\beta}. \tag{3.41}$$

We have also seen that under reparametrizations and Weyl rescaling the changes in the metric can be decomposed as

$$\delta h_{\alpha\beta} = (P\xi)_{\alpha\beta} + 2\tilde{\Lambda}h_{\alpha\beta} \tag{3.42}$$

where the operator P maps vectors into symmetric traceless tensors. The covariant derivatives in above expressions are with respect to the metric $h_{\alpha\beta} = e^{2\phi}\hat{h}_{\alpha\beta}$ which is also used to raise and lower indices. The integration measure can now be written as

$$\mathcal{D}h = \mathcal{D}(P\xi)\mathcal{D}\tilde{\Lambda} = \mathcal{D}\xi\mathcal{D}\Lambda\left|\frac{\partial(P\xi,\tilde{\Lambda})}{\partial(\xi,\Lambda)}\right| \tag{3.43}$$

The Jacobian is easy to evaluate formally:

$$\left|\frac{\partial(P\xi,\tilde{\Lambda})}{\partial(\xi,\Lambda)}\right| = \left|\det\begin{pmatrix} P & 0 \\ * & 1 \end{pmatrix}\right| = |\det P| = (\det PP^\dagger)^{1/2} \tag{3.44}$$

where $*$ is some operator which does not enter the determinant. The integral over reparametrizations simply gives the volume of the diffeomorphism group (more precisely, the volume of the component connected to the identity). This volume does depend on the Weyl degree of freedom as the measure $\mathcal{D}\xi$ does. We do however assume that all dependence on the conformal factor will eventually drop out in the critical dimension. We thus ignore it and drop the integral over Λ. We then have

$$Z = \int \mathcal{D}X^\mu (\det PP^\dagger)^{1/2} e^{iS_p[e^\phi \hat{h}_{\alpha\beta}, X^\mu]}. \tag{3.45}$$

The last step of the Faddeev-Popov procedure is to rewrite the determinant by introducing anticommuting ghost fields c^α and $b_{\alpha\beta}$, where $b_{\alpha\beta}$ is symmetric and traceless. We then get

$$(\det PP^\dagger)^{1/2} = \int \mathcal{D}c\mathcal{D}b \, \exp\left(-\frac{i}{2\pi}\int d\sigma^2 \sqrt{h}\, h^{\alpha\beta} b_{\beta\gamma} \nabla_\alpha c^\gamma \right) \tag{3.46}$$

where $h_{\alpha\beta} = e^{2\phi}\hat{h}_{\alpha\beta}$ is the gauge fixed metric. Both b and c are hermitian. Note that the c^α corresponds to infinitesimal reparametrizations and $b_{\alpha\beta}$ to variations perpendicular to the gauge slice. One often refers to $b_{\alpha\beta}$ as the antighost. If we insert eq.(3.46) in the partition function we get

$$Z = \int \mathcal{D}X^\mu(\sigma,\tau)\mathcal{D}c(\sigma,\tau)\mathcal{D}b(\sigma,\tau) e^{iS[X,\hat{h},b,c]} \tag{3.47}$$

where

$$S = -\frac{1}{8\pi}\int d^2\sigma \sqrt{\hat{h}}\, \hat{h}^{\alpha\beta}\{\partial_\alpha X^\mu \partial_\beta X_\mu + 4i b_{\beta\gamma}\hat{\nabla}_\alpha c^\gamma\}. \tag{3.48}$$

If we now choose $\hat{h}_{\alpha\beta} = \eta_{\alpha\beta}$, i.e. go to conformal gauge, we find

$$S = S[X] + S_{\text{ghost}}[b,c]$$

where

$$S[X] = \frac{1}{2\pi}\int d^2\sigma \, \partial_+ X^\mu \partial_- X_\mu$$

$$S_{\text{ghost}}[b,c] = \frac{i}{\pi}\int d^2\sigma \, (c^+\partial_- b_{++} + c^-\partial_+ b_{--}). \tag{3.49}$$

Since $b_{\alpha\beta}$ is traceless symmetric its only non-vanishing components are b_{++} and b_{--}.

We have to point out that our treatment above left unmentioned some subtle points. One, having to do with the conformal anomaly, was already touched upon and will be taken up shortly. Another issue has to do with reparametrizations which satisfy $(P\xi)_{\alpha\beta} = 0$, i.e. with the possible existence of conformal Killing fields. The equation of motion for the c ghosts is just the conformal Killing equation so that the c zero modes correspond to diffeomorphisms which can be absorbed by a Weyl rescaling. In the functional integration we are to integrate over each metric deviation only once. Since the ones corresponding to conformal Killing vectors are already taken care of by the integration over the conformal factor, we have to omit the zero modes from the integration over c.

Another problem has to do with the question whether all symmetric traceless metric deformations can be generated by reparametrizations. As we know from Chapter 2, this is not the case if $P^\dagger$ has zero modes; they correspond to zero modes of the b ghosts. If present, they have to be treated separately to get a non-vanishing result, since $\int d\theta\, 1 = 0$ for θ a Grassmann variable. We will come back to the issue of ghost zero modes in Chapter 6.

The energy momentum tensor of the ghosts fields can be derived from eq.(3.48) using eq.(2.19). Dropping hats from now on, we find

$$T_{\alpha\beta} = i\Big(b_{\alpha\gamma}\nabla_\beta c^\gamma + b_{\beta\gamma}\nabla_\alpha c^\gamma - c^\gamma \nabla_\gamma b_{\alpha\beta} - h_{\alpha\beta}b_{\gamma\delta}\nabla^\gamma c^\delta\Big). \qquad (3.50)$$

The last term vanishes on-shell, as does the trace $T^\alpha{}_\alpha$. In the derivation one also has to vary the metric dependence of the covariant derivatives and has to take into account the tracelessness of $b_{\alpha\beta}$. One can verify that $\nabla^\alpha T_{\alpha\beta} = 0$ if one uses the equations of motion. In light-cone gauge, the non-vanishing components are

$$T_{++} = i\left(2b_{++}\partial_+c^+ + (\partial_+b_{++})c^+\right)$$
$$T_{--} = i\left(2b_{--}\partial_-c^- + (\partial_-b_{--})c^-\right)$$

(3.51)

and energy-momentum conservation is

$$\partial_-T_{++} = \partial_+T_{--} = 0.$$

(3.52)

The equations of motion are

$$\partial_-b_{++} = \partial_+b_{--} = 0$$
$$\partial_+c^- = \partial_-c^+ = 0.$$

(3.53)

They have to be supplemented by periodicity (closed string) and boundary conditions (open string). The periodicity condition is simply $b(\sigma + 2\pi) = b(\sigma)$ and likewise for c. In the closed string case the equations of motion imply that b_{++} and c^+ are purely left-moving whereas b_{--} and c^- are purely right-moving. Left- and right-movers know about each other only through the constraints. For the open string, the boundary terms which arise in the derivation of the equations of motion vanish if we require $b_{++} = b_{--}$ and $c^+ = c^-$ at the ends of the string.

The ghost system, being anti-commuting, is quantized by the following canonical anti-commutation relations:

$$\{b_{++}(\sigma,\tau), c^+(\sigma',\tau)\} = 2\pi\delta(\sigma - \sigma')$$
$$\{b_{--}(\sigma,\tau), c^-(\sigma',\tau)\} = 2\pi\delta(\sigma - \sigma')$$

(3.54)

with all others vanishing.

We can now solve the equations of motion and express the canonical brackets in terms of the Fourier modes. We then define the Virasoro operators of the b, c system as the moments of the constraints $T_{++} = T_{--} = 0$. We will do this only for the closed string.

The solutions to the equations of motion, periodic in σ with period 2π are

$$c^+(\sigma,\tau) = \sum_{n=-\infty}^{+\infty} \bar{c}_n e^{-in(\tau+\sigma)}$$
$$c^-(\sigma,\tau) = \sum_{n=-\infty}^{+\infty} c_n e^{-in(\tau-\sigma)}$$
(3.55)

and

$$b_{++}(\sigma,\tau) = \sum_{n=-\infty}^{+\infty} \bar{b}_n e^{-in(\tau+\sigma)}$$
$$b_{--}(\sigma,\tau) = \sum_{n=-\infty}^{+\infty} b_n e^{-in(\tau-\sigma)}$$
(3.56)

and the canonical anti-commutators become

$$\{b_m, c_n\} = \delta_{m+n}$$
$$\{b_m, b_n\} = \{c_m, c_n\} = 0$$
(3.57)

and likewise for the barred oscillators. Left-moving modes anticommute with right-moving modes. The Virasoro operators are defined as in eq.(2.66) and we get

$$L_m = \sum_{n=-\infty}^{+\infty} (m-n) :b_{m+n}c_{-n}:$$
$$\bar{L}_m = \sum_{n=-\infty}^{+\infty} (m-n) :\bar{b}_{m+n}\bar{c}_{-n}:$$
(3.58)

Hermiticity of b and c entails

$$c_n = c^\dagger_{-n}, \qquad b_n = b^\dagger_{-n}$$
(3.59)

from which we get

$$L_m = L^\dagger_{-m}$$
(3.60)

with identical relations for the left-movers.

Again, L_0 and $\bar{L}_0$ are ambiguous due to operator ordering. We have defined them as normal ordered expressions.

We can now compute the commutator of the Virasoro operators and find the algebra they satisfy. In the same way as it was done in the previous chapter, we obtain

$$[L_m, L_n] = (m - n)L_{m+n} + A(m)\delta_{m+n} \tag{3.61}$$

where the anomaly is

$$A(m) = \frac{1}{6}(-13m^3 + m). \tag{3.62}$$

Let us now look at the combined matter-ghost system. We define the Virasoro operators as the sum of the Virasoro operators for the X^μ fields and the conformal ghost system

$$L_m = L_m^X + L_m^{gh} - a\delta_m \tag{3.63}$$

where the last term accounts for the normal ordering constant in L_0^X and L_0^{gh}. We then get

$$[L_m, L_n] = (m - n)L_{m+n} + A(m)\delta_{m+n} \tag{3.64}$$

with

$$A(m) = \frac{d}{12}m(m^2 - 1) + \frac{1}{6}(m - 13m^3) + 2am. \tag{3.65}$$

The first term is due to the X^μ fields ($\mu = 1, \ldots, d$), the second is due to the ghosts and the last arises from the shift in L_0. $A(m)$ vanishes if and only if $d = 26$ and $a = 1$. These are precisely the values we got from requiring Lorentz invariance of the theory quantized in light cone gauge. Here they arose from requiring that the total (ghost plus matter) anomaly of the Virasoro algebra vanishes. This in turn is the condition for conformal symmetry to be preserved in the transition from the classical theory to the quantum theory.

Recall that the anomaly in light-cone gauge was 24, the number of transverse dimensions. Even though light-cone quantization is completely consistent we cannot expect the anomaly to vanish since in making this gauge choice we have completely fixed the gauge and the light cone action is no longer invariant under the transformations generated by T_{++} and T_{--}.

Appendix A. The Virasoro algebra

In this appendix we want to derive the algebra satisfied by the Virasoro operators. The fact that in the quantum theory the L_n's are normal ordered expressions requires some care.

The following commutator will be useful:

$$[\alpha_m^i, L_n] = \frac{1}{2} \sum_{p=-\infty}^{+\infty} [\alpha_m^i, : \alpha_p^j \alpha_{n-p}^j :]. \tag{A.1}$$

Here we can drop the normal ordering symbol since α_m^i commutes with c-numbers. Using $[A, BC] = [A, B]C + B[A, C]$ we get

$$\begin{aligned}
[\alpha_m^i, L_n] &= \frac{1}{2} \sum_{p=-\infty}^{+\infty} \left\{ [\alpha_m^i, \alpha_p^j]\alpha_{n-p}^j + \alpha_p^j[\alpha_m^i, \alpha_{n-p}^j] \right\} \\
&= \frac{1}{2} \sum_{p=-\infty}^{+\infty} \left\{ \delta_{m+p}\alpha_{n-p}^j + \delta_{m+n-p}\alpha_p^j \right\} m\delta^{ij} = m\alpha_{m+n}^i.
\end{aligned} \tag{A.2}$$

Next we write

$$[L_m, L_n] = \frac{1}{2} \sum_{p=-\infty}^{+\infty} [: \alpha_p^i \alpha_{m-p}^i :, L_n], \tag{A.3}$$

break up the sum to eliminate normal ordering and use eq.(A.2):

$$\begin{aligned}
[L_m, L_n] &= \frac{1}{2} \sum_{p=-\infty}^{0} [\alpha_p^i \alpha_{m-p}^i, L_n] + \frac{1}{2} \sum_{p=1}^{+\infty} [\alpha_{m-p}^i \alpha_p^i, L_n] \\
&= \frac{1}{2} \sum_{p=-\infty}^{0} \left\{ (m - p)\alpha_p^i \alpha_{m+n-p}^i + p\alpha_{n+p}^i \alpha_{m-p}^i \right\} \\
&\quad + \frac{1}{2} \sum_{p=1}^{+\infty} \left\{ (m - p)\alpha_{m+n-p}^i \alpha_p^i + p\alpha_{m-p}^i \alpha_{n+p}^i \right\}.
\end{aligned} \tag{A.4}$$

Now change the summation variable in the second and fourth term to $q = p + n$ and get

$$[L_m, L_n] = \frac{1}{2}\Big\{ \sum_{p=-\infty}^{0} (m - p)\alpha_p^i \alpha_{m+n-p}^i + \sum_{q=-\infty}^{n} (q - n)\alpha_q^i \alpha_{m+n-q}^i$$
$$+ \sum_{p=1}^{+\infty} (m - p)\alpha_{m+n-p}^i \alpha_p^i + \sum_{q=n+1}^{+\infty} (q - n)\alpha_{m+n-q}^i \alpha_q^i \Big\}.$$
$$(A.5)$$

Let us now assume that $n > 0$ (the case $n \leq 0$ is treated similarly). We then get

$$[L_m, L_n] = \frac{1}{2}\Big\{ \sum_{q=-\infty}^{0} (m - n)\alpha_q^i \alpha_{m+n-q}^i + \sum_{q=1}^{n} (q - n)\alpha_q^i \alpha_{m+n-q}^i$$
$$+ \sum_{q=n+1}^{+\infty} (m - n)\alpha_{m+n-q}^i \alpha_q^i + \sum_{q=1}^{n} (m - q)\alpha_{m+n-q}^i \alpha_q^i \Big\}.$$
$$(A.6)$$

We now notice that except for the second term all terms are already normal ordered (the only critical case is when $m + n = 0$). The second term can be rewritten as

$$\sum_{q=1}^{n} (q - n)\alpha_q^i \alpha_{m+n-q}^i = \sum_{q=1}^{n} (q - n)\alpha_{m+n-q}^i \alpha_q^i + \sum_{q=1}^{n} (q - n)q d\, \delta_{m+n}$$

where $d = \delta_i^i$. Using this we get

$$[L_m, L_n] = \frac{1}{2} \sum_{q=-\infty}^{+\infty} (m - n) : \alpha_q^i \alpha_{m+n-q}^i : + \frac{1}{2} d \sum_{q=1}^{n} (q^2 - nq)\delta_{m+n}. \quad (A.7)$$

If we now use

$$\sum_{q=1}^{n} q^2 = \frac{1}{6} n(n + 1)(2n + 1) \qquad \text{and} \qquad \sum_{q=1}^{n} q = \frac{1}{2} n(n + 1) \qquad (A.8)$$

we finally get the Virasoro algebra

$$[L_m, L_n] = (m - n)L_{m+n} + \frac{d}{12} m(m^2 - 1)\delta_{m+n}. \qquad (A.9)$$

In Chapter 4 we will see how conformal field theory provides a simple tool to rederive this algebra.

REFERENCES

[1] C. Lovelace, *"Pomeron form factors and dual Regge cuts"*,
Phys. Lett. **34B** (1971) 500.

[2] P. Goddard, J. Goldstone, C. Rebbi and C.B. Thorn, *"Quantum dynamics of a massless relativistic string"*, *Nucl. Phys.* **B56** (1973) 109.

[3] R.C. Brower, *"Spectrum-generating algebra and no-ghost theorem for the dual model"*, *Phys. Rev.* **D6** (1972) 1655.

[4] P. Goddard and C.B. Thorn, *"Compatibility of the dual Pomeron with unitarity and the absence of ghosts in the dual resonance model"*, *Phys. Lett.* **40B** (1972) 235.

[5] A.M. Polyakov, *"Quantum geometry of bosonic strings"*,
Phys. Lett. **103B** (1981) 207.

Chapter 4

Introduction to Conformal Field Theory

This chapter is an introduction to conformal field theory. The basic reference on two-dimensional conformal field theory is the work by Belavin, Polyakov and Zamolodchikov [1]. Some review articles which were used for the preparation of this chapter are refs. [2, 3, 4].

4.1 General introduction

In distinction to higher dimensions the conformal group[1] in two dimensions is infinite dimensional: it is the group generated by analytic and anti-analytic vector fields. Associated with the infinity of generators is an infinity of conserved charges. That imposes important restrictions on the structure of two dimensional conformally invariant theories.

One class of physical systems which are described by conformal field theory are two dimensional statistical systems at the critical point (i.e. at $T = T_C$) where they are conformally invariant. The representation theory of the conformal group places constraints on the critical exponents. We will however not have much to say about these systems.

The second important application of conformal field theory is to string theory. We have already seen in Chapter 2 that the string action in conformal gauge is still invariant under conformal transformations with the

---

[1]The conformal group is the subgroup of those general coordinate transformations which preserve the angle between any two vectors. They leave the metric invariant up to a scale transformation.

associated infinite dimensional Virasoro algebra. The classical solutions of string theory are conformally invariant two-dimensional field theories. A particular choice corresponds to a particular vacuum which determines e.g. the number of space-time dimensions, the gauge group etc. There are of course constraints that a conformal field theory has to satisfy in order to be an acceptable string vacuum. One obvious condition that we have already encountered is the vanishing of the conformal anomaly. Others, coming from modular invariance, spin-statistics etc. will be discussed in subsequent chapters. We can then use methods of conformal field theory to determined the string spectrum and to compute string scattering amplitudes.

In order for conformal field theory to be applicable to string theory we have to continue the signature of the world-sheet metric from Minkowskian to Euclidean. Consider the world-sheet of a closed string – the cylinder – parametrized by $\sigma \in [0, 2\pi]$ and $\tau \in [-\infty, +\infty]$. We now make a Wick rotation, i.e. go to imaginary τ : $\tau \to -i\tau$ or

$$\sigma^{\pm} = \tau \pm \sigma \to -i(\tau \pm i\sigma). \tag{4.1}$$

We then define complex coordinates on the cylinder

$$\begin{aligned} z' &= \tau - i\sigma \\ \bar{z}' &= \tau + i\sigma. \end{aligned} \tag{4.2}$$

We can now map the cylinder to the complex plane via the conformal transformation

$$\begin{aligned} z &= e^{z'} = e^{\tau - i\sigma} \\ \bar{z} &= e^{\bar{z}'} = e^{\tau + i\sigma}. \end{aligned} \tag{4.3}$$

This is illustrated in figure 4.1.

The conformal map from the cylinder to the plane will not change the theory if it is conformally invariant. This will be the case for string theory in the critical dimension.

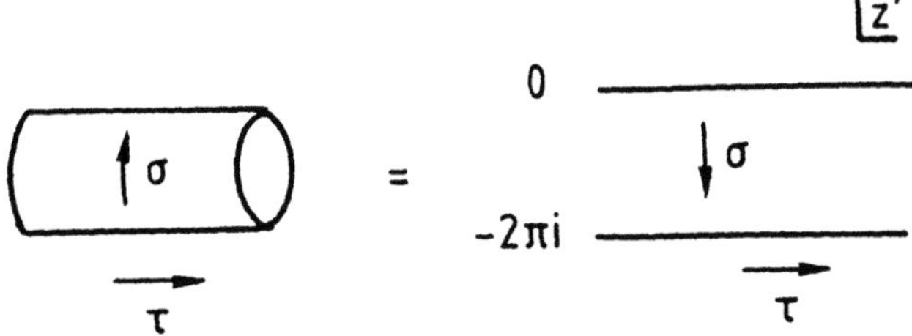

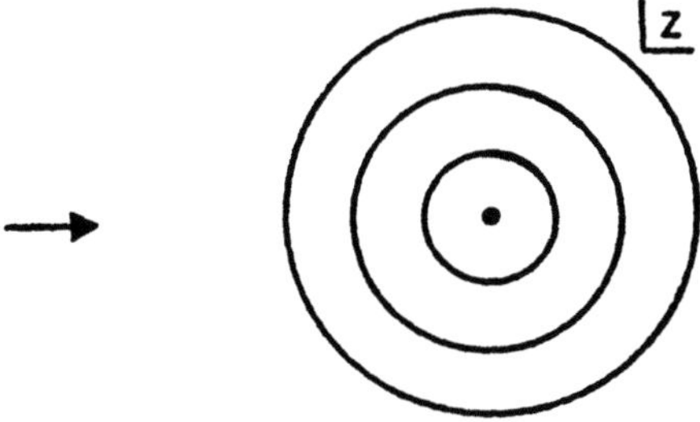

Fig.4.1. Conformal map from the cylinder to the complex plane

Having defined the theory on the complex plane we can now use all the powerful techniques of complex analysis. Lines of equal time τ are mapped into circles around the origin. Integrals over σ will be replaced by contour integrals around the origin. The infinite past becomes $z = 0$ and the infinite future $z = \infty$. σ translations become rotations: $\sigma \to \sigma + \theta \Rightarrow z \to e^{-i\theta} z$ and time translations become dilatations: $\tau \to \tau + a \Rightarrow z \to e^a z$. In the quantized theory the generator of dilatations will take the role of the Hamiltonian and time ordering will be replaced by radial ordering. Equal time commutators will be equal radius commutators. This is known as radial quantization. Products of fields are only defined if we put them in radial order. Radial order is defined in analogy with time order in ordinary field theory:

$$R\big(\phi_1(z)\phi_2(w)\big) = \begin{cases} \phi_1(z)\phi_2(w) & \text{for } |z| > |w| \\ \phi_2(w)\phi_1(z) & \text{for } |w| > |z|. \end{cases} \tag{4.4}$$

There will be a relative minus sign for the case of two anticommuting fields. Products of operators will always be assumed to be R-ordered and we drop the ordering symbol. The necessity to put operators in radial order will be illustrated below. The equal radius commutator is then defined by

$$[\phi_1(z), \phi_2(w)]_{|z|=|w|} = \lim_{\delta \to 0} \left\{ \big(\phi_1(z)\phi_2(w)\big)_{|z|=|w|+\delta} - \big(\phi_2(w)\phi_1(z)\big)_{|z|=|w|-\delta} \right\}$$
$$\tag{4.5}$$

After Wick rotation eq.(4.1) and the map eq.(4.3) right- and left-moving is replaced by holomorphic in z and $\bar{z}$ respectively. We will use both terminologies interchangeably. Also, we will call fields holomorphic in $\bar{z}$ antiholomorphic. Most expressions of Chapters 2 and 3 which were expressed in isothermal coordinates are unchanged if we replace σ^- by z and σ^+ by $\bar{z}$ and include the Jacobian factors from the map eq.(4.3). For instance, the nonvanishing components of the energy momentum tensor are now T_{zz} and $T_{\bar{z}\bar{z}}$.

The basic objects of a conformal field theory are the conformal fields (also called primary fields) $\phi(z, \bar{z})$. Consider a conformal transformation $z \to z' = f(z)$, $\bar{z} \to \bar{z}' = \bar{f}(\bar{z})$. Primary fields transform as tensors under conformal transformations:

$$\phi(z, \bar{z}) \to \phi'(z, \bar{z}) = \left(\frac{\partial z'}{\partial z}\right)^h \left(\frac{\partial \bar{z}'}{\partial \bar{z}}\right)^{\bar{h}} \phi(z'(z), \bar{z}'(\bar{z})). \tag{4.6}$$

In this chapter we will only consider single-valued fields, which requires $h = \bar{h} \in \mathbf{Z}$. Under infinitesimal transformations

$$z' = z + \xi(z) \quad , \qquad \bar{z}' = \bar{z} + \bar{\xi}(\bar{z}) \tag{4.7}$$

we get

$$\phi'(z, \bar{z}) = \phi(z, \bar{z}) + \delta_{\xi, \bar{\xi}} \phi(z, \bar{z}) \tag{4.8}$$

with

$$\delta_{\xi,\bar{\xi}}\phi(z,\bar{z}) = \left(h\partial\xi + \bar{h}\bar{\partial}\bar{\xi} + \xi\partial + \bar{\xi}\bar{\partial}\right)\phi(z,\bar{z}) \tag{4.9}$$

where we have introduced the notation $\partial = \frac{\partial}{\partial z}$ and $\bar{\partial} = \frac{\partial}{\partial \bar{z}}$. h and $\bar{h}$ are called the conformal weights of ϕ under analytic and anti-analytic transformations. Holomorphic and anti-holomorphic tensors have $\bar{h} = 0$ and $h = 0$ respectively. ($\bar{h}$ does not denote the complex conjugate of h.) Purely left- or right-moving fields are called chiral. Under a rescaling (dilatation) $z \rightarrow \lambda z$, λ real we have $\phi \rightarrow \lambda^{h+\bar{h}}\phi$ and $h + \bar{h}$ is called the scaling dimension. The generator of dilatations plays the role of the Hamiltonian and the scaling dimension is related to the energy. Under rotations $z \rightarrow e^{-i\theta}z$ we get $\phi \rightarrow e^{-i(h-\bar{h})\theta}\phi$; hence $h - \bar{h}$ is referred to as the conformal spin.

Consider the map eq.(4.3) from the cylinder to the complex plane. Applying eq.(4.6), the fields on the cylinder and plane are related as follows $(z' = \ln z)$:

$$\phi(z)_{\text{plane}} = \left(\frac{1}{z}\right)^h \phi(z'(z))_{\text{cylinder}}. \tag{4.10}$$

If $\phi(z')_{\text{cylinder}}$ has a mode expansion

$$\phi(z')_{\text{cylinder}} = \sum_{n\in\mathbf{Z}} \phi_n e^{-nz'} = \sum_n \phi_n z^{-n}; \tag{4.11}$$

then the mode expansion on the complex plane is

$$\phi_{\text{plane}}(z) = \sum_{n\in\mathbf{Z}} z^{-n-h}\phi_n \tag{4.12}$$

with the same coefficients ϕ_n. From now on, unless stated otherwise, all fields will be on the complex plane. The inverse of eq.(4.12) is

$$\phi_n = \oint_{C_0} \frac{dz}{2\pi i}\, \phi(z)z^{n+h-1} \tag{4.13}$$

where the integration is counterclockwise around the origin. The value of the integral is independent of the contour around the origin. For fields single valued on the complex plane the mode numbers n have to be such that $n+h \in \mathbf{Z}$. They will however always be integer spaced. Note that single valuedness on the plane does not mean single valuedness on the cylinder due to the Jacobian factor of the map eq.(4.3).

We already know from Chapter 2 that in a conformally invariant theory the energy momentum tensor is traceless, i.e. $T_\alpha^\alpha = 0$. Expressed in conformal coordinates it reads:

$$T_{z\bar{z}} = 0 \, . \tag{4.14}$$

This, together with energy-momentum conservation

$$\partial_{\bar{z}} T_{zz} + \partial_z T_{\bar{z}z} = 0 \quad , \qquad \partial_z T_{\bar{z}\bar{z}} + \partial_{\bar{z}} T_{z\bar{z}} = 0 \tag{4.15}$$

shows that in a conformally invariant theory we have

$$\partial_{\bar{z}} T_{zz} = 0, \qquad \partial_z T_{\bar{z}\bar{z}} = 0. \tag{4.16}$$

The two non-vanishing components of the energy-momentum tensor of a conformally invariant theory are analytic and anti-analytic functions respectively. We will use the notation $T(z) = T_{zz}(z)$ and $\bar{T}(\bar{z}) = T_{\bar{z}\bar{z}}(\bar{z})$. From the conservation law eq.(4.16) we immediately find that if $T(z)$ is conserved so is $\xi(z)T(z)$ if ξ depends only analytically on its argument. This infinity of conserved currents is equivalent to our statement at the beginning of this chapter that the conformal group in two dimensions is infinite dimensional. With each current we associate a conserved charge

$$T_\xi = \oint_{C_0} \frac{dz}{2\pi i} \, \xi(z)T(z) \tag{4.17}$$

which generates infinitesimal conformal transformations

$$z \to z' = z + \xi(z) \tag{4.18}$$

with similar expression for the anti-analytic component $\bar{T}$. From now on we will restrict our attention to chiral fields, say right-moving ones.

The transformation eq.(4.9) is implemented by the commutator of $\phi(z)$ and $T_\xi(w)$

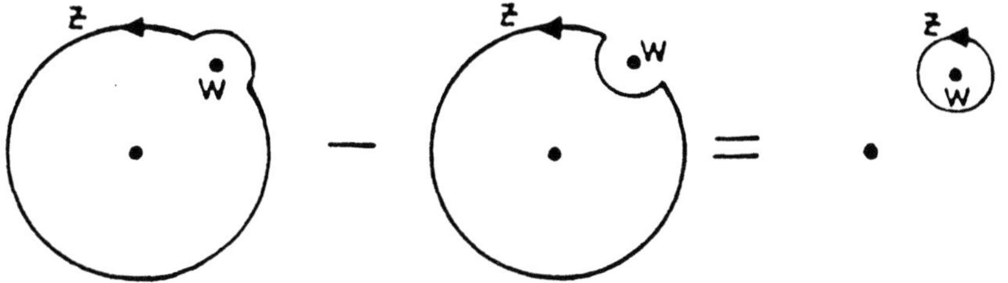

Fig.4.2. Integration contours in eq.(4.20)

$$\delta_\xi \phi(w) = [T_\xi, \phi(w)]. \tag{4.19}$$

Using the prescription of radial ordering this gives:

$$\delta_\xi \phi(w) = \oint_{\substack{C_0 \\ |z|>|w|}} \frac{dz}{2\pi i}\, \xi(z)T(z)\phi(w) - \oint_{\substack{C_0 \\ |z|<|w|}} \frac{dz}{2\pi i}\, \xi(z)T(z)\phi(w)$$

$$= \oint_{C_w} \frac{dz}{2\pi i}\, \xi(z)T(z)\phi(w). \tag{4.20}$$

The contours are shown in figure 4.2.

Recall that all operator products are assumed to be radially ordered. Comparing this with eq.(4.9) for $\bar{h} = 0$ and $\bar{\partial}\phi = 0$ we find with the help of the Cauchy-Riemann formula

$$\oint_{C_w} \frac{dz}{2\pi i} \frac{f(z)}{(z-w)^n} = \frac{1}{(n-1)!} f^{(n-1)}(w), \tag{4.21}$$

that any conformal field must have the following (R-ordered) operator product with $T(z)$:

$$T(z)\phi(w) = \frac{h\phi(w)}{(z-w)^2} + \frac{\partial\phi(w)}{(z-w)} \quad + \quad \text{finite terms.} \tag{4.22}$$

Therefore, instead of eq.(4.6), the operator product with the energy-momentum tensor can serve as the definition of a conformal field of weight h.

Eq.(4.22) is our first example of an operator product expansion of two fields. The basic idea is that if $\{O_i\}$ is a complete set of local operators with definite scaling dimensions, then the product of two operators can be expanded as[2]

$$O_i(z)O_j(w) = \sum_k C_{ijk}\big((z-w)\big)O_k(w).\tag{4.23}$$

Invariance under rescaling specifies the structure functions up to numerical constants:

$$C_{ijk}\big((z-w)\big) = (z-w)^{h_k-h_i-h_j}C_{ijk}\tag{4.24}$$

where h_i are the scaling dimensions of the fields which are not necessarily primary. Operator products should always be thought as inserted into correlation functions (cf. below). The radius of convergence of the operator product is restricted by the positions of the other operators in the correlation function. Completeness of the set of operators $\{O_i\}$ means that any state can be generated by their linear action.

Let us make one remark about the evaluation of commutators. For the contour deformation of figure 4.2 it is crucial that the fields in the integrand commute. Otherwise we would get the anticommutator. Also, we see that (anti-)commutators depend only on the singularities of the operator product expansion.

We can now examine the conformal transformation properties of the energy momentum tensor. Using the commutation properties of infinitesimal conformal transformations

$$[\delta_{\xi_1}, \delta_{\xi_2}] = \delta_{(\xi_1\partial\xi_2 - \xi_2\partial\xi_1)}\tag{4.25}$$

we find

$$T(z)T(w) = \frac{c/2}{(z-w)^4} + \frac{2T(w)}{(z-w)^2} + \frac{\partial T(w)}{(z-w)} \quad + \quad \text{finite terms.}\tag{4.26}$$

---

[2] Here we consider chiral fields only.

The first term is allowed by eq.(4.25) and is consistent with Bose symmetry and scale invariance. We can rewrite eq.(4.26) in an equivalent way as

$$\delta_\xi T(z) = \frac{c}{12}\partial^3\xi(z) + 2\partial\xi(z)T(z) + \xi(z)\partial T(z). \qquad (4.27)$$

We see that $T(z)$ transforms as a tensor of weight two under those transformations for which $\partial^3\xi(z) = 0$ but fails to do so for general conformal transformations if $c \neq 0$. Classically c is zero, and $c \neq 0$ represents a conformal anomaly, a purely quantum mechanical effect. Note that the scaling dimension of $T(z)$ does not get modified by quantum effects. This can be understood from the fact that since $T(z)$ is a symmetric traceless tensor it has spin two and for $\bar{h} = 0$ the spin is the same as the scaling dimension. One should compare eq.(4.26) with eq.(3.15). The $T(z)T(w)$ operator product is of course equivalent to the Virasoro algebra and c is its central charge. We expand $T(z)$ in modes

$$T(z) = \sum_n z^{-n-2}L_n \qquad (4.28)$$

which in turn gives

$$L_n = \oint \frac{dz}{2\pi i} z^{n+1}T(z) \qquad (4.29)$$

where the L_n's are the Virasoro generators. They satisfy the hermiticity relation

$$L_n^\dagger = L_{-n} \qquad (4.30)$$

which follows from the reality of the energy-momentum tensor in Minkowski space (c.f. Chapter 2). In general, the hermitian conjugate of a field of weight h is defined by

$$\left[\phi(z)\right]^\dagger = \phi^\dagger\left(\frac{1}{\bar{z}}\right)\frac{1}{\bar{z}^{2h}}. \qquad (4.31)$$

For the modes this means

$$\left(\phi^\dagger\right)_{-n} = \left(\phi_n\right)^\dagger. \qquad (4.32)$$

A hermitian field satisfies $\phi^\dagger = \phi$. A word of explanation is in order. Consider the continuation back to the Minkowski space cylinder. The missing factors of i in Euclidean space-time evolution, $\phi(\sigma,\tau) = e^{H\tau}\phi(\sigma,0)e^{-H\tau}$, must be compensated in the definition of adjoint by an explicit time reversal $\tau \to -\tau$. This corresponds on the complex plane to $z \to 1/\bar{z}$.

The Virasoro algebra is then easily obtained, using eq.(4.26), as the difference of a double contour integral:

$$[L_n, L_m] = \oint_{C_0} \frac{dw}{2\pi i} \oint_{C_w} \frac{dz}{2\pi i} z^{n+1} w^{m+1} \left[\frac{c/2}{(z-w)^4} + \frac{2T(w)}{(z-w)^2} + \frac{\partial T(w)}{z-w} \right]$$

$$= \frac{c}{12} n(n-1)(n+1)\delta_{m+n} + (n-m)L_{n+m} .$$

$$(4.33)$$

$\bar{T}(\bar{z})$ or, equivalently, the $\bar{L}_n$ satisfy identical algebras as $T(z)$ and the L_n. The two algebras commute, i.e. $[L_n, \bar{L}_m] = 0$. The central charges of the left- and right-moving algebras are the same since $T + \bar{T}$ is real. Then the $(T+\bar{T})(T+\bar{T})$ operator product is real only if $c = \bar{c}$.

The L_n's act as the generators of all possible conformal transformations. A primary field is defined via the L_n's as

$$[L_n, \phi(z)] = z^n [z\partial + (n+1)h]\phi(z) \tag{4.34}$$

or, in terms of the modes of ϕ:

$$[L_n, \phi_m] = [n(h-1) - m]\phi_{n+m}. \tag{4.35}$$

Comparing eq.(4.34) with eq.(4.9) we see that the Virasoro generator L_n is associated with the infinitesimal transformation $\epsilon(z) = z^{n+1}$. In particular, L_0, L_1, L_{-1}, generate infinitesimal transformations $\delta z = \alpha + \beta z + \gamma z^2$; they are the generators of $SL(2, \mathbf{R})$, the maximal closed subalgebra of the conformal group. The finite transformations are

$$z \to z' = \frac{az+b}{cz+d} \tag{4.36a}$$

with

$$\begin{pmatrix} a & b \\ c & d \end{pmatrix} \in SL(2,\mathbf{R}), \qquad \text{i.e.} \qquad \begin{array}{l} a,b,c,d \in \mathbf{R}, \\ ad - bc = 1. \end{array} \tag{4.36b}$$

Indeed, if we expand eq.(4.36) around $a = d = 1, b = c = 0$ we get $\delta z = \delta b + (\delta a - \delta c - \delta d)z - \delta c z^2$. Adding $\bar{L}_0, \bar{L}_{+1}, \bar{L}_{-1}$, we generate $SL(2,\mathbf{C})$. $SL(2,\mathbf{C})$ transformations are the only globally defined invertible conformal mappings of the Riemann sphere $(\mathbf{C} \cup \infty)$ one-to-one onto itself. This is in fact easy to see. The transformations δ_ϵ are generated by the vector fields $\epsilon(z)\partial_z$; only those which satisfy $\partial^3 \epsilon(z) = 0$ are defined at both $z = 0$ and $z = \infty$ (for $z = \infty$ we use the map $w = 1/z$). The transformations eq.(4.36) are called fractional linear or Moebius transformations. If we define the generators $X_0 = L_0$, $X_1 = \frac{1}{2}(L_1 + L_{-1})$ and $X_2 = \frac{i}{2}(L_1 - L_{-1})$ we easily verify that they satisfy the three dimensional Lorentz algebra $[X_i, X_j] = i\epsilon_{ijk}X^k$ $(X^0 = -X_0)$.

Eq.(4.27) can be integrated to give the behavior of $T(z)$ under finite transformations $z \to f(z)$. One finds

$$T(z) \to T'(z) = (\partial f(z))^2 T(f(z)) + \frac{c}{12} D(f)_z \tag{4.37}$$

where

$$D(f)_z = \frac{\partial f(z)\partial^3 f(z) - \frac{3}{2}(\partial^2 f(z))^2}{(\partial f)^2} \tag{4.38}$$

is the Schwarzian derivative. It has the following properties

$$D(f)_z = 0 \iff f = \frac{az + b}{cz + d}$$

$$D\left(\frac{af + b}{cf + d}\right)_z = D(f)_z \tag{4.39}$$

$$D(f)_z = (\partial_z g)^2 D(f)_g + D(g)_z.$$

The Schwarzian derivative is in fact the only weight two object with these properties. For the map from the cylinder to the plane, eq.(4.37) gives

$$T_{\text{cyl}}(z') = z^2 T_{\text{plane}}(z) - \frac{c}{24}. \tag{4.40}$$

In particular for L_0 this gives

$$(L_0)_{\text{cyl}} = (L_0)_{\text{plane}} - \frac{c}{24}. \tag{4.41}$$

Now, let us consider the Hilbert space and also some of the representation theory of a conformal field theory. Denote the in-vacuum by $|0\rangle$. Regularity of the energy-momentum tensor at $z = 0$ ($\tau = -\infty$) requires that

$$L_n|0\rangle = 0 \quad \text{for } n \geq -1. \tag{4.42}$$

The L_n's with $n \geq -1$ generate the conformal transformations which are regular at the origin. To get the conditions on the out-vacuum $\langle 0|$ following from regularity at $z = \infty$ ($\tau = +\infty$) we map the point at infinity to the origin via $w = -1/z$. The mode expansion of T is then $T'(w) = \sum w^{n-2} L_n$ and we find

$$\langle 0|L_n = 0 \quad \text{for} \quad n \leq 1. \tag{4.43}$$

Here T' is the energy-momentum tensor expressed in the coordinates where $w \to 0$ corresponds to $z \to \infty$. Eqs.(4.42) and (4.43) are hermitian conjugates of each other. The generators of $SL(2, \mathbf{C})$ annihilate both the in- and the out-vacuum. We refer to this vacuum as the $SL(2, \mathbf{C})$ invariant vacuum. The requirement of regularity at $z = 0$ and $z = \infty$ leads for a primary field of weight h with mode expansion as in eq.(4.12) to

$$\begin{aligned}
\phi_n|0\rangle &= 0 \qquad \text{for } n \geq 1 - h, \\
\langle 0|\phi_n &= 0 \qquad \text{for } n \leq h - 1.
\end{aligned} \tag{4.44}$$

Note that for $h < 0$ there are modes of ϕ which annihilate neither the in- nor the out-vacuum. (The case $h = 0$ is trivial since in unitary theories the only conformal field with $h = 0$ is the identity.) We will however see below that unitarity restricts the conformal weights to $h \geq 0$. This is however avoided by the ghost system. The c-ghost has $h = -1$ and the three zero modes c_{-1}, c_0 and c_{+1} do not annihilate the vacuum.

Let us now construct the asymptotic in- and out-states of the conformal field theory. Since the time $\tau \to -\infty$ on the cylinder corresponds to the origin on the z-plane, it is natural to define in-states as

$$|\phi_{j\,\text{in}}\rangle = \lim_{z\to 0} \phi_j(z)|0\rangle = \phi_j(0)|0\rangle = \phi_{-h_j}|0\rangle \qquad (4.45)$$

where

$$\phi_{-h_j} = \oint_{C_0} \frac{dz}{2\pi i}\, \frac{1}{z}\phi(z). \qquad (4.46)$$

To define the out-states $\langle\phi_{j\,\text{out}}|$ we have to construct the analogous objects for $z \to \infty$. We want, of course, that $\langle\phi_{j\,\text{out}}| = |\phi_{j\,\text{in}}\rangle^\dagger$. Using eqs.(4.31) and (4.32) leads to the following definition for $\langle\phi_{j\,\text{out}}|$:

$$\langle\phi_{j\,\text{out}}| = \lim_{z\to\infty} \langle 0|\phi_j^\dagger(z)z^{2h_j} = \langle 0|(\phi^\dagger)_{h_j}\,. \qquad (4.47)$$

Since $\phi(z)$ is primary, one derives from eq.(4.34) (we have dropped the subscript 'in')

$$\begin{aligned}
L_0|\phi_j\rangle &= h_j|\phi_j\rangle\,, \\
L_n|\phi_j\rangle &= 0, \quad n > 0\,.
\end{aligned} \qquad (4.48)$$

Also,

$$L_0\big(L_{-n}|\phi_j\rangle\big) = (n + h_j)\big(L_{-n}|\phi_j\rangle\big) \qquad \text{for } n \geq 0, \qquad (4.49)$$

i.e. the L_{-n} $(n \geq 0)$ raise the eigenvalue of L_0. States satisfying eq.(4.48) should be called 'lowest weight states'; however, in analogy to the terminology used in the representation theory of Lie algebras they are called highest weight states of the Virasoro algebra. We have thus established a correspondence between conformal fields and highest weight states. The vacuum $|0\rangle$ is itself a highest weight state; in a unitary theory ($h_j \geq 0$, cf. below) it has the lowest eigenvalue of the 'Hamiltonian' L_0. Highest weight states with different L_0 eigenvalue are orthogonal.

The complete Hilbert space is obtained by acting with the raising operators L_{-n} $(n > 0)$ on highest weight states. The new states obtained in this way are called descendant states. Each highest weight state $|\phi_j\rangle$ determines a representation of the Virasoro algebra labelled by h_j. This representation is called a Verma module, consisting of all fields of the form

$$|\phi_j^{k_1\dots k_m}\rangle = L_{-k_1}\cdots L_{-k_m}|\phi_j\rangle, \qquad k_i > 0 \tag{4.50}$$

with L_0 eigenvalue $h_j + \sum_i k_i$. States in different Verma modules are easily seen to be orthogonal to each other. Descendant states are created from the vacuum by descendant fields which are not primary but rather secondary operators. They are contained in the operator product of the primary field with the energy-momentum tensor:

$$T(z)\phi_i(w) = \sum_{k=0}^{\infty}(z-w)^{-2+k}\phi_i^{(-k)}(w) \tag{4.51}$$

i.e.

$$\phi_i^{(-k)}(w) = \oint \frac{\mathrm{d}z}{2\pi i}(z-w)^{1-k}T(z)\phi_i(w) \equiv \hat{L}_{-k}\phi_i(w). \tag{4.52}$$

Especially

$$\phi_i^{(0)}(z) = \hat{L}_0\phi_i(z) = h_i\phi_i(z)$$
$$\phi_i^{(-1)}(z) = \hat{L}_{-1}\phi_i(z) = \partial\phi_i(z). \tag{4.53}$$

The other descendants for $k \geq 2$ appear in the regular terms of the operator product eq.(4.51). The fields $\phi_i^{(-k)}$ do not exhaust the descendants of the primary $\phi_i(z)$. The operator product $T(z)\phi_i^{(-k_1)}(w)$ contains the fields $\phi_i^{(-k_1,-k_2)}(w)$ and so on. For the descendant field that creates the state eq.(4.50) we get

$$\phi_i^{\{k\}}(z) = \hat{L}_{-k_1}\cdots\hat{L}_{-k_m}\phi_i(z).$$

These fields constitute the conformal family $[\phi_j]$. We have already encountered one example of a secondary field, namely the energy-momentum tensor. It is in the conformal family of the identity operator $[I]$ which is present in any conformal field theory. Indeed

$$I^{(-2)}(z) = \hat{L}_{-2}I(z) = \oint_{C_z}\frac{\mathrm{d}w}{2\pi i}\frac{T(w)}{w-z}I(z) = T(z). \tag{4.54}$$

Note that the states eq.(4.50) are not all independent due to the relation between the L_n's given by the Virasoro algebra. A basis is given by those states for which $k_1 \geq \ldots \geq k_m > 0$[3]. A state is defined to be in the n'th level of the Virasoro algebra if its L_0 eigenvalue is $h_j + n$. Thus the n'th level is spanned by the vectors of eq.(4.50) with $\sum k_i = n$. There are $P(n)$ such states, where $P(n)$ is the number of ways of writing n as a sum of positive integers. One easily convinces oneself that the generating function for $P(n)$ is given by

$$\sum_{n=0}^{\infty} P(n)q^n = (1 + q + q^2 + \ldots)(1 + q^2 + q^4 + \ldots)(1 + q^3 + \ldots) \cdots$$

$$= \prod_{n=1}^{\infty} \frac{1}{(1 - q^n)}$$

$$(4.55)$$

where we have defined $P(0) = 1$.

The partition function, also called character of a conformal family, contains the information of the number of states at each "energy" level; it is defined as

$$Ch_j(\tau) = \mathrm{Tr}\ q^{L_0 - \frac{c}{24}} = \sum_{n=0}^{\infty} P(n)q^{h_j + n - \frac{c}{24}}, \qquad q = e^{2\pi i \tau}$$

$$= q^{h_j - \frac{c}{24}} \prod_{n=1}^{\infty} (1 - q^n)^{-1}$$

$$(4.56)$$

where the trace is over all members of the conformal family $[\phi_j]$. In Chapter 6 the identification of τ with the complex modulus of the world-sheet torus, relevant for one-loop string calculations, will be explained. In a statistical mechanics context it is related to the inverse temperature.

Let us now take up our discussion of operator product expansions, eq.(4.23). In general, the product of two operators will contain primary and descendant fields. In particular, if ϕ_i and ϕ_j are primary, we get

---

[3]This is not quite true if there are null states, a complication which we will not discuss.

$$\phi_i(z)\phi_j(w) = \sum_k \sum_{\{l\}} C_{ijk}^{\{l\}} \, (z-w)^{h_k + \sum_n l_n - h_i - h_j} \phi_k^{\{l\}}(w). \qquad (4.57)$$

It can be shown that $C_{ijk}^{\{l\}} = C_{ijk}\beta_{ijk}^{\{l\}}$ where the $\beta_{ijk}^{\{l\}}$ are uniquely determined by conformal invariance in terms of the dimensions h_i, h_j and h_k. Then the spectrum of primary fields, their operator product coefficients and the central charge of the Virasoro algebra completely specify a conformal field theory. These parameters cannot be determined from conformal symmetry. One needs extra dynamical principles such as associativity of the operator algebra. Not any set of parameters $\{c, h_i, C_{ijk}\}$ defines a conformal field theory. Their classification is still an open problem.

The information which conformal families are contained in the operator product of two primary fields is encoded in the fusion rules

$$\phi_i \times \phi_j = \sum_k N_{ij}{}^k \phi_k, \qquad N_{ij}{}^k \in \mathbf{N}_0. \qquad (4.58)$$

$N_{ij}{}^k > 1$ means that there is more than one way the primary field ϕ_k is contained in the product of ϕ_i and ϕ_j. This is similar to the situation in the theory of finite dimensional groups where a representation can appear more than once in the product of two representations.

Let us briefly investigate the constraints of unitarity for representations of the Virasoro algebra. Unitarity means that the inner product in the Hilbert space is positive definite. The inner product of any two states can be computed from eqs.(4.33) and (4.48)

$$\langle \phi_j | L_n L_{-n} | \phi_j \rangle = [2nh_j + \frac{c}{12}(n^3 - n)]\langle \phi_j | \phi_j \rangle. \qquad (4.59)$$

Taking n sufficiently large implies that $c > 0$, while for $n = 1$ we find that $h_j \geq 0$; i.e. the vacuum is the state of lowest energy. A more detailed analysis shows that unitarity places no further constraints if $c \geq 1$. Then one has in general an infinite number of primary fields and c and h can take continuous values. On the other hand, if $c < 1$, both the value for h and c are quantized. c is given by [5]

$$c = 1 - \frac{6}{m(m+1)} \quad m = 2, 3, 4, \ldots \tag{4.60}$$

and h is limited to the values:

$$h_{p,q} = \frac{[(m+1)p - mq]^2 - 1}{4m(m+1)} \tag{4.61}$$

$$p = 1, 2, \ldots, m-1, \quad q = 1, 2, \ldots p.$$

The conformal theories with c and h given by eqs.(4.60) and (4.61) are called minimal models. c and h are rational and there is only a finite number of primary fields. Conformal field theories with these properties are also called rational. For the first few values of m the minimal models have been identified with statistical systems. The first non-trivial one is obtained for $m = 3$ and describes the continuum limit of the two-dimensional Ising model at the critical point.

Let us now make some general remarks about correlation functions. Recall that they are vacuum expectation values of R-ordered products. Their general structure is severely restricted in a conformal field theory as we will demonstrate. Since the vacuum is invariant under $SL(2, \mathbf{C})$ (however not under the full conformal group) correlation functions have to satisfy

$$\langle \phi_1'(z_1) \cdots \phi_n'(z_n) \rangle = \langle \phi_1(z_1) \cdots \phi_n(z_n) \rangle \tag{4.62}$$

where $\phi'(z)$ is the $SL(2, \mathbf{C})$ transformed of $\phi(z)$. For primary fields we know how they transform under conformal transformations (cf. eq.(4.6)). The following discussion is however valid for a less restricted class of fields, the so-called quasi-primary fields, which transform as tensors under $SL(2, \mathbf{C})$ but not necessarily under the full conformal group. Many secondary fields, as e.g. the energy-momentum tensor, are of this type. The generators of $SL(2, \mathbf{C})$ act as

$$
\begin{array}{llll}
L_{-1}: & \text{translations} & z' = z + b & \phi'_i(z) = \phi_i(z+b) \\[2ex]
L_0: & \text{dilatations and rotations} & z' = az & \phi'_i(z) = a^{h_i}\phi_i(az) \\[2ex]
L_{+1}: & \text{special conformal transformations} & z' = \frac{z}{cz+1} & \phi'_i(z) = \left(\frac{1}{cz+1}\right)^{2h_i}\phi_i\left(\frac{z}{cz+1}\right)
\end{array}
\tag{4.63}
$$

Invariance under translations tells us that a general n-point correlation can only depend on $z_{ij} = z_i - z_j$. This means in particular that the one-point function must be a constant. Then, from dilatation invariance it follows that they vanish; i.e.

$$
\langle \phi_i(z) \rangle = 0 \tag{4.64}
$$

for all quasi-primary fields. For two-point functions invariance under dilatations and rotations means that they can only be $\langle \phi_i(z)\phi_j(w)\rangle = C_{ij}(z-w)^{-(h_i+h_j)}$ where C_{ij} is a constant which can not be determined from $SL(2,\mathbf{C})$ invariance. Invariance under special conformal transformations restricts this further to $h_i = h_j$ so that finally

$$
\langle \phi_i(z)\phi_j(w)\rangle = \begin{cases} \dfrac{C_{ij}}{(z-w)^{2h_i}} & \text{for } h_i = h_j, \\[2ex] 0 & \text{for } h_i \neq h_j. \end{cases} \tag{4.65}
$$

We can normalize the primary fields of a given theory such the only non-vanishing two-point functions are

$$
\langle \phi_i(z)\phi_j(w)\rangle = \frac{\delta_{ij}}{(z-w)^{2h_i}}. \tag{4.66}
$$

Three-point functions are constrained by dilatations and rotations to be of the form $f(z_{12}z_{13}z_{23})$ where f is a homogeneous functions of degree $h_1+h_2+h_3$. This function is completely determined, up to a constant, by invariance under special conformal transformations. We easily find

$$
\langle \phi_i(z_1)\phi_j(z_2)\phi_k(z_3)\rangle = \frac{C_{ijk}}{z_{12}^{h_i+h_j-h_k}\, z_{13}^{h_i+h_k-h_j}\, z_{23}^{h_j+h_k-h_i}}. \tag{4.67}
$$

The C_{ijk} are just the operator product coefficients (cf. eq.(4.24)). Using this equation we can also evaluate the expectation value of a field $\phi_j(z)$ between an asymptotic in-state $|\phi_{k\,\text{in}}\rangle$ and an asymptotic out-state $\langle\phi_{i\,\text{out}}|$. This simply amounts to taking the limit $z_1 \to \infty$ and $z_3 \to 0$ in eq.(4.67):

$$\langle\phi_{i\,\text{out}}|\phi_j(z)|\phi_{k\,\text{in}}\rangle = \frac{C_{ijk}}{z^{h_j+h_k-h_i}}\,. \tag{4.68}$$

For correlation functions of four and more quasi-primary fields the situation becomes more complicated. They are no longer determined up to a constant. The reason is that out of four points z_i we can form so-called anharmonic quotients or cross ratios

$$X_{ij}^{kl} = \frac{(z_i - z_j)(z_k - z_l)}{(z_i - z_l)(z_k - z_j)} \tag{4.69}$$

which are invariant under $SL(2,\mathbf{C})$ transformations of the z_i. It is easy to see that for the four-point function there is only one independent cross ratio. For n-point functions there are $n-3$. Then, repeating above reasoning, we find the following general structure for an arbitrary correlation function of n quasi-primary fields:

$$\langle\phi_1(z_1)\cdots\phi_n(z_n)\rangle = \prod_{i<j} z_{ij}^{-\gamma_{ij}} f(X_{ij}^{kl}) \tag{4.70}$$

where the $\gamma_{ij} = \gamma_{ji}$ are any solution of the set of $\frac{1}{2}n(n-1)$ equations

$$\sum_{j\neq i} \gamma_{ij} = 2h_i \tag{4.71}$$

and f is an undetermined function of the $(n-3)$ independent cross ratios; it cannot be determined from $SL(2,\mathbf{C})$ invariance. (Note that $\prod_{i<j} z_{ij}^{-(\gamma_{ij}-\bar\gamma_{ij})}$ for γ_{ij} and $\bar\gamma_{ij}$ two different solutions of eq.(4.71) is always a function of the X_{ij}^{kl}.) One solution for $n = 4$ is

$$\langle\phi_1(z_1)\phi_2(z_2)\phi_3(z_3)\phi_4(z_4)\rangle = \frac{z_{13}^{h_2+h_4}\,z_{24}^{h_1+h_3}}{z_{12}^{h_1+h_2}\,z_{23}^{h_2+h_3}\,z_{34}^{h_3+h_4}\,z_{14}^{h_1+h_4}} f\!\left(\frac{z_{12}z_{34}}{z_{13}z_{24}}\right). \tag{4.72}$$

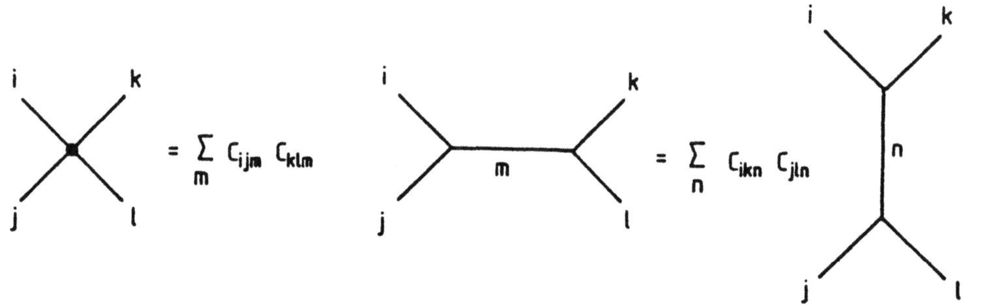

Fig.4.3. Crossing symmetry of the four point amplitude

The four-point amplitudes can be used to obtain some constraints on the operator product coefficients C_{ijk}. One evaluates the four-point function in two ways as shown schematically in figure 4.3. Associativity of the operator algebra implies that the two ways give the same result. This is known as crossing symmetry or duality of the four-point amplitude. In this way we obtain an infinite number of equations that the C_{ijk}'s have to satisfy. The procedure of solving these relations is known as conformal bootstrap; in general this is very difficult to do in practice.

We complete the discussion about general properties of amplitudes in conformal field theory by writing down the conformal Ward identities satisfied by correlation functions of primary fields $\phi_i(z)$. Ward identities among correlation functions generally reflect the symmetries of a theory. We want to investigate the constraints of the local conformal algebra on the correlation functions of the primary fields. Therefore consider the action of the generator of infinitesimal conformal transformations, $\oint_{C_0} \frac{dz}{2\pi i} \xi(z) T(z)$, on the correlation function of n primary fields $\phi_i(w_i)$ $(i = 1, \ldots, n)$ where the z-contour surrounds all points w_i. Analyticity allows to deform the contour to a sum over contours encircling each of the points w_i:

$$\langle \oint_{C_0} \frac{dz}{2\pi i} \xi(z) T(z) \phi_1(w_1) \ldots \phi_n(w_n) \rangle$$

$$= \sum_{j=1}^{n} \langle \phi_1(w_1) \ldots (\oint_{C_{w_j}} \frac{dz}{2\pi i} \xi(z) T(z) \phi_j(w_j)) \ldots \phi_n(w_n) \rangle \qquad (4.73)$$

$$= \sum_{j=1}^{n} \langle \phi_1(w_1) \ldots \delta_\xi \phi_j(w_j) \ldots \phi_n(w_n) \rangle.$$

This must hold for arbitrary ξ leading to

$$\langle T(z) \phi_1(w_1) \ldots \phi_n(w_n) \rangle$$

$$= \sum_{j=1}^{n} \left(\frac{h_j}{(z - w_j)^2} + \frac{1}{z - w_j} \partial_{w_j} \right) \langle \phi_1(w_1) \ldots \phi_n(w_n) \rangle. \qquad (4.74)$$

This is the unintegrated form of the conformal Ward identity.

4.2 Application to string theory

Now let us return to the closed bosonic string theory and study the simplest example of a conformal field theory, the massless free scalar field $X(z, \bar{z})$. In Euclidean space, the action for such a field is

$$S = \frac{1}{4\pi} \int d^2 z \, \partial X(z, \bar{z}) \bar{\partial} X(z, \bar{z}) \qquad (4.75)$$

which, up to the index μ, is the Euclidean action of the bosonic string in units where $\alpha' = 2$. It leads to the equation of motion

$$\partial \bar{\partial} X(z, \bar{z}) = 0 \qquad (4.76)$$

with general solution

$$X(z, \bar{z}) = X(z) + \bar{X}(\bar{z}). \qquad (4.77)$$

The fields $X(z)$ and $\bar{X}(\bar{z})$ correspond to the right- and left-moving coordinates of the closed bosonic string respectively. The propagator for the free boson $X(z, \bar{z})$ following from the action eq.(4.75) is

$$\langle X(z, \bar{z}) X(w, \bar{w}) \rangle = -2 \log |z - w| \qquad , \quad |z| > |w|. \qquad (4.78)$$

It satisfies the equation

$$\partial\bar{\partial}\langle X(z,\bar{z})X(w,\bar{w})\rangle = -2\pi\delta^{(2)}(z-w) \qquad (4.79)$$

which follows from[4]

$$\partial\bar{\partial}\log|z|^2 = 2\pi\delta^{(2)}(z). \qquad (4.80)$$

Making the split into left- and right-movers, eq.(4.77), we get

$$\begin{aligned}
\langle X(z)X(w)\rangle &= -\log(z-w)\,, \\
\langle \bar{X}(\bar{z})\bar{X}(\bar{w})\rangle &= -\log(\bar{z}-\bar{w})\,.
\end{aligned} \qquad (4.81)$$

Here we have treated X and $\bar{X}$ as completely independent fields (cf. the discussion in Chapter 3). From its two-point function we see that the field X does not have a definite scaling dimension. However, we will only need its derivatives and exponentials, both of which have definite scaling dimension and are good fields of the conformal field theory. This will be demonstrated below.

The energy-momentum tensor following from the action eq.(4.75) is

$$T(z) = -\frac{1}{2}:\partial X(z)\partial X(z): \qquad (4.82)$$

and likewise for $\bar{T}(\bar{z})$. This can be found with reference to eq.(2.41). However, the easiest way to derive the energy-momentum tensor which does not require a metric, is to compute the change of the action under infinitesimal coordinate transformations $\delta z = \xi$ and $\delta\bar{z} = \bar{\xi}$:

$$\delta S = -\frac{1}{2\pi}\int d^2z\left(\partial\bar{\xi}\,\bar{T} + \bar{\partial}\xi\,T\right). \qquad (4.83)$$

Eq.(4.82) then follows with $\delta X = \xi\partial X + \bar{\xi}\bar{\partial}X$. In eq.(4.82) normal ordering is defined by

---

[4] One way to see this is as follows: $\partial\bar{\partial}\log|z| = \frac{1}{2}\partial\frac{1}{\bar{z}} = \frac{1}{2}\partial\frac{z}{|z|^2} = \frac{1}{2}\lim_{\epsilon\to 0}\partial\frac{z}{|z|^2+\epsilon^2} = \frac{1}{2}\lim_{\epsilon\to 0}\frac{\epsilon^2}{(|z|^2+\epsilon^2)^2} = \pi\delta^{(2)}(z).$ An alternative way is to integrate with a test function.

$$: \phi_i(z)\phi_j(z): = \lim_{w \to z} \Big(\phi_i(w)\phi_j(z) - \text{poles} \Big) \tag{4.84}$$

where the pole terms to be subtracted are those arising in the operator product expansion of $\phi_i(w)\phi_j(z)$.

It is now straightforward to compute the operator product of the energy-momentum tensor with itself. Since we are dealing with free fields we can use Wick's theorem. Remembering that no contractions are to be made within are normal ordered expression, we find with the help of the basic contraction eq.(4.81)

$$T(z)T(w) = \frac{\frac{1}{2}}{(z-w)^4} + \frac{2T(w)}{(z-w)^2} + \frac{\partial T(w)}{(z-w)} + \ldots \tag{4.85}$$

This shows that we have a conformal field theory with $c = 1$.

Finally, we have to specify the conformal fields of this model. As already mentioned, $X(z)$ is not a conformal field due to the logarithmic z-dependence of its propagator. Computing the operator product of $T(z)$ with $\partial X(z)$,

$$T(z)\partial X(w) = \frac{\partial X(w)}{(z-w)^2} + \frac{\partial(\partial X(w))}{(z-w)} + \ldots, \tag{4.86}$$

shows that $\partial X(z)$ is a conformal field with dimension $h = 1$. We can expand it in modes as

$$i\partial X(z) = \sum_{n \in \mathbf{Z}} \alpha_n z^{-n-1} \ , \ \alpha_0 = p \, . \tag{4.87}$$

(Cf. Chapter 3.) Higher derivatives $\partial^{(n)} X(z)$ are not primary but descendant fields with $h = n$. For example, $\partial^2 X(z) = L_{-1}\partial X(z)$. The only other conformal fields in the free scalar model are normal ordered exponentials of $X(z)$:

$$T(z) :e^{i\alpha X(w)}: \ = \ \Big[\frac{\frac{1}{2}\alpha^2}{(z-w)^2} + \frac{1}{(z-w)}\partial_w \Big] :e^{i\alpha X(w)}: + \ldots \tag{4.88}$$

This is again shown using a Wick expansion. We find that the operator $: e^{i\alpha X(z)} :$ has conformal dimension $h = \frac{\alpha^2}{2}$ with α being a continuous variable.

The complete operator product algebra among the conformal fields has the following form:

$$\partial X(z)\,\partial X(w) = -\frac{1}{(z-w)^2} + \text{finite} \tag{4.89a}$$

$$:e^{i\alpha X(z)}: \; :e^{i\beta X(w)}: \; = (z-w)^{\alpha\beta} \; :e^{i(\alpha X(z)+\beta X(w))}:$$

$$= (z-w)^{\alpha\beta} \; :e^{i(\alpha+\beta)X(w)}: \tag{4.89b}$$

$$+ \, i\alpha(z-w)^{\alpha\beta+1} \; :\partial X(w)\, e^{i(\alpha+\beta)X(w)}: + \ldots$$

$$\partial X(z) \; :e^{i\alpha X(w)}: \; = -\frac{i\alpha}{z-w}\, e^{i\alpha X(w)} + \text{finite} \tag{4.89c}$$

From eq.(4.89a) we find

$$\{\alpha_m, \alpha_n\} = m\delta_{m+n} . \tag{4.90}$$

$\alpha^\dagger_{-n} = \alpha_n$ follows from Hermiticity of $i\partial X$. We can now calculate its two-point function:

$$\langle \partial X(z)\partial X(w)\rangle = -\sum_{m,n}\langle 0|\alpha_n\alpha_m|0\rangle z^{-n-1}w^{-m-1}$$

$$= -\sum_{n>0} nz^{-n-1}w^{+n-1} \tag{4.91}$$

$$= -\frac{1}{(z-w)^2}, \qquad \text{for } |z| > |w| .$$

where we have used eq.(4.44). This result also follows from eq.(4.78) upon taking derivatives. Above derivation however demonstrates that the operators have to be radially ordered for the sum to converge.

In the bosonic string theory one deals with d $(d = 26)$ identical free bosonic fields $X^\mu(z, \bar{z})$ $(\mu = 0, \ldots d-1)$ and their contribution to the central charge is $c = \bar{c} = d$. Physical string states must satisfy the following conditions:

$$L_n|\phi\rangle = \bar{L}_n|\phi\rangle = 0, \qquad\qquad n > 0$$

$$(L_0 - 1)|\phi\rangle = (\bar{L}_0 - 1)|\phi\rangle = 0, \tag{4.92}$$

$$(L_0 - \bar{L}_0)|\phi\rangle = 0.$$

This means that physical string states correspond to primary fields of the conformal field theory:

$$|\phi\rangle = \phi(0)|0\rangle = \lim_{z,\bar{z}\to 0} \phi(z,\bar{z})|0\rangle. \tag{4.93}$$

$\phi(z,\bar{z})$ are conformal fields and create asymptotic states. They are called vertex operators. String scattering amplitudes are then simply correlation functions of vertex operators (cf. Chapters 6 and 15). Eq.(4.92) also implies that the conformal dimension of vertex operators is $(h,\bar{h}) = (1,1)$.

Let us briefly discuss the spectrum of the closed bosonic string in the context of conformal field theory. The lowest state is the tachyon; it is a space-time scalar with momentum k_μ.

$$|k\rangle = \lim_{z,\bar{z}\to 0} :e^{ik_\rho X^\rho(z,\bar{z})}: |0\rangle. \tag{4.94}$$

The physical state condition eq.(4.92) requires that $:e^{ik_\rho X^\rho(z,\bar{z})}:$ has conformal dimension $h = \bar{h} = 1$. With $X^\mu(z,\bar{z}) = X^\mu(z) + \bar{X}^\mu(\bar{z})$ this leads to $k^2 = 2$ which is the mass shell condition for a tachyon: $m^2 = -k^2 = -2$. Let us verify that the state $|k\rangle$ carries momentum k_μ. We restrict our attention to one, say the right-moving sector. Then $|k\rangle =:e^{ik_\rho X^\rho(0)}: |0\rangle$ and the momentum operator is $\alpha_0^\mu = i\oint \frac{dz}{2\pi i}\partial X^\mu(z)$. Then

$$\alpha_0^\mu|k\rangle = i\oint \frac{dz}{2\pi i}\partial X^\mu(z) :e^{ik_\rho X^\rho(z)}: |0\rangle = k^\mu|k\rangle \tag{4.95}$$

where we have used the operator product eq.(4.89c).

The states at the next level have the form:

$$|k,\epsilon\rangle = \epsilon_{\mu\nu} \lim_{z,\bar{z}\to 0} :\partial X^\mu(z)\partial\bar{X}^\nu(\bar{z})e^{ik_\rho X^\rho(z,\bar{z})}: |0\rangle$$

$$= \alpha_{-1}^\mu\bar{\alpha}_{-1}^\nu|k\rangle\epsilon_{\mu\nu}. \tag{4.96}$$

$\epsilon_{\mu\nu}$ is a polarization tensor. These states are just those discussed in eq.(3.36). Since $\partial X^\mu(z)\partial\bar{X}^\nu(\bar{z})$ has $h = \bar{h} = 1$, we find $k^2 = -m^2 = 0$ for these states; they must be massless. We still have to check whether this vertex operator is a conformal field. We therefore take its operator product with the energy-momentum tensor $T(z)$; we can do this for the holomorphic and anti-holomorphic parts separately:

$$\epsilon_{\mu\nu}T(z) :\partial X^\mu(w)\bar{\partial}X^\nu(\bar{w})e^{ik_\rho X^\rho(w,\bar{w})}: = \frac{k^\mu \epsilon_{\mu\nu}}{(z-w)^3} : \bar{\partial}X^\nu(\bar{w})e^{ik_\rho X^\rho(w,\bar{w})}:$$

$$+ \left[\frac{\frac{1}{2}k^2 + 1}{(z-w)^2} + \frac{\partial_w}{(z-w)} \right] \epsilon_{\mu\nu} :\partial X^\mu(w)\bar{\partial}X^\nu(\bar{w})e^{ik_\rho X^\rho(w,\bar{w})}: + \ldots \tag{4.97}$$

Thus, we recognize that in order to get rid of the unwanted cubic singularity one has to demand that

$$k^\mu \epsilon_{\mu\nu} = 0 \tag{4.98}$$

which, together with $k^2 = 0$, is nothing else than the on-shell condition for a massless tensor particle which we identify with the graviton, antisymmetric tensor and dilaton, depending on whether $\epsilon_{\mu\nu}$ is symmetric traceless, antisymmetric or pure trace.

We close this chapter with an observation concerning string scattering amplitudes. Scattering amplitudes of asymptotic string states are correlation functions of the corresponding vertex operators[5]. We have already said that the vertex operators are primary fields of the Virasoro algebra of weight $(h, \bar{h}) = (1, 1)$. It involves, however, also an integration over the positions of the vertex operators. (This will be discussed in more detail in the last chapter where we explicitly compute string scattering amplitudes.) We then get expressions of the form

---

[5] The discussion here is limited to string tree level amplitudes since there the world-sheet is the (Riemann) sphere which is conformally equivalent to $\mathbf{C} \cup \infty$.

$$\int \prod_{i=1}^{n} \mathrm{d}^2 z_i \, \langle V(z_1, \bar{z}_2) \cdots V(z_n, \bar{z}_n) \rangle. \tag{4.99}$$

Since the V's have weight $(\bar{h}, h) = (1, 1)$ and the integration measure transforms under $SL(2, \mathrm{C})$ as $\mathrm{d}^2 z_i \to \frac{\mathrm{d}^2 z_i}{|cz+d|^4}$ we find, using eq.(4.6), that string amplitudes are $SL(2, \mathrm{C})$ invariant.

Let us now demonstrate that secondary states decouple from string scattering amplitudes. Consider the correlation function

$$A = \langle \phi_N^{(-k)}(z) \phi_1(z_1) \cdots \phi_n(z_n) \rangle \tag{4.100}$$

where the ϕ_i, $i = 1, \cdots, n$ are primary and

$$\phi_N^{(-k)}(z) = \hat{L}_{-k} = \oint_{C_z} \frac{\mathrm{d}w}{2\pi i} \frac{T(w)}{(w-z)^{k-1}} \phi_N(z) \tag{4.101}$$

is secondary. We now insert eq.(4.101) into eq.(4.100) and deform the contour to enclose the points $z_i, \ldots, z_n$ instead of z. (This is easy to visualize for the sphere which is conformally equivalent to $\mathrm{C} \cup \infty$.) Then, expanding the operator products $T(w)\phi_i(z_i)$, we get

$$A = \sum_{i=1}^{n} \left\{ -\frac{(k-1)h_i}{(z_i - z)^k} + \frac{\partial_{z_i}}{(z_i - z)^{k-1}} \right\} \langle \phi_N(z) \phi_1(z_1) \cdots \phi_n(z_n) \rangle. \tag{4.102}$$

If the ϕ_i are vertex operators, we have $h_i = 1$, $\forall i$ and this can be written as

$$A = \sum_{i=1}^{n} \partial_{z_i} \left\{ \frac{1}{(z_i - z)^{k-1}} \langle \phi_N(z) \phi_1(z_1) \cdots \phi_n(z_n) \rangle \right\} \tag{4.103}$$

which vanishes upon integration over the positions of the of the vertex operators. This argument can be generalized to the case of several descendant fields.

REFERENCES

[1] A.A. Belavin, A.M. Polyakov and A.B. Zamolodchikov, *"Infinite conformal symmetry in two-dimensional quantum field theory"*, Nucl. Phys. **B241** (1984) 333.

[2] P. Ginsparg, *"Applied conformal field theory"*, lectures presented at Les Houches summer school 1988, preprint HUTP-88/A054, in proceedings Les Houches, Session XLIX, 1988.

[3] M. Peskin, *"Introduction to string and superstring theory II"*, lectures presented at the 1986 TASI School, preprint SLAC-PUB-4251, in Santa Cruz TASI proceedings.

[4] T. Banks, *"Lectures on conformal field theory"*, lectures presented at the 1987 TASI School, preprint SCIPP 87/111, in Santa Fe TASI proceedings.

[5] D. Friedan, Z. Qiu and S. Shenker, *"Conformal invariance, unitarity and critical exponents in two dimensions"*, Phys. Rev. Lett. **52** (1984) 1575.

Chapter 5

Reparametrization Ghosts and BRST Quantization

As a second application of conformal field theory we want to examine the reparametrization ghosts which we introduced within the path integral quantization of the bosonic string in Chapter 3. In the second part of this chapter we study the related issue of BRST [1, 2] quantization, using the formalism of ref. [3]. The BRST treatment of string theory was first discussed in [4]. References [5, 6] provide an overview over the material presented in this chapter and contain an exhaustive list of literature.

5.1 The ghost system as a conformal field theory

In conformal coordinates the ghost action is

$$S = \frac{1}{2\pi} \int \mathrm{d}^2 z (b_{zz} \partial_{\bar{z}} c^z + b_{\bar{z}\bar{z}} \partial_z c^{\bar{z}}) \tag{5.1}$$

and the solutions of the equations of motion are

$$\begin{aligned} c^z = c(z) \quad , \quad b_{zz} = b(z) \\ c^{\bar{z}} = \bar{c}(\bar{z}) \quad , \quad b_{\bar{z}\bar{z}} = \bar{b}(\bar{z}). \end{aligned} \tag{5.2}$$

The ghost fields are effectively free fermions with integer spin. $b(z)$ is an analytic conformal field of dimension (spin) $(h, \bar{h}) = (2, 0)$, $c(z)$ is a field of dimension $(h, \bar{h}) = (-1, 0)$; $\bar{b}(z)$ and $\bar{c}(\bar{z})$ are antianalytic conformal tensors with $(h, \bar{h}) = (0, 2)$ and $(0, -1)$ respectively. In the following we will restrict the discussion to the analytic fields $b(z)$ and $c(z)$. Their propagator, following from the action eq.(5.1) is

$$\langle b(z)c(w)\rangle = \langle c(z)b(w)\rangle = \frac{1}{z-w}\,. \tag{5.3}$$

It satisfies

$$\partial_{\bar{z}}\langle b(z)c(w)\rangle = 2\pi\delta^{(2)}(z-w) \tag{5.4}$$

where we have used

$$\partial_{\bar{z}}\frac{1}{z-w} = 2\pi\delta^{(2)}(z-w). \tag{5.5}$$

From the propagator we deduce the following operator products

$$c(z)b(w) = b(z)c(w) = \frac{1}{z-w} + \dots \tag{5.6}$$

Next we expand the ghost fields into modes:

$$\begin{aligned}
c(z) &= \sum_n c_n z^{-n+1} \\
b(z) &= \sum_n b_n z^{-n-2}
\end{aligned} \tag{5.7}$$

with hermiticity conditions $b_n^\dagger = b_{-n}$ and $c_n^\dagger = c_{-n}$. On the SL_2 invariant ghost vacuum $|0\rangle_{b,c}$ the oscillators act as

$$\begin{aligned}
b_n|0\rangle_{b,c} &= 0 \quad \text{for} \quad n \geq -1, \\
c_n|0\rangle_{b,c} &= 0 \quad \text{for} \quad n \geq 2.
\end{aligned} \tag{5.8}$$

From the operator products eq.(5.6) we derive the following anticommutation relations:

$$\begin{aligned}
\{b_m, c_n\} &= \delta_{n+m}, \\
\{c_n, c_m\} &= \{b_n, b_m\} = 0.
\end{aligned} \tag{5.9}$$

Note that since we are dealing with anticommuting fields, the contour trick of Chapter 4 leads to anticommutators. The energy momentum tensor of the b,c system is (use eq.(4.83))

$$T^{b,c}(z) = -2b(z)\partial c(z) - \partial b(z)c(z). \tag{5.10}$$

This is equivalent to the following expression for the Virasoro generators (cf. eq.(3.58)):

$$L_n^{b,c} = \sum_{m=-\infty}^{\infty} (n-m) : b_{n+m} c_{-m} : \tag{5.11}$$

Normal ordering means again that we put negative frequency modes to the right of positive frequency modes, taking due care of minus signs arising due to the Grassmann property of the ghosts. It is now straightforward to work out the operator product of the stress tensors with itself:

$$T^{b,c}(z)\, T^{b,c}(w) = \frac{-26/2}{(z-w)^4} + \frac{2T^{b,c}(w)}{(z-w)^2} + \frac{\partial T^{b,c}(w)}{(z-w)} + \text{finite terms.} \tag{5.12}$$

This shows that the central charge of the b,c system is $c^{b,c} = -26$. Since $c < 0$ the conformal field theory of the ghost system is non-unitary as expected. This already follows from the negative conformal weight of $c(z)$. Adding the contribution to the central charge from d bosonic fields $X^\mu(z)$ and of the ghost fields,

$$c^{\text{tot}} = c^X + c^{b,c} = d - 26 \tag{5.13}$$

we find again that the conformal anomaly vanishes if $d = 26$. The operator product of $T^{b,c}$ with the ghost fields can be easily worked out and shows that b and c are primary fields.

Another important operator of the b,c system is the $U(1)$ ghost number current $j(z)$,

$$j(z) = - :b(z)c(z): = \sum_n z^{-n-1} j_n \tag{5.14}$$

where

$$j_n = \sum_m :c_{n-m} b_m: \tag{5.15}$$

$j(z)$ is a conformal field of dimension $h = 1$. Classically the ghost number current is conserved. In the quantum theory there is however an anomaly. This will be discussed in Chapter 13. The ghost charge is given by the contour integral of $j(z)$:

$$N_g = \oint_{C_0} \frac{\mathrm{d}z}{2\pi} j(z) = j_0 = \sum_m :c_{-m} b_m: \tag{5.16}$$

Thus, the ghost charge N_g of a particular conformal field $\phi(z)$ is given by the singular part of its operator product with j.

$$j(z)\,\phi(w) = \frac{N_g}{(z-w)} + \text{finite terms.} \qquad (5.17)$$

It follows that $c(z)$ and $b(z)$ have $N_g = +1$ and -1 respectively.

5.2 BRST quantization

Let us now turn to the question of how to identify physical states. We have seen in Chapter 3 that in light-cone quantization the longitudinal and timelike components of X^μ are not independent degrees of freedom and can be eliminated. All states can be built as excitations of the transverse oscillators only. In covariant Polyakov quantization we keep all components of X^μ and, in addition, have the ghost fields b and c. The excitation of the longitudinal and timelike components of X^μ and the ghosts will now become part of the spectrum of string theory and we need some way to distinguish physical from unphysical states. The tool to do this is the BRST charge. So let us briefly review some general aspects of BRST quantization and then apply it to the bosonic string.

BRST quantization was introduced to quantize systems with a local gauge symmetry G. After gauge fixing, BRST symmetry is a remnant of the local gauge symmetry. Let us first review the general strategy. Consider a system with gauge invariances generated by charges K_i which form a closed finite dimensional Lie algebra[1]

$$[K_i, K_j] = f_{ij}{}^k K_k \quad , \qquad i,j,k = 1,\ldots,\dim G \qquad (5.18)$$

with $f_{ij}{}^k$ being the structure constants of G. One now defines a hermitian nilpotent operator which commutes with the Hamiltonian and which acts

---

[1] For instance, in a Yang-Mills theory the K_i are the non-abelian generalizations of Gauss law.

on all fields like a fermionic gauge transformation. The gauge parameter is replaced by an anticommuting variable c^i, called a ghost. This operator is the BRST charge Q. An explicit expression for the BRST charge is given by

$$\begin{aligned}
Q &= c^i (K_i - \frac{1}{2} f_{ij}{}^k \, c^j b_k) \\
&= c^i (K_i + \frac{1}{2} K_i^{\text{ghost}})
\end{aligned}$$

(5.19)

where we have introduces so-called anti-ghosts b_i which obey the following commutation relations with the c^i:

$$\{c^i, b_j\} = \delta^i_j.$$

(5.20)

In the following, both b_i and c^i will be collectively referred to as ghosts. The first part of Q clearly acts in the required way on the fields. The second part is needed to make Q nilpotent. It acts like a gauge transformation on the ghost fields. Nilpotency of Q is easy to verify. Using the symmetry algebra eq.(5.18), the anticommutation relations eq.(5.20) and antisymmetry of the structure constants, we find without difficulty

$$Q^2 = \frac{1}{4} f_{[ij}{}^k f_{l]k}{}^m \, (c^j c^i c^l b_m) = 0.$$

(5.21)

The Jacobi identity was used in the last step. The BRST transformation acts on the ghosts as

$$\begin{aligned}
\delta c^i &= \{Q, c^i\} = -\frac{1}{2} f_{kl}{}^i c^k c^l \\
\delta b_i &= \{Q, b_i\} = K_i - f_{ij}{}^k \, c^j b_k = K_i + K_i^{\text{ghosts}} \equiv \tilde{K}_i.
\end{aligned}$$

(5.22)

One now shows that

$$[\tilde{K}_i, \tilde{K}_j] = f_{ij}{}^k \tilde{K}_k$$

(5.23)

i.e. the $\tilde{K}_i$ also satisfy the symmetry algebra but, in contrast to the K_i, also incorporate the ghost degrees of freedom. The ghost contribution to the action can be written to be the BRST transformed of a term of the form $b^i F_i$ where F_i is called the gauge fixing function; i.e. $\mathcal{L}^{\text{ghost}} \sim \delta(b^i F_i)$.

This guarantees the BRST invariance of the total action. BRST invariance is thus a symmetry of the gauge fixed action. Finally, one also introduces a ghost number operator

$$N_g = - \sum_{i=1}^{\dim G} b_i \, c^i. \tag{5.24}$$

c^i and b_i have ghost number $+1$ and -1 respectively.

Let us now consider the Hilbert space of the theory. Eigenstates of the Hamiltonian are said to be BRST invariant if they are annihilated by Q:

$$Q|\phi\rangle = 0. \tag{5.25}$$

States which satisfy eq.(5.25) are gauge independent, a necessary requirement for physical states. There are two types of BRST invariant states. First, any state of the form

$$|\phi\rangle = Q|\lambda\rangle \tag{5.26}$$

is trivially BRST invariant due to the nilpotency of Q. The states $|\phi\rangle$ and $|\lambda\rangle$ form a BRST doublet. They differ in ghost charge by one unit. $|\phi\rangle$ has zero norm, due to the hermiticity and nilpotency of the BRST charge: $\langle\lambda|QQ|\lambda\rangle = 0$. These states decouple in S-matrix elements. (Recall that Q commutes with the Hamiltonian.) Therefore we have to look for states of the form

$$Q|\phi\rangle = 0, \quad |\phi\rangle \neq Q|\lambda\rangle. \tag{5.27}$$

They are BRST singlets. These states will henceforth be referred to as physical states. Two states $|\phi\rangle$ and $|\phi'\rangle$ are said to be equivalent if

$$|\phi\rangle - |\phi'\rangle = Q|\lambda\rangle. \tag{5.28}$$

The equivalence classes are called BRST cohomology classes. Clearly, all states within one given cohomology class have the same ghost number. S-matrix elements are independent of which representative of a cohomology class one uses: $\langle\phi_1|S|\phi_2\rangle = \langle\phi_1'|S|\phi_2'\rangle$, for ϕ and ϕ' related as in eq.(5.28).

If $|\phi\rangle = \phi|0\rangle$ is a physical BRST singlet, then $[Q, \phi] = 0$. For states without ghost excitation this implies $[K_i, \phi] = 0$. Those are the states identified with physical particles.

Let us now apply the BRST formalism to the bosonic string. In distinction to the case of gauge theories with finite dimensional symmetry groups, we are now dealing with the infinite dimensional Virasoro algebra. Expression such as the BRST charge and the ghost number must be normal ordered and a normal ordering constant will appear. Finally, nilpotency of the BRST charge, which contains the symmetry generators, might be anomalous. Let us start with the BRST charge. The generalization of eq.(5.19) to the case of the Virasoro algebra is

$$
\begin{aligned}
Q &= \sum_{m=-\infty}^{\infty} \left(c_{-m} L_m^X - \frac{1}{2} \sum_{n=-\infty}^{\infty} (m-n) : c_{-m} c_{-n} b_{m+n} : \right) - a c_0 \\
&= \sum_m : c_{-m} \left[L_m^X + \frac{1}{2} L_m^{b,c} - a \delta_m \right] :
\end{aligned}
\tag{5.29}
$$

In the first line we have used the explicit form for the structure constants of the Virasoro algebra: $f_{mn}{}^p = (m-n)\delta_{p,m+n}$. $L_m^{(b,c)}$ was given in eq.(5.11). We already know from Chapter 3 that $L_m^{\text{tot}} = L_m^X + L_m^{b,c} - a \delta_m$ satisfy the Virasoro algebra and that the central charge vanishes if $a = 1$ and $d = 26$. L_m^{tot} corresponds to $\tilde{K}_i$ in eq.(5.22).

Q can be equivalently written as a contour integral:

$$
\begin{aligned}
Q &= \oint_{C_0} \frac{dz}{2\pi i} \, c(z)[T^X(z) + \frac{1}{2} T^{b,c}(z)] \\
&= \oint_{C_0} \frac{dz}{2\pi i} \, j_{\text{BRST}}(z).
\end{aligned}
\tag{5.30}
$$

The operator $j_{\text{BRST}}(z)$ is the BRST current. Eq.(5.30) defines it only up to a total derivative which must however be of dimension one and ghost number one. The most general form is then

$$
j_{\text{BRST}} = c T^{(X)} + \frac{1}{2} c T^{(b,c)} + \kappa \partial^2 c.
\tag{5.31}
$$

Requiring the BRST current to be a conformal field of weight one gives $\kappa = \frac{3}{2}$.

Let us now check the nilpotency of the BRST charge Q which is crucial for the identification of physical states. One finds ($L^{\text{tot}} = L^X + L^{b,c}$)

$$Q^2 = \frac{1}{2}\{Q,Q\} = \frac{1}{2}\sum_{m,n=-\infty}^{\infty}\left([L_m^{\text{tot}}, L_n^{\text{tot}}] - (m-n)L_{m+n}^{\text{tot}}\right)c_{-m}c_{-n} \quad (5.32)$$

which implies that $Q^2 = 0$ if the conformal anomaly vanishes, i.e. for $d = 26$ and $a = 1$. One can also show that $Q^2 = 0$ implies the vanishing of the conformal anomaly.

It is now easy to work out the BRST transformation properties of the various fields. We find

$$[Q, X^\mu(z)] = c\partial X^\mu(z)\,,$$

$$[Q, T^{\text{tot}}(z)] = \frac{1}{12}(d - 26)\partial^3 c(z)\,,$$

$$\{Q, c(z)\} = c\partial c(z)\,,$$

$$\{Q, b(z)\} = T^{\text{tot}}(z)\,, \qquad\qquad (5.33)$$

where $T^{\text{tot}} = T^X + T^{b,c}$. Expressed in terms of modes they are

$$[Q, \alpha_n] = -\sum_m nc_m\alpha_{n-m}\,,$$

$$[Q, L_n] = -\frac{1}{12}(d - 26)n(n^2 - 1)c_n\,,$$

$$\{Q, c_n\} = -\sum_m(2n + m)c_{-m}c_{m+n}\,, \qquad\qquad (5.34)$$

$$\{Q, b_n\} = L_n^{\text{tot}}\,.$$

It is now straightforward to verify that the total action

$$S = S^X + S^{b,c}$$

$$= \frac{1}{4\pi}\int d^2z\left(\partial X\bar\partial X + 2b\bar\partial c + 2\bar b\partial\bar c\right) \qquad\qquad (5.35)$$

is invariant under BRST transformations. Alternatively, we could have followed the procedure common in gauge theories and chosen a gauge fixing function. The gauge fixing function leading to the action eq.(5.35) would have been $F^{\alpha\beta} = \sqrt{h}h^{\alpha\beta} - \eta^{\alpha\beta}$.

Let us now turn to the problem of identifying physical states. We again demand that they are BRST singlet states, i.e. $Q|\phi\rangle = 0$ but $|\phi\rangle \neq Q|\lambda\rangle$. According to our discussion in Chapter 4, a state $|\phi\rangle$ is created from the SL_2 invariant vacuum by a local vertex operator: $|\phi\rangle = \phi(0)|0\rangle$. BRST invariance then implies that

$$[Q, \phi(z)] = \oint_{C_z} \frac{dw}{2\pi i} j_{\text{BRST}}(w)\phi(z) = \text{ total derivative} \qquad (5.36)$$

i.e. the operator product of j_{BRST} and ϕ must not have a pole of order one, unless the residue is a total derivative, in which case it vanishes upon integration over the insertion point of the vertex operator. Then correlation functions will be BRST invariant. Consider states without ghost excitations. Then

$$\begin{aligned}
\oint \frac{dw}{2\pi i} j_{\text{BRST}}(w)\phi(z) &= \oint \frac{dw}{2\pi i} c(w) T^\phi(w)\phi(z) \\
&= \oint \frac{dw}{2\pi i} c(w)\Big[\frac{h_\phi \phi(z)}{(w-z)^2} + \frac{\partial\phi(z)}{w-z} + \cdots\Big] \qquad (5.37) \\
&= h_\phi(\partial c)\phi(z) + c\partial\phi(z)
\end{aligned}$$

which is a total derivative whenever the conformal weight of ϕ is $h_\phi = 1$. We have thus found that primary fields of dimension one create asymptotic BRST invariant states.

Let us finally look at the ghost sector of the theory. Both b and c have zero frequency components which satisfy the anti-commutation relations $b_0^2 = c_0^2 = 0$ and $\{c_0, b_0\} = 1$. b_0 and c_0 commute with the Hamiltonian L_0 (cf. eq.(4.35)). There are then two degenerate states. One, denoted $|\uparrow\rangle$, is annihilated by c_0; the second is then defined by $|\downarrow\rangle = b_0|\uparrow\rangle$:

$$c_0|\!\uparrow\rangle = 0 \qquad\qquad b_0|\!\downarrow\rangle = 0$$
$$b_0|\!\uparrow\rangle = |\!\downarrow\rangle \qquad\qquad c_0|\!\downarrow\rangle = |\!\uparrow\rangle. \tag{5.38}$$

These states clearly have zero norm. They differ in ghost number by one unit. Since the normal ordering of the zero frequency term in eq.(5.16) is arbitrary, we can choose it to be symmetric; i.e.

$$N_g = \frac{1}{2}(c_0 b_0 - b_0 c_0) + \sum_{m>0}(c_{-n}b_n - b_{-n}c_n). \tag{5.39}$$

With this convention the states $|\!\uparrow\rangle$ and $|\!\downarrow\rangle$ have ghost charge $+\frac{1}{2}$ and $-\frac{1}{2}$ respectively.

On the other hand we notice that the SL_2 invariant ghost vacuum $|0\rangle_{b,c}$ obeys eq.(5.8). This however means that while being a highest weight state of the Virasoro algebra, it is not a highest weight state of the b,c algebra since it is not annihilated by all the negative frequency modes:

$$c_1|0\rangle_{b,c} = c(0)|0\rangle_{b,c} \neq 0. \tag{5.40}$$

Since $[L_0, c_1] = -c_1$, $|0\rangle_{b,c}$ is not the ground state of the ghost system. The state

$$c_0 c_1|0\rangle_{b,c} = -c\partial c(0)|0\rangle_{b,c} \neq 0 \tag{5.41}$$

has also L_0 eigenvalue -1; i.e. the two states $c_1|0\rangle_{b,c}$ and $c_0 c_1|0\rangle_{b,c}$ are degenerate. It is easy to see that if we identify $c_1|0\rangle_{b,c} = |\!\downarrow\rangle$ and $c_0 c_1|0\rangle_{b,c} = |\!\uparrow\rangle$, we can verify the relations eq.(5.38) and their ghost number assignments. In addition, we find that $\langle\uparrow\,|\,\uparrow\rangle = \langle\downarrow\,|\,\downarrow\rangle = 0$ and $\langle\uparrow\,|\,\downarrow\rangle = \langle\downarrow\,|\,\uparrow\rangle = {}_{b,c}\langle 0|c_{-1}c_0 c_1|0\rangle_{b,c} \neq 0$. We choose the normalization such that

$${}_{b,c}\langle 0|c_{-1}c_0 c_1|0\rangle_{b,c} = 1. \tag{5.42}$$

This shows that the SL_2 invariant vacuum carries three units of ghost number. They correspond to the three global diffeomorphisms of the sphere (compactified plane), generated by $L_0, L_{\pm 1}$. Since correlation functions are invariant under SL_2 (cf. Chapter 4), the gauge fixing is not complete which is reflected by the presence of ghost zero modes. For details we refer to the following chapter where we learn how to deal with the ghost zero modes in the computation of scattering amplitudes.

Physical states are characterized by BRST cohomology classes of some definite ghost number. We can build states on either of the two ghost ground states $|\uparrow\rangle$ and $|\downarrow\rangle$. Let us consider states of the form

$$|\psi\rangle = |\phi\rangle_X \otimes |\downarrow\rangle_{b,c}. \tag{5.43}$$

BRST invariance of this state then requires

$$Q|\psi\rangle = \Big(c_0(L_0^X - 1) + \sum_{n>0} c_{-n} L_n^X\Big)|\psi\rangle = 0 \tag{5.44}$$

which is equivalent to the physical state conditions

$$(L_0^X - 1)|\phi\rangle_X = 0 \qquad \text{and} \qquad L_n^X|\phi\rangle_X = 0, \quad \text{for } n > 0. \tag{5.45}$$

Had we instead taken the state $|\uparrow\rangle$, we could not have obtained the condition $(L_0^X - 1) = 0$ since $c_0|\uparrow\rangle = 0$. BRST invariant states are then of the form

$$|\psi\rangle = |\phi\rangle_X \otimes |\downarrow\rangle = |\phi\rangle_X \otimes (c_1|0\rangle_{b,c}) \tag{5.46}$$

where $|\phi\rangle_X$ is a highest weight state of the Virasoro algebra with L_0 eigenvalue $+1$. The corresponding vertex operators are

$$\psi(z) = \phi(z)c(z). \tag{5.47}$$

One easily shows that

$$[Q, \psi(z)] = (h_\phi - 1)(\partial c)c(z)\phi(z) \tag{5.48}$$

which vanishes for $h_\phi = 1$; i.e. if $\phi(z)$ with $h_\phi = 1$ satisfies eq.(5.37), $c\phi(z)$ commutes with Q without any derivative terms. The fields $c\phi(z)$ then have zero conformal weight.

As an example, consider the tachyon with

$$|\psi\rangle_X = |k\rangle = \lim_{z,\bar{z}\to 0} : \exp ik_\mu X^\mu(z,\bar{z}) : |0\rangle_X. \tag{5.49}$$

The mass operator is now $H = L_0^{\text{tot}} = L_0^X + L_0^{b,c}$. Since the ghost ground state gives $L_0^{b,c}|\downarrow\rangle = -|\downarrow\rangle$, we have $H = L_0^X - 1$. Therefore we can attribute the negative $(\text{mass})^2$ of the tachyon to the ghost contribution to the string spectrum.

REFERENCES

[1] C. Becchi, A. Rouet and R. Stora, *"The Abelian Higgs Kibble model and unitarity of the S-operator"*, Phys. Lett. **52B** (1974) 344; *"Renormalization of gauge theories"*, Ann. Phys. **98** (1976) 287.

[2] I.V. Tyupin, *"Gauge invariance in field theory and in statistical physics in the operator formalism"*, Lebedew institute preprint FIAN No.39 (1975), unpublished.

[3] E.S. Fradkin and G.A. Vilkovisky, *"Quantization of relativistic systems with constraints"*, Phys. Lett. **55B** (1975) 224.

[4] M. Kato and K. Ogawa, *"Covariant quantization of string based on BRS invariance"*, Nucl. Phys. **B212** (1983) 443.

[5] M. Peskin, *"Introduction to string and superstring theory II"*, lectures presented at the 1986 TASI School, preprint SLAC-PUB-4251, in Santa Cruz TASI proceedings.

[6] J.H. Schwarz, *"Faddeev-Popov Ghosts and BRS Symmetry in String Theories"*, Prog. Theor. Phys. **S86** (1986) 70.

Chapter 6

Global Aspects of String Perturbation Theory and Riemann Surfaces

In this chapter we want to study some issues which are of relevance for the perturbation theory of closed oriented bosonic[1] strings. Global aspects of the string world-sheet were addressed by Friedan [1] and Alvarez [2]. The role of modular invariance for closed string loop calculations was first discussed by Shapiro [3]. Some reviews which influenced the presentation in this chapter and further extensions of these topics are refs. [4, 5, 6]. In the following we will assume that a Wick rotation of the world-sheet and the embedding space has been performed, so that both the world-sheet and the d-dimensional space have Euclidean signature.

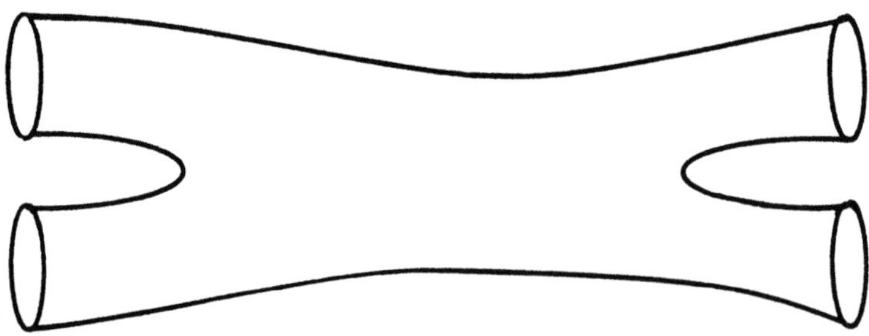

Fig. 6.1. Tree level scattering of four closed strings

---

[1] The restriction to oriented strings entails that all the surfaces we are considering below are orientable.

Consider as an example the tree level scattering amplitude of four closed strings shown in figure 6.1. The interactions of strings result from their splitting and joining. The corresponding world-sheet has tubes extending into the past and the future corresponding to incoming and outgoing strings. In the Polyakov formulation[2] the scattering amplitudes are given by a functional integral over oriented surfaces bounded by the position curves of the initial and final string configurations, weighted with the exponential of the free action (Polyakov action) and integrated with the string wave functions. The key observation is now that conformal invariance allows to consider compact world-sheets instead of surfaces with boundaries corresponding to incoming and outgoing strings. The incoming and outgoing strings can be conformally mapped to points of the two-dimensional surface (see figure 6.2).

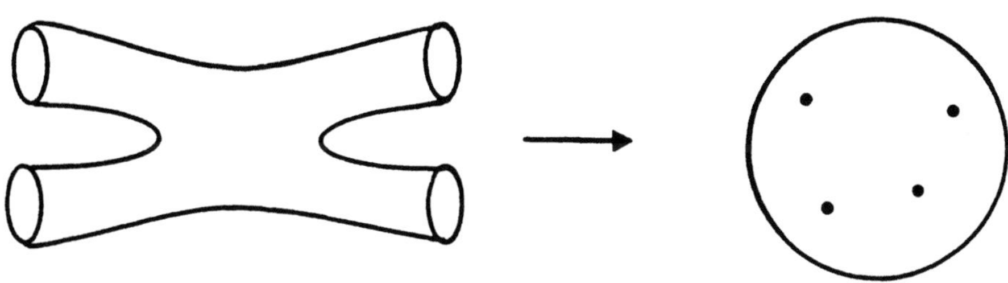

Fig. 6.2. Map of asymptotic string states to points on the sphere

Consider, for example, the case of a world-sheet with only one incoming and one outgoing string, described by a cylinder with metric $ds^2 = d\tau^2 + d\sigma^2$, $-\infty < \tau < \infty$, $0 \le \sigma < 2\pi$. Taking $\tau = \ln r$ this becomes $ds^2 = r^{-2}(dr^2 + r^2 d\sigma^2)$. The incoming string ($\tau = -\infty$) has been mapped to the point

---

[2] We will only discuss the Polyakov formulation for the calculation of string scattering amplitudes and not the older operator approach. They do lead to the same results.

$r = 0$ and the outgoing string ($\tau = +\infty$) to $r = \infty$. The string world-sheet has been mapped to the plane. A suitable choice of a conformal factor maps the plane to the sphere. We rescale the metric by $4r^2(1 + r^2)^{-2}$ and get $d\tilde{s}^2 = 4\frac{dr^2 + r^2 d\sigma^2}{(1+r^2)^2} = \frac{4 dz d\bar{z}}{(1+|z|^2)^2}$ where $z = re^{i\sigma}$. This is the standard round metric of the sphere, stereographically projected onto the plane. Indeed, with $z = \cot\left(\frac{\theta}{2}\right)e^{i\phi}$ we find $d\tilde{s}^2 = d\theta^2 + \sin^2\theta d\phi^2$. The incoming and outgoing strings are now finite points, namely the south and north pole of the sphere. For more complicated string diagrams with several incoming and outgoing strings, the conformal factor can always be chosen to map all of them to points on the sphere. This remains also true for loop diagrams (cf. below) where the external strings are mapped to points on spheres with g handles if g is the number of loops. The quantum numbers of the external string states are generated by local operators inserted at these points. These are the vertex operators introduced in Chapter 4. In summary, performing the conformal mapping, the world-sheet becomes a two-dimensional surface with the incoming and outgoing particles inserted by local vertex operators. In this sense vertex operators can be viewed as conformal projections of asymptotic states. It is however only known how to construct vertex operators for on-shell states.[3] Then string scattering amplitudes of on-shell particles are correlation functions of vertex operators.

Analogously, the one loop scattering amplitudes are described by world-sheets with one "hole" (handle) as shown in figure 6.3. A four-point multi-loop diagram is drawn in figure 6.4.

Two dimensional surfaces are topologically completely characterized by their number g of handles in terms of which the Euler number is given as $\chi = \frac{1}{4\pi} \int \sqrt{h} R \, d^2\sigma = 2(1-g)$. The number of handles is also called the genus of the surface. In summary, a n-point, g-loop amplitude is described by a

---

[3]In previous chapters we have seen how the requirement of conformal or BRST invariance puts them on shell.

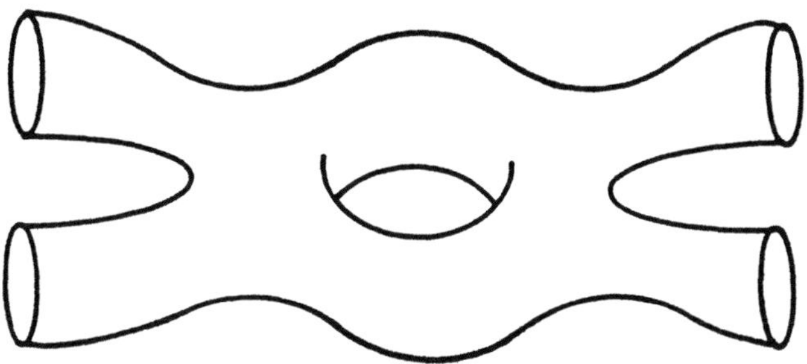

Fig.6.3 One loop scattering of four closed strings

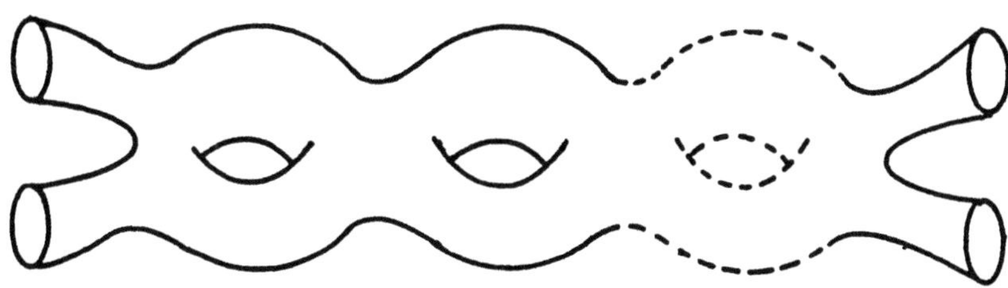

Fig.6.4 Multiloop scattering of four closed strings

two-dimensional surface with g handles and n vertex operator insertions.

Then, using the prescription of Polyakov, a general n-point amplitude can be computed as the following path integral:

$$
\begin{aligned}
A_n &= \sum_{g=0}^{\infty} A_n^{(g)} \\
&= \sum_{g=0}^{\infty} \int \mathcal{D}h \mathcal{D}X^{\mu} \int \mathrm{d}^2 z_1 \ldots \mathrm{d}^2 z_n \, V(z_1, \bar{z}_1) \ldots V(z_n, \bar{z}_n) e^{-S[X,h]} \quad (6.1) \\
&= \sum_{g=0}^{\infty} \int \mathrm{d}^2 z_1 \ldots \mathrm{d}^2 z_n \, \langle V(z_1, \bar{z}_1) \ldots V(z_n, \bar{z}_n) \rangle
\end{aligned}
$$

where we sum over all topologies of the world-sheet and integrate over the insertion points of the vertex operators. A_0 is called the partition function and is commonly denoted by Z.

Since the action S is invariant under conformal transformations and diffeomorphisms of the world-sheet, this integral is highly divergent; one integrates infinitely many times over gauge equivalent metric configurations. Therefore, in order to get rid of this overcounting one has to divide the measure in eq.(6.1) by the volume of the symmetry group which is generated by diffeomorphisms and conformal rescaling. Here and below we will always assume that we are in the critical dimension.

We have seen in Chapter 2 that the Polyakov action does not depend on the Weyl degree of freedom of the metric. This classical result was also found to hold in the quantum theory in the critical dimension. Locally we can choose isothermal (or conformal) coordinates, in which the metric takes the simple form $\mathrm{d}s^2 = 2e^\sigma\big((\mathrm{d}\sigma^1)^2 + (\mathrm{d}\sigma^2)^2\big)$ which, in terms of complex coordinates $z = \sigma^1 + i\sigma^2$, becomes $\mathrm{d}s^2 = 2e^\sigma \mathrm{d}z\mathrm{d}\bar{z}$. The factor 2 has been introduced for convenience to get $h_{z\bar{z}} = h_{\bar{z}z} = e^\sigma$. Conformal transformations $z \to f(z)$ and $\bar{z} \to \bar{f}(\bar{z})$ only change the conformal factor e^σ. However, we also know that if the operator $P^\dagger$ defined in Chapter 2 has zero modes, then there exist metrics on the world-sheet which are not conformally related, i.e. they cannot be obtained from each other by reparametrizations and Weyl rescaling. They are said to have different conformal structures. All metrics which are conformally related will take the form $\mathrm{d}s^2 = 2e^\sigma \mathrm{d}z\mathrm{d}\bar{z}$ in some fixed coordinate system. Conformally unrelated metrics will however take the form $\mathrm{d}s^2 = 2e^{\tilde{\sigma}}|\mathrm{d}z + \mu\mathrm{d}\bar{z}|^2$ in that same coordinate system. $\mu = \mu_{\bar{z}}{}^z(z,\bar{z})$ is called a Beltrami differential. If we make an infinitesimal change of coordinates to $z \to z + \delta z$ where δz is not a globally defined vector field, we find, to first order in δz, $\mathrm{d}s^2 = 2e^\sigma\big(1 + \frac{\mathrm{d}\delta z}{\mathrm{d}z} + \frac{\mathrm{d}\delta \bar{z}}{\mathrm{d}\bar{z}}\big)\big|\mathrm{d}z + \frac{\mathrm{d}\delta z}{\mathrm{d}\bar{z}}\mathrm{d}\bar{z}\big|^2$ and $\mu_{\bar{z}}{}^z = \frac{\mathrm{d}\delta z}{\mathrm{d}\bar{z}}$. Under these transformations the metric changes as $\delta h_{\bar{z}\bar{z}} = \frac{d\delta z}{d\bar{z}} h_{z\bar{z}} = \mu_{\bar{z}}{}^{z} h_{z\bar{z}}$ and we find

$$\mu_{\bar{z}}{}^{z} = h^{z\bar{z}} \delta h_{\bar{z}\bar{z}} . \tag{6.2}$$

The conformal structures are distinguished by a finite number of parameters called moduli, denoted by τ_i. They take value in the moduli space $\mathcal{M}_g$. The moduli space is the space of all metrics divided by all conformal rescalings and diffeomorphisms:

$$\mathcal{M}_g = \frac{\{\text{metrics}\}}{\{\text{Weyl rescalings}\} \times \{\text{diffeomorphisms}\}} . \tag{6.3}$$

We can then write the change of the metric as $h^{z\bar{z}}\delta h_{zz} = \sum_i \delta\tau_i h^{z\bar{z}} \partial_{\tau_i} h_{zz} = \sum_i \delta\tau_i \mu_{iz}{}^{\bar{z}}$. The number of moduli, i.e. the dimension of moduli space, depends on the genus of the surface in a well defined way as we will explain below. The integral over metrics will then reduce to a finite dimensional integral over the moduli. We will discuss the integration measure in more detail below.

If we cover the world-sheet by conformal coordinate patches U_α, then on the overlaps the metrics will be conformally related, i.e. the transition functions on the overlaps are analytic and the complex coordinates are globally defined. A system of analytic coordinate patches is called a complex structure, which, as we have now seen, is the same as a conformal structure. A two-dimensional (topological) manifold with a complex structure is called a Riemann surface Σ_g where the subscript denotes the genus. We have thus seen that the theory of Riemann surfaces will play an important role in string perturbation theory.

Since we have globally defined complex coordinates we can define on any Riemann surface vectors $V^z \partial_z$ and $V^{\bar{z}} \partial_{\bar{z}}$ and 1-forms $V_z dz$ and $V_{\bar{z}} d\bar{z}$ and in general tensors with components $V^{z...z\bar{z}...\bar{z}}{}_{z...z\bar{z}...\bar{z}}$. Since the indices z and $\bar{z}$ range only over one value all tensors are one component objects. Denote by

$T^{(n,\bar{n})}_{(m,\bar{m})}$ a tensor with $n(\bar{n})$ upper $z(\bar{z})$ and $m(\bar{m})$ lower $z(\bar{z})$ indices. Under conformal transformations it transforms with weight $(h = m-n, \bar{h} = \bar{m}-\bar{n})$. Its conformal spin is $h - \bar{h} = (m - n) - (\bar{m} - \bar{n})$. The non-vanishing metric component $h_{z\bar{z}}$ and its inverse $h^{z\bar{z}}$ allow one to convert upper $\bar{z}$ indices into lower z indices and vice versa. It follows that any tensor can be written with one type of indices only, say z. Such a tensor is called holomorphic. A holomorphic tensor with p lower and q upper indices is said to have rank $n = p - q$. Its rank is also equal to the conformal weight h and to the conformal spin (since $\bar{h} = 0$). A rank n holomorphic tensor transforms under analytic coordinate transformations as $T(z,\bar{z}) \rightarrow \left(\frac{\partial f(z)}{\partial z}\right)^n T(f(z), \bar{f}(\bar{z}))$. From now on we will only consider holomorphic tensors. We will call the space of holomorphic rank n tensors $\mathcal{T}^{(n)}$. Note that elements of $\mathcal{T}^{(n)}$ are in general functions of z and $\bar{z}$. An analytic tensor is a holomorphic tensor whose components depend only analytically on the coordinates in each local coordinate chart.

We now define a scalar product and norm on $\mathcal{T}^{(n)}$ by

$$(V^{(n)}|W^{(n)}) = \int \mathrm{d}^2z \sqrt{h}(h^{z\bar{z}})^n V^{(n)*}W^{(n)} \tag{6.4}$$

and

$$||V^{(n)}||^2 = (V^{(n)}|V^{(n)}) \tag{6.5}$$

where $V^{(n)}, W^{(n)} \in \mathcal{T}^{(n)}$. This is the only possible covariant local norm. We note that it is invariant under Weyl rescalings of the metric only for $n = 1$.

We can now define covariant derivatives. As connection we take the usual Levi-Civita connection with the Christoffel symbols as connection coefficients. In conformal coordinates only two of them are non-vanishing, namely

$$\Gamma_{zz}{}^z = \partial\sigma \quad , \quad \Gamma_{\bar{z}\bar{z}}{}^{\bar{z}} = \bar{\partial}\sigma \tag{6.6}$$

and the Riemann tensor for a conformally flat metric is (recall that in two dimensions the curvature tensor has only one independent component)

$$R_{z\bar{z}z\bar{z}} = -h_{z\bar{z}}R_{z\bar{z}} = -\frac{1}{2}(h_{z\bar{z}})^2 R$$

$$= \partial\bar{\partial}h_{z\bar{z}} - h^{z\bar{z}}\partial h_{z\bar{z}}\,\bar{\partial}h_{z\bar{z}} \qquad (6.7)$$

$$= e^{\sigma}\partial\bar{\partial}\sigma$$

where $R_{z\bar{z}}$ is the Ricci tensor and R the Ricci curvature scalar

$$R = -2h^{z\bar{z}}\partial\bar{\partial}\ln h_{z\bar{z}}$$

$$= -2e^{-\sigma}\partial\bar{\partial}\sigma. \qquad (6.8)$$

We now have the following covariant derivatives:

$$\nabla_z^{(n)}: \quad \mathcal{T}^{(n)} \to \mathcal{T}^{(n+1)}: \quad \nabla_z^{(n)}T^{(n)}(z,\bar{z}) = \big(\partial - n(\partial\sigma)\big)T^{(n)}(z,\bar{z}), \quad (6.9)$$

$$\nabla_{(n)}^z: \quad \mathcal{T}^{(n)} \to \mathcal{T}^{(n-1)}: \quad \nabla_{(n)}^z T^{(n)}(z,\bar{z}) = h^{z\bar{z}}\,\nabla_{\bar{z}}T^{(n)} = h^{z\bar{z}}\,\bar{\partial}T^{(n)}(z,\bar{z})$$

$$(6.10)$$

which are nothing but the ordinary covariant derivatives with connection coefficients eq.(6.6). They commute with holomorphic coordinate changes. ($\nabla_z^{(1)}$ is the operator P of Chapter 2.) The adjoint of $\nabla_z^{(n)}$ is defined as $(W^{(n+1)}|\nabla_z^{(n)}V^{(n)}) = (\nabla_z^{(n)\dagger}W^{(n+1)}|V^{(n)})$. We find

$$\left(\nabla_z^{(n)}\right)^{\dagger} = -\nabla_{(n+1)}^z \ . \qquad (6.11)$$

The Ricci identity is also easy to derive:

$$[\nabla_{(n+1)}^z\nabla_z^{(n)} - \nabla_z^{(n-1)}\nabla_{(n)}^z] = \frac{1}{2}nR \ . \qquad (6.12)$$

The complex geometry we have just introduced can now be used to get more information about the moduli space $\mathcal{M}_g$ associated with the Riemann surface Σ_g. Consider an arbitrary infinitesimal change of metric (cf. Chapter 2):

$$\delta h_{\alpha\beta} = \Lambda h_{\alpha\beta} + \nabla_\alpha V_\beta + \nabla_\beta V_\alpha + \sum_i \frac{\partial}{\partial\tau_i}h_{\alpha\beta}\delta\tau_i. \qquad (6.13)$$

The first term corresponds to a Weyl rescaling, the second term to a reparametrization, parametrized by a vector field V_α, and the last term to changes of the conformal structure parametrized by the variation of the moduli parameters τ_i. They can, by definition, not be compensated by Weyl rescalings and diffeomorphisms. The transformations eq.(6.13) are in the tangent space to the space of metrics at $h_{\alpha\beta}$; we will denote it by $T_{(h)}$. What we want is an orthogonal decomposition of $T_{(h)}$. We absorb the trace parts of reparametrizations and variations of the conformal structure in the Weyl rescaling by redefining $\Lambda \to \Lambda - \nabla^\gamma V_\gamma - \frac{1}{2} h^{\gamma\delta} \sum_i (\partial_{\tau_i} h_{\gamma\delta}) \delta\tau_i$ and get, in conformal coordinates

$$\delta h_{z\bar{z}} = \Lambda h_{z\bar{z}}$$
$$\delta h_{zz} = \nabla_z^{(+1)} V_z + \sum_i \delta\tau^i \mu^i{}_{zz} \tag{6.14}$$

where $\mu^i{}_{zz} = \partial_{\tau_i} h_{zz} = h_{z\bar{z}} \mu_z^{i\bar{z}}$. We see that the changes of $h_{z\bar{z}}$ can always be written as Weyl rescalings. Eq.(6.14) is not an orthogonal decomposition of $T_{(h)}$. Let us define a basis $\phi^i{}_{zz}$ which spans the orthogonal complement of $\nabla_z^{(+1)}$ in $T_{(h)}$. Using eq.(6.4) this means that

$$(\phi^i{}_{zz} \mid \nabla_z^{(+1)} V_z) = -(\nabla_{(+2)}^z \phi^i{}_{zz} | V_z) = 0 \tag{6.15}$$

for arbitrary V. This is the case if and only if

$$h_{z\bar{z}} \nabla_{(+2)}^z \phi^i{}_{zz} = \partial_{\bar{z}} \phi^i{}_{zz} = 0. \tag{6.16}$$

This means that the $\phi^i{}_{zz}$ are global analytic tensors of rank 2; they are called quadratic differentials. Therefore the dimension of moduli space is equal to the number of linear independent quadratic differentials on a given Σ_g. In other words, the quadratic differentials span the kernel or the space of zero modes of the operator $(\nabla_z^{(+1)})^\dagger = -\nabla_{(+2)}^z$. We then have the following orthogonal decomposition of $T_{(h)}$:

$$T_{(h)} = \{\Lambda h_{z\bar{z}}\} \oplus \{\mathrm{image} \nabla_z^{(+1)}\} \oplus \{\ker \nabla_{(+2)}^z\} \oplus c.c. \tag{6.17}$$

In contrast to the μ^i, the ϕ^i are not tangent to the gauge slice. We can write eq.(6.14) in this orthogonal basis. To do this we have to project the μ^i on the space spanned by the ϕ^i. The projection operator is $P = \sum_{ij} |\phi^i) M_{ij} (\phi^j|$ where $(M^{-1})_{ij} = (\phi_i|\phi_j)$. (The ϕ^i are not necessarily an orthonormal basis of ker $\nabla^z_{(+2)}$.) Using this we get

$$\sum_i \delta\tau^i \mu^i_{zz} = \sum_i \delta\tau^i (P\mu_{zz})^i + \sum_i \delta\tau^i ((1-P)\mu_{zz})^i$$
$$= \sum_i \delta\tau^i (P\mu_{zz})^i + \sum_i \delta\tau^i \partial_z \xi^i_z \qquad (6.18)$$

for some vector fields ξ^i and finally

$$\delta h_{zz} = \nabla^{(+1)}_z V_z + \sum_{ijk} \delta\tau^i \phi^k_{zz} M_{kj} (\phi^j|\mu^i) \qquad (6.19)$$

where we have shifted $V_z + \sum_i \delta\tau^i \xi^i_z \rightarrow V_z$ and used $(\mu^i|\phi^j) = \int d^2z\, \mu^{iz}_{\bar{z}} \phi^j_{zz}$ which does not depend on the metric but only on the conformal class.

The kernel of $\nabla^{(+1)}_z$ is spanned by tensors $\in \mathcal{T}^{(1)}$ satisfying

$$\nabla^{(+1)}_z V_z = h_{z\bar{z}} \partial V^{\bar{z}} = 0 \qquad (6.20)$$

which defines conformal Killing vectors $V^{\bar{z}}$. They are globally defined vector fields which span the kernel of $\nabla^{(+1)}_z$. They generate the conformal Killing group (CKG), the group of conformal isometries. The diffeomorphisms generated by them can be completely absorbed by Weyl rescalings.

The question of how many moduli parameters τ_i exist for a compact Riemann surface of genus g can be answered with the help on an index theorem, the Riemann-Roch theorem, which we will state without proof. If we define the index of $\nabla^{(n)}_z$ to be the number of its zero modes minus the number of zero modes of its adjoint $\nabla^z_{(n+1)}$, then the theorem states that

$$\mathrm{ind}\nabla^{(n)}_z = \dim \ker \nabla^{(n)}_z - \dim \ker \nabla^z_{(n+1)} = -(2n+1)(g-1). \qquad (6.21)$$

This tells us that the number of complex moduli parameters minus the number of conformal Killing vectors is $(n = +1)$ $3g - 3$. $\chi = 2(1 - g)$ is the Euler number of Σ_g.

It is not hard to find the number of conformal Killing vectors for any compact Riemann surface. They have to be globally defined analytic vector fields whose norm is finite:

$$\int_{\Sigma_g} \mathrm{d}^2 z \sqrt{h} h_{z\bar{z}} V^z V^{\bar{z}} = \text{finite} \tag{6.22}$$

where $V^z = \sum_n V_n z^n$. On the sphere $(g=0)$ the metric is $\mathrm{d}s^2 = \frac{4\mathrm{d}z\mathrm{d}\bar{z}}{(1+|z|^2)^2}$. It then follows that there are three independent conformal Killing vectors: ∂_z, $z\partial_z$ and $z^2\partial_z$. To show these fields are also well behaved at ∞ we study their behavior at $w \to 0$ where $w = 1/z$: $-w^2\partial_w$, $-w\partial_w$, $-\partial_w$. They are the only holomorphic vector fields which are well-behaved at the origin and at infinity. They correspond to the transformations generated by L_0 and $L_{\pm 1}$. The conformal Killing group is thus $SL(2, \mathbf{C})$, as follows from our discussion in Chapter 4. From the Riemann-Roch theorem we get for the dimension of moduli space $\dim \mathcal{M}_0 = 0$; i.e. there are no moduli parameters. All metrics on the sphere are conformally equivalent and there is a unique Riemann surface at genus zero. In the same way that we have shown $\dim \ker \nabla_z^{(+1)} = 3$, we can show that $\dim \ker \nabla_{(2)}^z = 0$. In fact we easily find that $\dim \ker \nabla_z^{(n)} = 2n + 1$ and $\dim \ker \nabla_{(n)}^z = 0$ (for $n > 0$) thus verifying the Riemann-Roch theorem explicitly for the case $g = 0$. For $g > 0$ we use the Ricci identity eq.(6.12). Then for $V^{(n)} \in \ker \nabla_z^{(n)}$

$$0 = (\nabla_z^{(n)} V^{(n)} | \nabla_z^{(n)} V^{(n)})$$

$$= -(V^{(n)} | \nabla_{(n+1)}^z \nabla_z^{(n)} V^{(n)})$$

$$= -\frac{1}{2}(V^{(n)} | (\nabla_{(n+1)}^z \nabla_z^{(n)} + \nabla_z^{(n-1)} \nabla_{(n)}^z + \frac{1}{2} nR) V^{(n)}) \tag{6.23}$$

$$= +\frac{1}{2}\{(\nabla_z^{(n)} V^{(n)} | \nabla_z^{(n)} V^{(n)}) + (\nabla_{(n)}^z V^{(n)} | \nabla_{(n)}^z V^{(n)}) - \frac{1}{2} nR(V^{(n)} | V^{(n)})\}.$$

The torus ($g = 1$) admits globally a flat metric $ds^2 = dz\,d\bar{z}$, i.e. $R = 0$ and we find that $\partial_z V^{(n)} = \partial_{\bar{z}} V^{(n)} = 0$, i.e. $V^{(n)} = \text{const.}$ and $\dim \ker \nabla_z^{(n)} = 1$. For $n = 1$ this is just the generator of complex translations generating the conformal Killing group $U(1) \times U(1)$ of the torus. This group has one complex generator and therefore, by the Riemann-Roch theorem, the torus is described by one complex modulus τ. To get information about higher genus surfaces we use a result from the theory of Riemann surfaces which states that any Riemann surface with $g > 1$ admits a metric with constant negative curvature. We conclude that $\dim \ker \nabla_z^{(n)} = 0$ for $g > 1$, $n > 0$ and $\dim \ker \nabla_{(n+1)}^z = (2n + 1)(g - 1)$. For $n = 0$ $\dim \ker \nabla_z^{(0)}$ is spanned by constant functions. We can then complete the following table, valid for $n \geq 0$. To get the results for $n < 0$ we use $\dim \ker \nabla_z^{(-n)} = \dim \ker \nabla_{(n)}^z$.

Table 6.1:

g	$\dim \ker \nabla_z^{(n)}$		$\dim \ker \nabla_{(n+1)}^z$
0	$2n + 1$		0
1	1		1
> 1	1	for $n = 0$	g
	0	for $n > 0$	$(2n + 1)(g - 1)$

Let us now investigate the difference between two conformally inequivalent tori more carefully. Heuristically, the fat and the thin torus, depicted in figure 6.5 are conformally inequivalent; roughly speaking, the modulus is given by the ratio of the two radii of the torus. More precisely, consider the complex z-plane and pick two complex numbers λ_1 and λ_2 as shown in figure 6.6. The torus is defined by making the following identifications on the complex plane:

$$z \approx z + n\lambda_1 + m\lambda_2, \quad n, m \in \mathbf{Z}, , \quad \lambda_1, \lambda_2 \in \mathbf{C}. \tag{6.24}$$

Since λ_1, λ_2 are rescaled and rotated by the conformal transformation $z' = \alpha z$ it is clear that only their ratio $\tau = \frac{\lambda_2}{\lambda_1}$ can be a conformal invariant. We

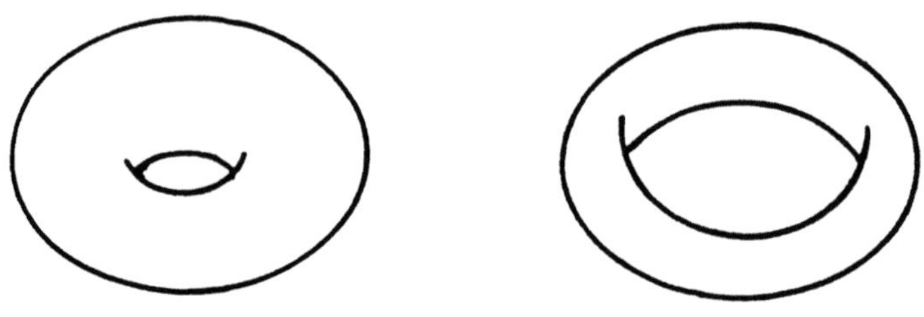

Fig.6.5. Two conformally inequivalent tori

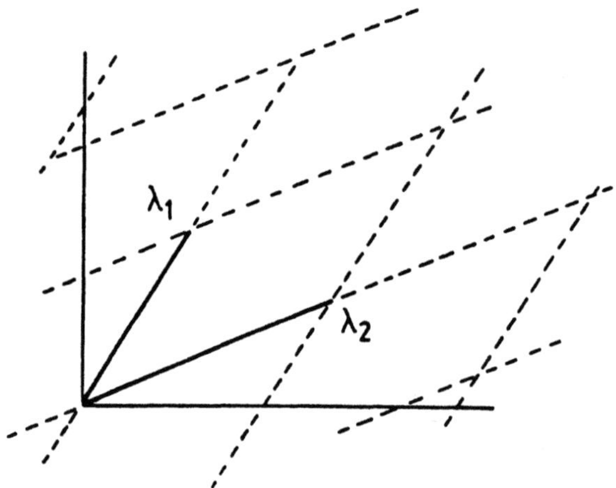

Fig.6.6 Definition of the two-dimensional torus by the complex numbers λ_1 and λ_2

can therefore set $\lambda_1 = 1$ and also, because of the freedom of interchanging λ_2 and λ_1, we may restrict Im $\tau > 0$. The tori are thus characterized by points τ in the upper half plane as illustrated in figure 6.7. where opposite sides of the parallelograms are identified:

$$z \approx z + n + m\tau \qquad n, m \in \mathbf{Z}. \tag{6.25}$$

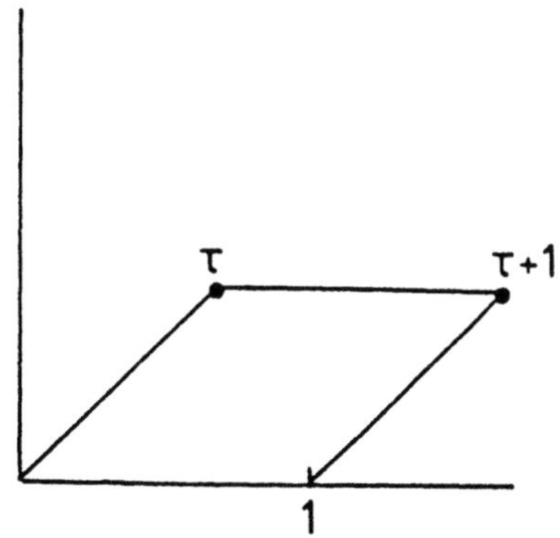

Fig.6.7. Definition of the two-dimensional torus by the complex number τ

τ is called Teichmüller parameter and describes points in Teichmüller space which for the case of the torus is the upper-half plane. Teichmüller space is the space of classes of conformally inequivalent Riemann surfaces and has the same dimension as the moduli space we are eventually looking for. It is, however, not quite true that τ is a conformal invariant that cannot be changed by rescalings and diffeomorphisms. The reason is that we must also consider global diffeomorphisms which cannot be smoothly connected to the identity. They leave the torus invariant but change the Teichmüller parameter τ. The global diffeomorphisms are the following operations on the torus. Cut the torus along the cycle a indicated in figure 6.8, twist one of its ends by 2π and glue them back together. Points that were in a neighborhood of each other before the twist will be so after the twist. Yet this twist is not connected to the identity. The same can now also be done along the cycle b. These operations are called Dehn twists and generate all global diffeomorphisms of the torus. The action on τ of a Dehn twist around the a cycle is shown in figure 6.9. In terms of λ_1, λ_2 it corresponds to $\lambda_1 \to \lambda_1$, $\lambda_2 \to \lambda_1 + \lambda_2$, which means $\tau \to \tau + 1$. A Dehn twist around the b cycle is shown in figure 6.10. To bring the transformed parallelogram into standard form we have to rotate and rescale it. Under the combined

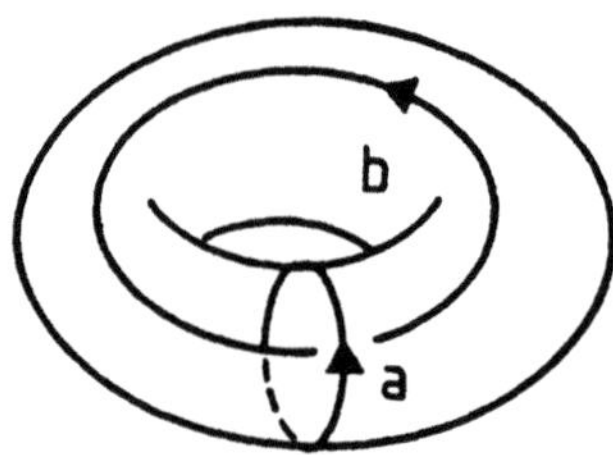

Fig.6.8. The two independent cycles on the torus

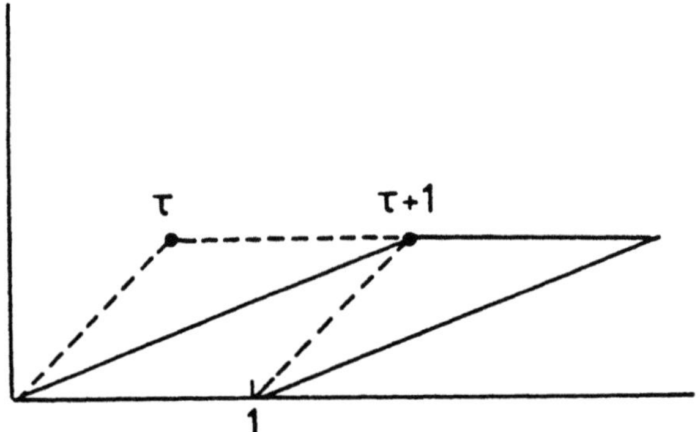

Fig.6.9 Action on τ of the Dehn twist around the a cycle

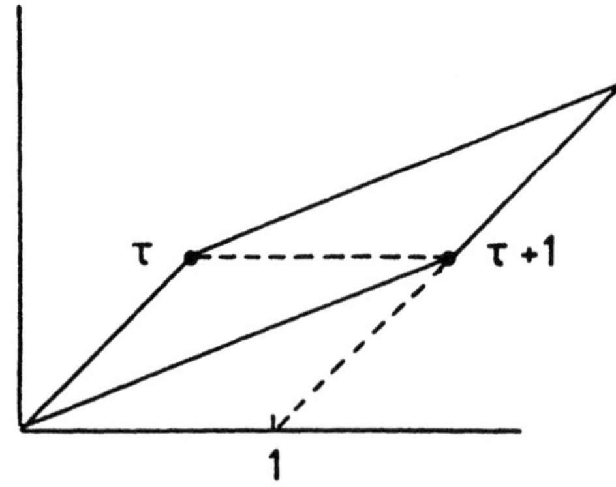

Fig.6.10 Action on τ of the Dehn twist around the b cycle

transformation we have $\tau \to \frac{\tau}{\tau+1}$. This again follows easily from the action on λ_1, λ_2 : $\lambda_1 \to \lambda_1 + \lambda_2$, $\lambda_2 \to \lambda_2$. The two transformations $\tau \to \tau + 1$ and $\tau \to \frac{\tau}{\tau+1}$ generate the group $SL(2,\mathbf{Z})$:

$$\tau \to \tau' = \frac{a\tau + b}{c\tau + d}, \qquad \begin{aligned} a,b,c,d &\in \mathbf{Z}, \\ ad - bc &= 1. \end{aligned} \tag{6.26}$$

Indeed, a general transformation is $\lambda_1 \to d\lambda_1 + c\lambda_2$, $\lambda_2 \to b\lambda_1 + a\lambda_2$ and the condition $ad - bc = 1$ preserves the area of the parallelogram. Since the two $SL(2,\mathbf{Z})$ matrices $\pm \begin{pmatrix} a & b \\ c & d \end{pmatrix}$ generate the same transformation of τ, the group of global diffeomorphisms, called the modular group for the torus, is $SL(2,\mathbf{Z})/\mathbf{Z}_2 = PSL(2,\mathbf{Z})$. We have thus learned that the parameter τ, subject to the equivalence relation eq.(6.26), describes conformally inequivalent tori. Therefore, the moduli space of the torus is the quotient of Teichmüller space and the modular group:

$$\mathcal{M}_1 = \frac{\text{Teichmüller space}}{\text{modular group}}. \tag{6.27}$$

The Dehn twists correspond to the following $SL(2,\mathbf{Z})$ matrices: $D_a = \begin{pmatrix} 1 & 1 \\ 0 & 1 \end{pmatrix}$ and $D_b = \begin{pmatrix} 1 & 0 \\ 1 & 1 \end{pmatrix}$. Instead of the two Dehn twists, one often uses the following transformations as the generators of the modular group:

$$\begin{aligned} T: & \quad \tau \to \tau + 1 \\ S: & \quad \tau \to -\frac{1}{\tau} \end{aligned} \tag{6.28}$$

We note that $TST : \tau \to \frac{\tau}{\tau+1}$. Any element of $SL(2,\mathbf{Z})$ can then be composed of S and T transformations. Any point in the upper half plane is, via a $SL(2,\mathbf{Z})$ transformation, related to a point in the so-called fundamental region $\mathcal{F}$ of the modular group. It is given by $\mathcal{F} = \left\{ -\frac{1}{2} \le \mathrm{Re}\,\tau \le 0, |\tau|^2 \ge 1 \cup 0 < \mathrm{Re}\,\tau < \frac{1}{2}, |\tau|^2 > 1 \right\}$. $\mathcal{F}$ is shown in figure 6.11.

It is the moduli space of the torus and points in $\mathcal{F}$ describe inequivalent tori. Any non-trivial modular transformation takes τ out of the fundamental

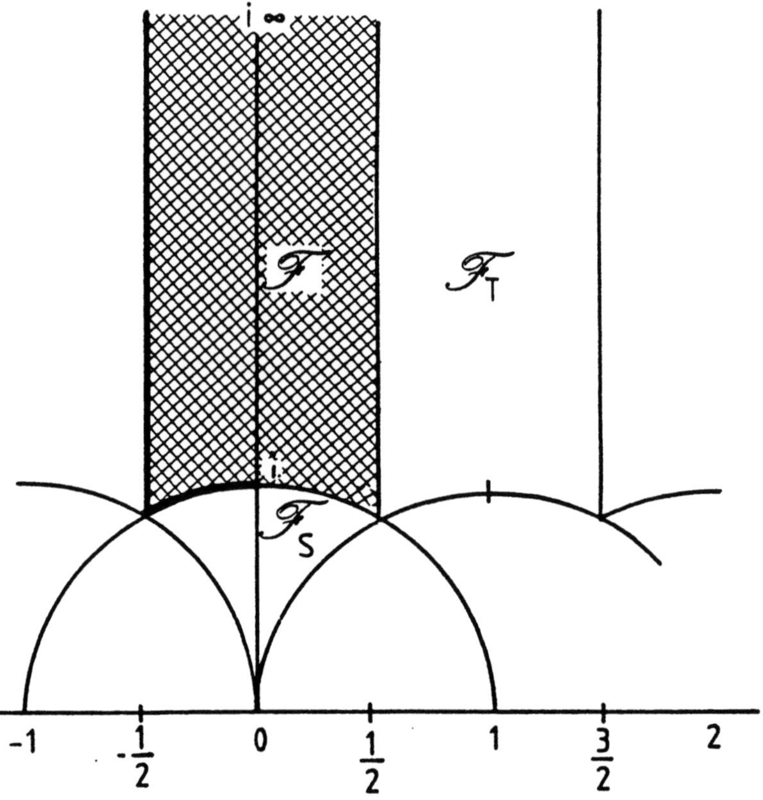

Fig.6.11. Fundamental region $\mathcal{F}$ in Teichmüller space and its images under S and T

region. The transformation S maps the fundamental region onto $\mathcal{F}_S$ shown in figure 6.11. T maps $\mathcal{F}$ onto $\mathcal{F}_T$. Of course, any image of $\mathcal{F}$ can serve equally well to parametrize the moduli space $\mathcal{M}_1$. Note that the modular group does not act freely on modular space: it has fixed points. $\tau = i$ is a fixed point of $S : \tau \rightarrow -\frac{1}{\tau}$, $S^2 = 1$ and $\tau = e^{2i\pi/3}$ of $ST : \tau \rightarrow -\frac{1}{\tau+1}$, $(ST)^3 = 1$. Because of these fixed points $\mathcal{M}_1$ is not a smooth manifold but rather a so-called orbifold with singularities at the fixed points.

It is obvious, since it is irrelevant which fundamental integration region we choose, that the integrand of one-loop string amplitudes must be invari-

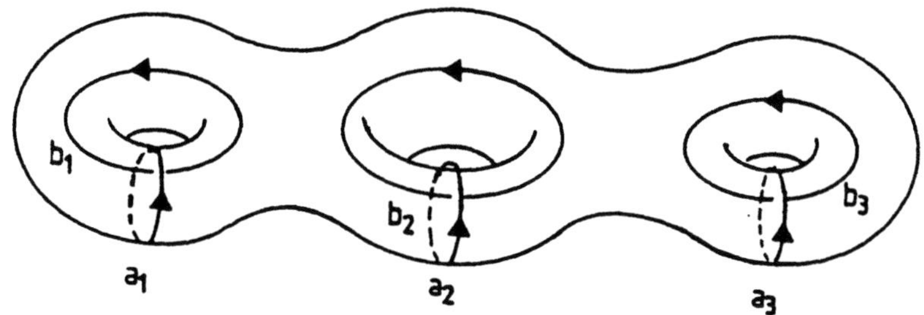

Fig.6.12. Cycles a_i, b_i as basis of first homology group $H_1(\Sigma_g, \mathbf{Z})$

ant under modular transformations eq.(6.26). This requirement of modular invariance plays an important role in string theory and has far reaching implications. We will encounter it in the construction of the heterotic string where it leads to strong restrictions on the possible gauge groups.

Let us now turn to the higher genus case. We will do this mainly to introduce some commonly used language and to point out some of the difficulties one encounters when going to higher loops. We have seen that for $g \geq 2$ the corresponding Riemann surface has no longer isometries but $3g-3$ complex moduli parameters, their number being identical to the complex dimension of moduli space. Choose $2g$ linear independent cycles a_i, b_i ($i = 1, \ldots, g$) on Σ_g which build a basis of the first homology group $H_1(\Sigma_g, \mathbf{Z}) = \mathbf{Z}^{2g}$. They are shown in figure 6.12. This basis has the property that the intersection pairings of cycles satisfy (including orientation)

$$\begin{aligned}
(a_i, a_j) &= \ (b_i, b_j) = 0 \\
(a_i, b_j) &= -(b_i, a_j) = \delta_{ij}.
\end{aligned} \tag{6.29}$$

Any such basis is called canonical. Now one can also find a set of g holomorphic and g antiholomorphic closed one forms $\omega_i, \bar{\omega}_i$ which are called Abelian differentials. A standard way of normalizing the ω_i's is to require:

$$\int_{a_i} \omega_j = \delta_{ij} . \tag{6.30}$$

Then the periods over the b cycles are completely determined as

$$\int_{b_i} \omega_j = \Omega_{ij}. \tag{6.31}$$

Ω_{ij} is the so-called period matrix of the Riemann surface; it can be shown to be a symmetric matrix with positive definite imaginary part. The space of all period matrices is a complex $\frac{g(g+1)}{2}$-dimensional space known as Siegel's upper-half plane H_g. In fact, the Ω_{ij} can be used to parametrize conformally inequivalent Riemann surfaces. However it is a highly redundant description, since the same surface will have in general many different matrices Ω corresponding to different canonical bases (cf. below). Remember that the dimension of moduli space is 0 for $g = 0$, 1 for $g = 1$ and $3g - 3$ for $g \geq 2$. On the other hand, the dimension of Siegel's upper half plane is $\frac{g(g+1)}{2}$. This coincides only for $g = 0, 1, 2, 3$. E.g. for $g = 1$ the period matrix Ω_{ij} is just the Teichmüller parameter τ – Siegel's upper half plane and Teichmüller space are identical. The abelian differentials are just constant one-forms. However, for $g \geq 4$ Teichmüller space T_g is embedded in a complicated way in Siegel's upper half plane H_g – not every symmetric $g \times g$ matrix corresponds to a point in Teichmüller space. This embedding problem and its formal solution can be phrased as the solutions to complicated differential equations (the so-called KP equations). We will not discuss this problem any further.

The second source of redundancy has to do, analogous to the one loop case, with the reduction of Teichmüller space T_g to the moduli space M_g, i.e. to find a fundamental region in T_g. In general, moduli space is obtained by dividing Teichmüller space by the group $\Omega(\Sigma)$ of disconnected diffeomorphisms of Σ_g. This group is known as mapping class group (MCG) so that we have the following relations:

$$T_g = \frac{\mathcal{M}_{g_h}}{\text{Weyl} \times \text{Diff}_0}$$

$$M_g = \frac{\mathcal{M}_{g_h}}{\text{Weyl} \times \text{Diff}} = \frac{T_g}{\text{MCG}} \qquad (6.32)$$

$$\text{MCG} = \frac{\text{Diff}}{\text{Diff}_0}$$

Here $\mathcal{M}_{g_h}$ is the space of all metrics on Σ_g and Diff_0 the diffeomorphisms connected to the identity.

A subclass of the mapping class group is the group of modular transformations which act non-trivially on a given homology basis. Suppose two canonical bases of the same Riemann surface are related by

$$\begin{pmatrix} a' \\ b' \end{pmatrix} = \begin{pmatrix} D & C \\ B & A \end{pmatrix} \begin{pmatrix} a \\ b \end{pmatrix} \qquad (6.33)$$

where A, B, C, D are $g \times g$ matrices. To preserve eq.(6.29) the matrix in eq.(6.33) must be a symplectic modular matrix with integer coefficients, i.e. an element of $Sp(2g, \mathbf{Z})$. These transformations are the analogue of the one loop modular transformation; indeed, for $g = 1$ $Sp(2, \mathbf{Z}) = SL(2, \mathbf{Z})$. We can now compute the transformation of Ω_{ij} made by the change of homology basis such that

$$\int_{a'_i} \omega'_j = \delta_{ij} \,. \qquad (6.34)$$

It then follows that $\omega'_j = \omega_k (C\Omega + D)^{-1}_{kj}$ and the new period matrix is then

$$\Omega' = (A\Omega + B)(C\Omega + D)^{-1} \,. \qquad (6.35)$$

The generators of modular transformations are the Dehn twists along the homologically non-trivial curves of figure 6.13. We have two generators for each handle and one generator for each curve linking the holes of two consecutive handles.

As in the torus case, there is a way of representing Dehn twists in terms of matrices. A Dehn twist around a non-trivial curve acts non-trivially

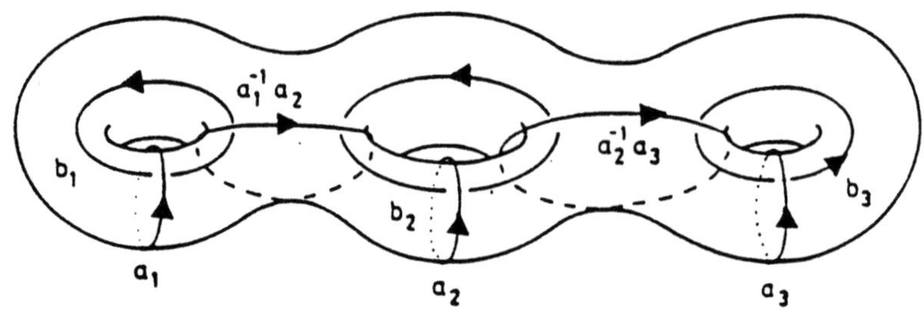

Fig.6.13. Homologically non-trivial curves on Σ_g

on the homology basis. For example, a Dehn twist around a_1 induces the following transformation on the homology basis: $a_1 \to a_1, b_1 \to b_1 + a_1$. Let D_γ be the modular transformation defined by a twist around γ. Then one can show that the matrices $D_{a_1}, D_{b_1}, D_{a_1^{-1}a_2}, \ldots, D_{a_g}, D_{b_g}$ generate in fact all matrices of $Sp(2g, \mathbf{Z})$. For the same reason as for the one loop case the integrand of the higher loop string amplitudes must be invariant under these modular transformations. E.g. for $g = 2$, the generators of $Sp(4, \mathbf{Z})$ are given by the following 4×4 matrices:

$$D_{a_1} = \begin{pmatrix} 1 & 0 & 0 & 0 \\ 0 & 1 & 0 & 0 \\ 1 & 0 & 1 & 0 \\ 0 & 0 & 0 & 1 \end{pmatrix}, \quad D_{b_1} = \begin{pmatrix} 1 & 0 & 1 & 0 \\ 0 & 1 & 0 & 0 \\ 0 & 0 & 1 & 0 \\ 0 & 0 & 0 & 1 \end{pmatrix},$$

$$D_{a_2} = \begin{pmatrix} 1 & 0 & 0 & 0 \\ 0 & 1 & 0 & 0 \\ 0 & 0 & 1 & 0 \\ 0 & 1 & 0 & 1 \end{pmatrix}, \quad D_{b_2} = \begin{pmatrix} 1 & 0 & 0 & 0 \\ 0 & 1 & 0 & 1 \\ 0 & 0 & 1 & 0 \\ 0 & 0 & 0 & 1 \end{pmatrix}, \quad D_{a_1^{-1}a_2} = \begin{pmatrix} 1 & 0 & 0 & 0 \\ 0 & 1 & 0 & 0 \\ -1 & 1 & 1 & 0 \\ 1 & -1 & 0 & 1 \end{pmatrix}.$$

$$\tag{6.36}$$

However, it is important to note that the modular transformations, i.e. the Dehn twists around the homologically non-trivial cycles do not generate the whole mapping class group. There are also twists around trivial cycles so that they do not affect the homology basis but nevertheless correspond to non-trivial diffeomorphisms. These form a subgroup of the mapping class group called the Torelli group. The quotient of the mapping class group and the Torelli group is precisely the symplectic modular group $Sp(2g, \mathbf{Z})$. We will however not consider this subtlety since for $g = 1$ (which we will be mainly interested in) the Torelli group is trivial so that the mapping class group is identical to the modular group $SL(2, \mathbf{Z})$. This concludes our considerations about the modular transformations.

We are now ready to discuss the integration measure. Our starting point is

$$\int_{\mathcal{M}_{g_h}} \frac{\mathcal{D}h\,\mathcal{D}X}{\mathrm{Vol}(\mathrm{Diff})\mathrm{Vol}(\mathrm{Weyl})} \tag{6.37}$$

where we have divided by the volume of the symmetry group. We can now replace the integral over the space of all metrics by an integral over moduli space, the conformal factor and the diffeomorphisms generated by vector fields V^z. This involves a non-trivial Jacobian determinant which we have already calculated in Chapter 3 where we have however neglected the presence of quadratic differentials and conformal Killing vectors. The Jacobian can be read off from eq.(6.19) and we find

$$\int_{\mathcal{M}_g} \mathrm{d}\tau_i^2 \int \frac{\mathcal{D}X\mathcal{D}'V\mathcal{D}\Lambda}{\mathrm{Vol}(\mathrm{Diff})\mathrm{Vol}(\mathrm{Weyl})} \frac{\det(\phi|\mu)\det(\mu|\phi)}{\det(\phi|\phi)} \mathrm{det}'\nabla_z^{(+1)}\mathrm{det}'\nabla_{(-1)}^z \tag{6.38}$$

where the prime indicates that we do not integrate over the diffeomorphisms generated by the conformal Killing vectors as they do not change h_{zz}. We treat the τ_i as complex parameters. If we now define $|\mathrm{MCG}| = \frac{\mathrm{Vol}(\mathrm{Diff})}{\mathrm{Vol}(\mathrm{Diff}_0)}$ and make an orthogonal decomposition of Diff_0 with

$$\mathrm{Vol}(\mathrm{Diff}_0) = \mathrm{Vol}(\mathrm{Diff}_0^\perp)\,\mathrm{Vol}(\mathrm{CKG}) \tag{6.39}$$

we can write the measure as

$$\frac{1}{|\mathrm{MCG}|} \int_{\mathcal{M}_g} \mathrm{d}\tau_i^2 \int \frac{\mathcal{D}X\mathcal{D}'V\mathcal{D}\Lambda}{\mathrm{Vol}(\mathrm{Diff}_0^\perp)\mathrm{Vol}(\mathrm{CKG})\mathrm{Vol}(\mathrm{Weyl})}$$

$$\times \frac{\det(\phi|\mu)\det(\mu|\phi)}{\det(\phi|\phi)}\mathrm{det}'\nabla_z^{(+1)}\mathrm{det}'\nabla_{(-1)}^z \,. \tag{6.40}$$

In the critical dimension, i.e. in the absence of a conformal anomaly, we can cancel $\int \mathcal{D}V'\mathcal{D}\Lambda$ against $\mathrm{Vol}(\mathrm{Diff}_0^\perp)\mathrm{Vol}(\mathrm{Weyl})$. Furthermore, we can replace $\frac{1}{|\mathrm{MCG}|}\int_{\mathcal{M}_g}\mathrm{d}\tau_i$ by an integral over a fundamental region. We then get the following expression for scattering amplitudes:

$$A_n = \sum_{g=0}^\infty \int \mathrm{d}^2 z_1 \ldots \mathrm{d}^2 z_n \int_{\mathcal{F}_g} \mathrm{d}\tau_i \int \frac{\mathcal{D}X}{\mathrm{Vol}(\mathrm{CKG})} \frac{\det(\phi|\mu)\det(\mu|\phi)}{\det(\phi|\phi)}$$

$$\times \mathrm{det}'\nabla_z^{(+1)}\mathrm{det}'\nabla_{(-1)}^z \, V(z_1,\bar z_1,\tau_i)\cdots V(z_n,\bar z_n,\tau_i)e^{-S[X,\tau_i]} \,. \tag{6.41}$$

At tree level there are no quadratic differentials and the corresponding factor in the measure is absent. At two and higher loop order there are no conformal Killing vectors. At one loop there is one of each, both being constants on the torus.

As we did in Chapter 3, we can replace the Jacobian determinant by an integral over anti-commuting Faddeev-Popov ghosts:

$$\mathrm{det}'\nabla_z^{(+1)}\mathrm{det}'\nabla_{(-1)}^z = \int \mathcal{D}'(b\bar b c\bar c)e^{-S[b,c,\tau_i]} \tag{6.42}$$

where the ghost action is (cf. Chapters 3 and 5)

$$S[b,c] = \frac{1}{2\pi}\int \mathrm{d}^2 z\big(b\bar\partial c + \bar b\partial\bar c\big)\,. \tag{6.43}$$

In eq.(6.42) we have excluded the integration over the ghost zero modes. The integral would vanish otherwise. We will show in the following that the remaining factors $\frac{1}{\mathrm{Vol}(\mathrm{CKG})}$ and $\frac{\det(\phi|\mu)\det(\mu|\phi)}{\det(\phi|\phi)}$ can be attributed to the c- and b-ghost zero modes respectively. This is in fact easy to see. The ghost zero modes satisfy the equations

$$\bar{\partial}c = h_{z\bar{z}}\nabla^{z}_{(-1)}c = 0$$
$$\bar{\partial}b = h_{z\bar{z}}\nabla^{z}_{(+2)}b = 0 \tag{6.44}$$

which tells us that the c zero modes correspond to the conformal Killing vectors and the b zero modes to the quadratic differentials. By the Riemann-Roch theorem, eq.(6.21), we then get

$$N_b - N_c = 3g - 3 \tag{6.45}$$

where $N_{b,c}$ denotes the number of zero modes. (Note that $\dim \ker\nabla^{z}_{(-1)} = \dim \ker\nabla^{(+1)}_{z}$ as $\nabla^{(n)}_{z}$ is the complex conjugate of $\nabla^{z}_{(-n)}$.) The presence of ghost zero modes means that the ghost number current is not conserved. We will give a more detailed and general description in Chapter 13.

Due to the anti-commutativity of the ghosts, integration over their zero modes will give a vanishing answer if we do not insert them into the integrand. To illustrate this, consider an anti-commuting variable ψ and split it into its zero-mode part ψ_0 and the remainder ψ':

$$\psi(z,\bar{z}) = \psi_0(z) + \psi'(z,\bar{z}) = \sum_{i=1}^{N} \psi_0^i \phi^i(z) + \psi'(z,\bar{z}) \tag{6.46}$$

where the zero mode wave functions ϕ^i satisfy $\bar{\partial}\phi^i = 0$. (E.g. for the b ghosts they are the quadratic differentials.) The ψ_0^i are constant anti-commuting parameters satisfying $\int d\psi_0^i\,\psi_0^i = 1$, $\int d\psi_0^i = 0$. Since the action for ψ does not depend on ϕ_0, $\int \mathcal{D}(\psi\bar{\psi})e^{-S}$ will vanish unless we restrict the integration to the non-zero modes or absorb them by inserting $\prod_{i=1}^{N} \psi(z_i)$ into the integrand:

$$\int \mathcal{D}(\psi\bar{\psi}) \prod_{i=1}^{n} \psi(z_i)\bar{\psi}(z_i)e^{-S[\psi']} = \frac{|\det(\phi^i(z_j))|^2}{\det(\phi^i|\phi^j)} \int \mathcal{D}'(\psi\bar{\psi})e^{-S[\psi']} \tag{6.47}$$

where the factor $\det(\phi^i|\phi^j)$ is relevant if the ϕ^i do not form an orthonormal basis. It renders the zero mode contribution basis independent.

Using this it is now easy to rewrite the string measure including the integration over the ghost zero modes. Let us first, for simplicity, restrict ourselves to the case $g \geq 2$ where there are no conformal Killing vectors. Using eq.(6.47) we find the following simple expression for the partition function:

$$Z_{g \geq 2} = \int_{\mathcal{F}_g} \mathrm{d}^2 \tau_i \int \mathcal{D}X \mathcal{D}(b\bar{b}c\bar{c}) \prod_{i=1}^{3g-3} |(\mu^i|b)|^2 e^{-S[X,b,c,\tau_i]}. \tag{6.48}$$

Needless to say, we won't evaluate it.

Let us now turn to the cases $g = 0$ and $g = 1$. At tree level we have no b zero modes (no moduli) but instead three complex c zero modes, corresponding to the conformal Killing vectors that generate the group $PSL(2, \mathbf{C})$. The zero mode wave functions were found to be 1, z and z^2. $PSL(2, \mathbf{C})$ acts freely on the insertion points z_i of the vertex operators and one can fix three of them, say z_1, z_2 and z_3, at arbitrary points with a unique $PSL(2, \mathbf{C})$ transformation. They are generated by vector fields $(\alpha + \beta z + \gamma z^2)\partial_z$ and we can trade the integrations over z_1, z_2, z_3 for an integration over α, β, γ. The Jacobian is

$$\left| \frac{\partial(z_1, z_2, z_3)}{\partial(\alpha, \beta, \gamma)} \right|^2 = |(z_1 - z_2)(z_2 - z_3)(z_1 - z_3)|^2. \tag{6.49}$$

The integration over α, β, γ then cancels the Vol(CKG) factor. Note that the Jacobian is also just $\det(V^i(z_j))$ where $\{V^i\} = \{1, z, z^2\}$ are the conformal Killing vectors (which are orthogonal to each other). More importantly, it can be written as

$$|\langle 0|c(z_1)c(z_2)c(z_3)|0\rangle|^2 = |\langle 0|c_{-1}c_0 c_1|0\rangle|^2 \det \begin{vmatrix} 1 & 1 & 1 \\ z_3 & z_2 & z_1 \\ z_3^2 & z_2^2 & z_1^2 \end{vmatrix}^2 \tag{6.50}$$

$$= |(z_1 - z_2)(z_1 - z_3)(z_2 - z_3)|^2$$

where we have used results from Chapter 5. For tree level amplitudes this means that the presence of conformal Killing vectors is taken care of if we drop the integration over the positions of three of the vertex operators and multiply each of them by $c(z_i)\bar{c}(\bar{z}_i)$. We know from Chapter 5 that if $\int V$ is BRST invariant, then so is $c\bar{c}V$. Tree level scattering amplitudes then take the form:

$$A_n^{g=0} = \int \mathcal{D}X\mathcal{D}(b\bar{b}c\bar{c})c\bar{c}V(z_1,\bar{z}_1)\,c\bar{c}V(z_2,\bar{z}_2)\,c\bar{c}V(z_3,\bar{z}_3)$$

$$\times \prod_{i=4}^{n}\int \mathrm{d}^2z_i\,V(z_i,\bar{z}_i)\,e^{-S[X,b,c]} \qquad (6.51)$$

$$= \langle V(z_1,\bar{z}_1)\,c\bar{c}V(z_2,\bar{z}_2)\,c\bar{c}V(z_3,\bar{z}_3)\prod_{i=4}^{n}\int \mathrm{d}^2z_i\,V(z_i,\bar{z}_i)\rangle.$$

If we have less than three vertex operator insertions we cannot completely factor out the $PSL(2,\mathbf{C})$ volume and the correlation functions vanish upon dividing by the infinite factor Vol(CKG). So 0,1 and 2 point functions vanish at tree level. There is no tree level cosmological constant, no tree level tadpoles and no tree level mass or wave-function renormalization.

Let us now turn to the torus. Here the conformal Killing vector V and the quadratic differential ϕ are (complex) constants, which we set to one for simplicity. We can parametrize the torus by two real variables ξ^1, ξ^2 with $0 \leq \xi^1,\xi^2 \leq 1$, in terms of which the complex coordinates become $z = \xi^1 + \tau\xi^2$ and we can use Weyl invariance to set $\mathrm{d}s^2 = |\mathrm{d}z|^2$. The area of the torus is $\int\sqrt{h}\mathrm{d}^2\xi = \mathrm{Im}\,\tau$. If we change τ to $\tau \to \tau + \delta\tau$, we find (up to a rescaling) $\mathrm{d}s^2 \to |\mathrm{d}z + \delta\tau\frac{i}{2\mathrm{Im}\,\tau}\mathrm{d}\bar{z}|^2$ and $\mu_{\bar{z}}{}^{z} = \frac{i}{2\mathrm{Im}\,\tau}$. We now easily compute $(\phi|\phi) = \mathrm{Im}\,\tau$, $(\mu|\phi) = 1$ and Vol(CKG)$= (V,V) = (\mathrm{Im}\,\tau)^2$ where we have neglected constant factors independent of τ. Consequently, the total contribution from the ghost zero modes is $(\frac{1}{\mathrm{Im}\,\tau})^3$. Performing the integral over X we get for the one-loop partition function

$$Z^{g=1} \sim \int_{\mathcal{F}_1}\frac{\mathrm{d}^2\tau}{(\mathrm{Im}\,\tau)^2}\frac{1}{\mathrm{Im}\,\tau}(\mathrm{Im}\,\tau)^{+13}(\det{}'\Box)^{-13}\det{}'\nabla_z^{(+1)}\det{}'\nabla_{(-1)}^{z} \qquad (6.52)$$

We have dropped a factor of the volume of 26-dimensional space-time and other numerical factors. $(\mathrm{Im}\,\tau)^{13}$ is the contribution from the X^μ zero modes. We will not go into details of how it arises, neither will we compute the determinants as we will rederive the partition function using the Hamiltonian formalism in light-cone gauge below. So let us simply state the results. For the determinants one gets

$$\det{}'\nabla_z^{(+1)}\det{}'\nabla_{(-1)}^z = \det{}'\square = (\operatorname{Im}\tau)^2|\eta(\tau)|^4 . \tag{6.53}$$

$\eta(\tau)$ is the famous Dedekind eta-function defined as

$$\eta(\tau) = q^{1/24} \prod_{n=1}^{\infty} (1 - q^n), \quad q = e^{2\pi i\tau} . \tag{6.54}$$

Then the final expression for the one-loop vacuum amplitude is

$$Z' = \int_{\mathcal{F}} \frac{\mathrm{d}^2\tau}{(\operatorname{Im}\tau)^2}\chi(\bar{\tau},\tau)$$

with

$$\chi(\bar{\tau},\tau) \sim (\operatorname{Im}\tau)^{-12}|\eta(\tau)|^{-48} . \tag{6.55}$$

We will often refer to $\chi(\tau,\bar{\tau})$ as the partition function. Eq.(6.53) indicates that the effect of the ghosts is to cancel the contribution of two coordinate degrees of freedom which correspond to the longitudinal and time like string excitations. Therefore, the partition function counts only the physical transverse string excitations.

Let us now check modular invariance of the one loop partition function. First note that the measure $\frac{\mathrm{d}^2\tau}{(\operatorname{Im}\tau)^2}$ is invariant by itself. This follows from

$$\begin{aligned} \mathrm{d}^2\tau &\rightarrow |c\tau + d|^{-4}\,\mathrm{d}^2\tau , \\ \operatorname{Im}\tau &\rightarrow |c\tau + d|^{-2}\operatorname{Im}\tau . \end{aligned} \tag{6.56}$$

To check modular invariance of $\chi(\tau,\bar{\tau})$ it suffices to do so for the two generators S and T of the modular group. With the well known transformation properties of the eta-function

$$\begin{aligned} \eta(\tau + 1) &= e^{i\pi/12}\eta(\tau) \\ \eta(-\frac{1}{\tau}) &= \sqrt{-i\tau}\,\eta(\tau) \end{aligned} \tag{6.57}$$

modular invariance follows straightforwardly.

Let us now evaluate the partition function by making use of the connection between the Euclidean path integral and the Hamiltonian formalism. Write the torus modular parameter as $\tau = \mathrm{Re}\,\tau + i\mathrm{Im}\,\tau$. $\mathrm{Im}\,\tau$ plays the role of an Euclidean time variable or, in statistical mechanics language, of the inverse temperature. If $\mathrm{Re}\,\tau = 0$, we obtain the functional integral as $\mathrm{Tr}\,\exp(-\mathrm{Im}\,\tau H)$, where the time evolution operator in the $\mathrm{Im}\,\tau$ direction is given by $H = H_L + H_R$ (H_L, H_R are the left- and right-moving Hamiltonians respectively). The partition function counts the number of states which propagate around the torus in the $\mathrm{Im}\,\tau$ direction and weighs them with a factor $e^{-\mathrm{Im}\,\tau H}$. If one thinks of the torus as a cylinder of length $\mathrm{Im}\,\tau$ whose ends are identified, one can twist the two ends relative to each other by an angle $2\pi\mathrm{Re}\,\tau$ before joining them. The operator which generates these twists is $P = H_R - H_L$. (Note that in the picture of Chapter 2, $\mathrm{Im}\,\tau$ corresponds to the world-sheet coordinate τ and the twist to a shift in σ.) The complete one-loop string partition function $\chi(\bar{\tau}, \tau)$ is then given by

$$\chi(\bar{\tau}, \tau) \sim \mathrm{Tr}\, e^{+2\pi i\mathrm{Re}\,\tau(H_R - H_L)} e^{-2\pi\mathrm{Im}\,\tau(H_L + H_R)}$$

$$= \mathrm{Tr}\, \bar{q}^{H_L} q^{H_R}\,. \tag{6.58}$$

We have rescaled $\mathrm{Im}\,\tau$ by a factor of 2π. In light-cone gauge only the physical states contribute and H_L and H_R are ($i = 1, \dots, 24$):

$$H_L = \frac{1}{2}p_i^2 + \sum_{n=1}^{\infty} \bar{\alpha}^i_{-n}\bar{\alpha}^i_n - 1 = \frac{1}{2}p_i^2 + N_L - 1\,,$$

$$H_R = \frac{1}{2}p_i^2 + \sum_{n=1}^{\infty} \alpha^i_{-n}\alpha^i_n - 1 = \frac{1}{2}p_i^2 + N_R - 1\,. \tag{6.59}$$

Inserting H_L, H_R into eq.(6.58) leads to[4]

---

[4] The calculation is completely analogous to the evaluation of the grand partition

$$\chi(\bar{\tau}, \tau) \sim \int \frac{d^{24}p}{(2\pi)^{24}} e^{-2\pi p^2 \operatorname{Im}\tau} e^{4\pi \operatorname{Im}\tau} \operatorname{Tr} \bar{q}^{N_L} q^{N_R}$$

$$\sim \frac{1}{(\operatorname{Im}\tau)^{12}} e^{4\pi \operatorname{Im}\tau} \prod_{n=1}^{\infty} (1 - \bar{q}^n)^{-24} (1 - q^n)^{-24} \tag{6.60}$$

$$= \frac{1}{(\operatorname{Im}\tau)^{12}} |\eta(\tau)|^{-48} .$$

Eqs.(6.60) and (6.55) agree. Deriving the partition function in this way it is evident that $\chi(\bar{\tau}, \tau)$ contains the information about the level density of string states, i.e. the number of states of each mass level. Expanding $\chi(\bar{\tau}, \tau)$ in powers of $q = e^{2\pi i \tau}$ one gets a power series of the form $\sum d_{mn} \bar{q}^m q^n$. d_{mn} is simply the number of states with $m_L^2 = m$ and $m_R^2 = n$. The first few terms of the expansion are

$$\chi(\bar{\tau}, \tau) \sim |\eta(\tau)|^{-48} = \frac{1}{|q|^2} + 24 q^{-1} + 24 \bar{q}^{-1} + 576 + \ldots \tag{6.61}$$

The first term corresponds to the negative $(\text{mass})^2$ tachyon and the constant term to the massless string states, namely to the on-shell graviton, antisymmetric tensor field and dilaton. Note however that $\chi(\bar{\tau}, \tau)$ contains also "unphysical" states that do not satisfy the reparametrization constraint $L_0 = \bar{L}_0$ and that they are not projected out when performing the integral over the modular parameter in the region of $\mathcal{F}$ where $\operatorname{Im}\tau < 1$. Due to the tachyon pole one finds that the one-loop cosmological constant of the closed bosonic string is infinite.

---

function of an ideal Bose gas with energy levels p. Each transverse degree of freedom contributes with

$$\sum_N \sum_{\substack{\{n_p\} \\ \sum n_p = N}} q^{\sum n_p p} = \sum_N \sum_{\substack{\{n_p\} \\ \sum n_p = N}} \prod_p q^{n_p p} = \prod_{n=1}^{\infty} \sum_{m=0}^{\infty} (q^n)^m = \prod_n (1 - q^n)^{-1} .$$

REFERENCES

[1] D. Friedan, *"Introduction to Polyakov's string theory"*, in proceedings of 1982 Les Houches Summer School (Elsevier), eds. J.B. Zuber and R. Stora.

[2] O. Alvarez, *"Theory of strings with boundary. Fluctuations, topology and quantum geometry"*, Nucl. Phys. **B216** (1983) 125.

[3] J.A. Shapiro, *"Loop graph in the dual-tube model"*, Phys. Rev. **D5** (1972) 1945.

[4] E. D'Hoker and D. Phong, *"The geometry of string perturbation theory"*, Rev. Mod. Phys. **60** (1986) 917.

[5] L. Alvarez-Gaumé and P. Nelson, *"Riemann surfaces and string theories"*, preprint CERN-TH.4615/86 (1986), in proceedings of 1986 Trieste School on strings.

[6] J. Bagger, *"Strings and Riemann surfaces"*, lectures presented at the 1987 TASI School; preprint HUTP-87/A079, in Santa Fe TASI proceedings.

Chapter 7

The Classical Closed Fermionic String

Up to now we have only discussed bosonic strings. That means that all physical degrees of freedom had been described by bosonic variables. We have treated the classical and the quantum theory, the algebra of the constraints (the Virasoro algebra) and we have found that at the quantum level the theory makes sense only in the critical dimension which was found to be 26. The spectrum of both the open and closed, oriented and unoriented theories was found to contain a tachyon, a fact which is at least alarming. Let us recall that its negative $(\text{mass})^2$ arose from the (regularized) zero point energy of an infinite set of bosonic harmonic oscillators. The problem with the tachyon may be cured if we introduce fermionic degrees of freedom which are quantized with anti-commutators. Then there is a chance that the zero point energies cancel and the tachyon is absent.

One basic symmetry principle that guarantees the absence of a tachyon in the string spectrum is space-time supersymmetry. It is important to keep in mind the distinction between world-sheet and space-time supersymmetry. The fermionic string theories that we will discuss all possess world-sheet supersymmetry but not necessarily space-time supersymmetry and are not all tachyon-free. Whether a particular string theory is space-time supersymmetric or not will manifest itself, for instance, in the spectrum. Especially, the existence of one or more massless gravitinos will signal space-time supersymmetry. The formulation of fermionic string theories which we will present is the Ramond [1], Neveu, Schwarz [2] spinning string. It has manifest world-sheet supersymmetry; space-time supersymmetry, if present, is however not manifest. A string spectrum with space-time supersymmetry is obtained after a suitable truncation as it was found by Gliozzi, Scherk and Olive [3]. We should mention that there exists also the so-called Green-Schwarz formalism [4] in which space-time supersymmetry is manifest at the cost of manifest world-sheet supersymmetry. It uses in a crucial way the triality property of $SO(8)$, the transverse Lorentz group in ten dimensions, which is, as we will see later in this chapter, the critical dimension for the fermionic string. Since we will also be interested in non space-time supersymmetric theories in less than ten space-time dimensions (see Chapter 14), the old formalism with manifest world-sheet supersymmetry is preferred. We will restrict our discussion in this and all the following chapters to closed oriented strings. Also, in this chapter we will discuss $n = 1$ supersymmetry on the world-sheet. Extended world-sheet supersymmetries will be covered in Chapter 12.

7.1 Superstring action and its symmetries

Let us now find the requirements on the field content coming from world-sheet supersymmetry and set up the supersymmetric extension of the Polyakov action. The bosonic string theory was described by the action for a collection of d scalar fields $X^\mu(\sigma, \tau)$ coupled to gravity $h_{\alpha\beta}$ in two dimensions. The purely gravitational part of the action was trivial, being a total derivative. This left us with

$$S_1 = -\frac{1}{8\pi} \int \mathrm{d}^2\sigma \sqrt{h} h^{\alpha\beta} \partial_\alpha X^\mu \partial_\beta X_\mu \qquad (7.1)$$

which is just the covariant kinetic energy for the "matter fields" X^μ. The supersymmetric extension of S_1 should be the coupling of supersymmetric "matter" to two-dimensional supergravity. With respect to the d-dimensional target space, which can be considered as an internal space from the world-sheet point of view, the fields $X^\mu(\sigma, \tau)$ transform as a vector. Hence, their supersymmetric partners should be world-sheet spinors with a target space vector index. We will denote them by $\psi^\mu(\sigma, \tau)$. Let us now see how the balance between bosonic and fermionic degrees of freedom works out. The fields X^μ, representing d real scalars, provide d bosonic degrees of freedom. If we impose on the d world-sheet fermions ψ^μ a Majorana condition, they provide $2d$ fermionic degrees of freedom. We have to introduce d real auxiliary scalar fields F^μ. Together (X^μ, ψ^μ, F^μ) form an off-shell scalar multiplet of two-dimensional $n = 1$ supersymmetry. On-shell (X^μ, ψ^μ) suffice.

Let us now turn to the gravity sector. The supergravity multiplet consists of the zweibein $e_\alpha{}^a$ (n-bein in n dimensions) and the gravitino χ_α. The zweibein has two different kinds of indices. a is a Lorentz index and takes part in local Lorentz transformations whereas α is called an Einstein index and takes part in coordinate transformations (reparametrizations). Einstein indices are raised and lowered with the world-sheet metric $h_{\alpha\beta}$ and Lorentz indices with the Lorentz metric η_{ab}. The zweibein allows to transform Lorentz into Einstein indices and vice versa. The introduction of the zweibein is necessary if we want to describe spinors on a curved manifold since the group $GL(n, R)$ does not have spinor representations[1] whereas the tangent space group, $SO(d - 1, 1)$, does. The inverse of $e^\alpha{}_a$, denoted by $e_\alpha{}^a$, is defined by

$$e_\alpha{}^a e^\alpha{}_b = \delta_b^a. \tag{7.2}$$

$e^\alpha{}_a$ defines an orthonormal set of basis vectors at each point, i.e. it satisfies

$$e^\alpha{}_a e^\beta{}_b h_{\alpha\beta} = \eta_{ab} \tag{7.3}$$

from which we derive

$$e^\alpha{}_a e^\beta{}_b \eta^{ab} = h^{\alpha\beta}. \tag{7.4}$$

The gravitino is a world-sheet vector and a world-sheet Majorana spinor.

---

[1] Under general coordinate transformations, tensor indices are acted on with elements of $GL(n, R)$.

In d dimensions the d-bein $e_\alpha{}^a$ has d^2 components. There are n reparametrizations and $\frac{1}{2}d(d-1)$ local Lorentz transformations as gauge symmetries, leaving $\frac{1}{2}d(d-1)$ degrees of freedom. The gravitino, being a Majorana spinor-vector has $2^{[\frac{d}{2}]}d$ components where $[\frac{d}{2}]$ denotes the integer part of $\frac{d}{2}$. For $n=1$ supersymmetry there are $2^{[\frac{d}{2}]}$ supersymmetry parameters leaving $(d-1)2^{[\frac{d}{2}]}$ degrees of freedom. For the case of interest, namely $d=2$, we find one bosonic and two fermionic degrees of freedom. To get a complete off-shell supergravity multiplet we have to introduce one auxiliary real scalar field A. The complete off-shell supergravity multiplet is then $(e_\alpha{}^a, \chi_\alpha, A)$. $(e_\alpha{}^a, \chi_\alpha)$ form the on-shell supergravity multiplet.

So far the discussion was independent of any particular action and only a statement of the field content of the two-dimensional supersymmetry multiplets. Let us now complete the string action. The kinetic energy term for the gravitino vanishes identically in two dimensions.[2] The kinetic energy for the matter fermions ψ^μ and the contribution of the auxiliary fields F^μ is

$$S_2 = \frac{1}{4\pi} \int \mathrm{d}^2\sigma\, e\Big\{ -i\bar{\psi}^\mu \rho^\alpha \partial_\alpha \psi_\mu + F^\mu F_\mu \Big\} \tag{7.5}$$

where $e = |\det e_\alpha{}^a| = \sqrt{h}$. Our notation is summarized in the appendix to this chapter. The fact that the derivative in S_2 is an ordinary derivative rather than a covariant derivative containing the spin connection is a consequence of the Majorana spin-flip property eq.(B.3). The action $S_1 + S_2$ is not yet locally supersymmetric. It is simply the covariantized form of the action of a scalar multiplet. Local supersymmetry requires the additional term

$$S_3 = \frac{i}{8\pi} \int \mathrm{d}^2\sigma \sqrt{h}\, \bar{\chi}_\alpha \rho^\beta \rho^\alpha \psi^\mu \Big(\partial_\beta X_\mu - \frac{i}{4}\bar{\chi}_\beta \psi_\mu\Big). \tag{7.6}$$

The auxiliary field A does not appear. The auxiliary matter scalars F^μ can

---

[2]In any number of dimensions it is given by $\bar{\chi}_\alpha \Gamma^{\alpha\beta\gamma} D_\beta \chi_\gamma$ where $\Gamma^{\alpha\beta\gamma}$ is the antisymmetrized product of three Dirac matrices which vanishes in two dimensions.

be eliminated via their equations of motion. This will be assumed to be
done from now on. The complete action

$$S = -\frac{1}{8\pi} \int d^2\sigma \sqrt{h} \Big\{ h^{\alpha\beta}\partial_\alpha X^\mu \partial_\beta X_\mu + 2i\bar\psi^\mu \rho^\alpha \partial_\alpha \psi_\mu$$
$$- i\bar\chi_\alpha \rho^\beta \rho^\alpha \psi^\mu (\partial_\beta X_\mu - \frac{i}{4}\bar\chi_\beta \psi_\mu) \Big\}$$

(7.7)

is invariant under the following local symmetries:

(i) supersymmetry

$$\delta_\epsilon X^\mu = i\bar\epsilon \psi^\mu,$$
$$\delta_\epsilon \psi^\mu = \frac{1}{2}\rho^\alpha (\partial_\alpha X^\mu - \frac{i}{2}\bar\chi_\alpha \psi^\mu)\epsilon,$$
$$\delta_\epsilon e_\alpha{}^a = \frac{i}{2}\bar\epsilon \rho^a \chi_\alpha,$$
$$\delta_\epsilon \chi_\alpha = 2D_\alpha \epsilon,$$

(7.8)

where $\epsilon(\sigma,\tau)$ is a Majorana spinor which parametrizes supersymmetry
transformations and D_α a covariant derivative with torsion:

$$D_\alpha \epsilon = \partial_\alpha \epsilon - \frac{1}{2}\omega_\alpha \bar\rho \epsilon,$$
$$\omega_\alpha = -\frac{1}{2}\epsilon^{ab}\omega_{\alpha ab} = \omega_\alpha(e) + \frac{i}{4}\bar\chi_\alpha \bar\rho \rho^\beta \chi_\beta,$$
$$\omega_\alpha(e) = -\frac{1}{e}e_{\alpha a}\epsilon^{\beta\gamma}\partial_\beta e_\gamma{}^a.$$

(7.9)

$\omega_\alpha(e)$ is the spin connection without torsion.

(ii) Weyl transformations

$$\delta_\Lambda X^\mu = 0,$$
$$\delta_\Lambda \psi^\mu = -\frac{1}{2}\Lambda \psi^\mu,$$
$$\delta_\Lambda e_\alpha{}^a = \Lambda e_\alpha{}^a,$$
$$\delta_\Lambda \chi_\alpha = \frac{1}{2}\Lambda \chi_\alpha.$$

(7.10)

(iii) super-Weyl transformations

$$\delta_\eta \chi_\alpha = \rho_\alpha \eta \,,$$
$$\delta_\eta (\text{others}) = 0 \,.$$

(7.11)

with $\eta(\sigma, \tau)$ being a Majorana spinor parameter.

(iv) two-dimensional Lorentz transformations

$$\delta_l X^\mu = 0 \,,$$
$$\delta_l \psi^\mu = \frac{1}{2} l \bar{\rho} \psi^\mu \,,$$
$$\delta_l e_\alpha{}^a = l \epsilon^a{}_b e_\alpha{}^b \,,$$
$$\delta_l \chi_\alpha = \frac{1}{2} l \bar{\rho} \chi_\alpha \,.$$

(7.12)

(v) reparametrizations

$$\delta_\xi X^\mu = \xi^\beta \partial_\beta X^\mu \,,$$
$$\delta_\xi \psi^\mu = \xi^\beta \partial_\beta \psi^\mu \,,$$
$$\delta_\xi e_\alpha{}^a = \xi^\beta \partial_\beta e_\alpha{}^a + e_\beta{}^a \partial_\alpha \xi^\beta \,,$$
$$\delta_\xi \chi_\alpha = \xi^\beta \partial_\beta \chi_\alpha + \chi_\beta \partial_\alpha \xi^\beta \,.$$

(7.13)

If we combine reparametrizations with a Lorentz transformation with parameter $l = -\xi^\alpha \omega_\alpha(e)$, they can be written in covariant form

$$e_{\beta a} \delta_\xi e_\alpha{}^a = \nabla_\alpha \xi_\beta = \frac{1}{2}(P\xi)_{\alpha\beta} - \frac{1}{2e^2}\epsilon_{\alpha\beta}(\epsilon^{\gamma\delta}\nabla_\gamma \xi_\delta) + \frac{1}{2} h_{\alpha\beta} \nabla \cdot \xi \,,$$

$$\delta_\xi \chi_\alpha = \xi^\beta \nabla_\beta \chi_\alpha + \chi_\beta \nabla_\alpha \xi^\beta \,,$$

$$\delta_\xi \psi^\mu = \xi^\alpha \nabla_\alpha \psi^\mu \,,$$

(7.14)

where ∇_α is a covariant derivative without torsion and the operator P has been defined in Chapter 2.

In eqs.(7.10)-(7.14) Λ, l and ξ are infinitesimal functions of (σ, τ).

There are several ways to get the complete action and the symmetry transformation rules. One possibility is to use the Noether method; another is to go to superspace. Either way, the procedure is analogous to the four dimensional case.

7.2 Superconformal gauge

We can now use local supersymmetry, reparametrizations and Lorentz transformations to gauge away two degrees of freedom of the zweibein and of the gravitino each. To do this we decompose the gravitino as

$$
\begin{aligned}
\chi_\alpha &= (h_\alpha{}^\beta - \tfrac{1}{2}\rho_\alpha\rho^\beta)\chi_\beta + \tfrac{1}{2}\rho_\alpha\rho^\beta\chi_\beta \\
&= \tfrac{1}{2}\rho^\beta\rho_\alpha\chi_\beta + \tfrac{1}{2}\rho_\alpha\rho^\beta\chi_\beta \\
&\equiv \tilde\chi_\alpha + \rho_\alpha\lambda
\end{aligned}
\tag{7.15}
$$

where $\tilde\chi_\alpha = \tfrac{1}{2}\rho^\beta\rho_\alpha\chi_\beta$ is ρ-traceless, i.e. $\rho\cdot\tilde\chi = 0$ and $\lambda = \tfrac{1}{2}\rho^\alpha\chi_\alpha$. This corresponds to a decomposition of the spin 3/2 gravitino into helicity $\pm3/2$ and helicity $\pm1/2$ components. It is orthogonal with respect to the inner product $(\phi|\psi) = \int \mathrm{d}^2\sigma\,\bar\phi^\alpha\psi_\alpha$. We can make the same decomposition for the supersymmetry transformation of the gravitino

$$
\begin{aligned}
\delta_\epsilon\chi_\alpha &= 2D_\alpha\epsilon \\
&\equiv 2(\Pi\epsilon)_\alpha + \rho_\alpha\rho^\beta D_\beta\epsilon
\end{aligned}
\tag{7.16}
$$

where we have defined the operator

$$
(\Pi\epsilon)_\alpha = (h_\alpha{}^\beta - \tfrac{1}{2}\rho_\alpha\rho^\beta)D_\beta\epsilon = \tfrac{1}{2}\rho^\beta\rho_\alpha D_\beta\epsilon
\tag{7.17}
$$

which maps spin 1/2 fields to ρ-traceless spin 3/2 fields. We can now write, at least locally, $\tilde\chi_\alpha = \rho^\beta\rho_\alpha D_\beta\kappa$ for some spinor κ where we have used the identity eq.(B.5). Comparing this with eq.(7.16) we see that κ can be eliminated by a supersymmetry transformation. We then use reparametrizations and local Lorentz transformations to transform the zweibein into the form $e_\alpha{}^a = e^\phi\delta_\alpha^a$ which we have demonstrated in Chapter 2 to be always possible locally. These transformations do not reintroduce trace parts into the gravitino since under reparametrizations it transforms as $\rho_\alpha(\sigma)\lambda(\sigma) \to \tilde{\rho}_\alpha(\sigma)\tilde{\lambda}(\sigma) = (\frac{\partial\tilde{\sigma}^\beta}{\partial\sigma^\alpha})\rho_\beta(\tilde{\sigma})\lambda(\tilde{\sigma})$. In this way we arrive at the so-called superconformal gauge [5, 6, 7] which is a generalization of the conformal gauge to the supersymmetric case:

$$e_\alpha{}^a = e^\phi \delta_\alpha^a \quad , \qquad \chi_\alpha = \rho_\alpha \lambda. \tag{7.18}$$

In the classical theory we can still use a Weyl rescaling and super-Weyl transformation to gauge away the remaining metric and gravitino degrees of freedom ϕ and λ, leaving only $e_\alpha{}^a = \delta_\alpha^a$ and $\chi_\alpha = 0$. In analogy to the bosonic case, these symmetries will be broken in the quantum theory except in the critical dimension.

Above arguments that were used to go to superconformal gauge were only true locally and one has to check under what conditions superconformal gauge can be reached globally. From our foregoing discussion it is clear that the condition is that there exists a globally defined spinor ϵ and a vector field ξ^α such that

$$(\Pi\epsilon)_\alpha = \tau_\alpha \qquad \text{and} \qquad (P\xi)_{\alpha\beta} = t_{\alpha\beta} \tag{7.19}$$

for arbitrary τ_α which satisfies $\rho \cdot \tau = 0$ and arbitrary symmetric traceless tensor $t_{\alpha\beta}$.

In Chapter 2 we have seen that the second condition is equivalent to the absence of zero modes of the operator $P^\dagger$. In the same way we can show that the absence of zero modes of the operator $\Pi^\dagger$, the adjoint of Π, allows to gauge away the trace part of the gravitino. $\Pi^\dagger$ maps ρ-traceless spin 3/2 fields to spin 1/2 fields via

$$(\Pi^\dagger \tau) = -2D^\alpha \tau_\alpha. \tag{7.20}$$

The zero modes of $P^\dagger$ were called moduli. In analogy we call the zero modes of $\Pi^\dagger$ supermoduli. We thus have

$$\# \text{ of moduli} = \dim \ker P^\dagger \,,$$
$$\# \text{ of supermoduli} = \dim \ker \Pi^\dagger \,. \qquad (7.21)$$

Also, zero modes of the operators P and Π mean that the gauge fixing is not complete. The zero modes of P are the conformal Killing vectors (CKV) (cf. Chapter 2); the zero modes of Π will be referred to as conformal Killing spinors (CKS); i.e.

$$\# \text{ of CKV} = \dim \ker P \,,$$
$$\# \text{ of CKS} = \dim \ker \Pi \,. \qquad (7.22)$$

We will compute the dimensions of the kernels of Π and $\Pi^\dagger$ in Chapter 9. P and $P^\dagger$ have been treated in Chapter 6.

In superconformal gauge the action simplifies to

$$S = -\frac{1}{8\pi} \int d^2\sigma \left\{ \partial_\alpha X^\mu \partial^\alpha X_\mu + 2i\bar{\psi}^\mu \rho^\alpha \partial_\alpha \psi_\mu \right\} \qquad (7.23)$$

which is nothing than the action of a free scalar superfield in two dimensions. To arrive at eq.(7.23) we have rescaled the matter fermions by $e^{\phi/2}\psi \to \psi$. World-sheet indices are now raised with the flat metric $\eta^{\alpha\beta}$ and $\rho^\alpha = \delta^\alpha_a \rho^a$. Also, the torsion piece in the spin connection now vanishes due to the identity eq.(B.5) and we have $\omega_\alpha = \epsilon_\alpha{}^\beta \partial_\beta \phi$. The action is still invariant under those local reparametrizations and supersymmetry transformations which satisfy $P\xi^\alpha = 0$ and $\Pi\epsilon = 0$. Under the supersymmetry transformations the fields transform as

$$\delta_\epsilon X^\mu = i\bar{\epsilon}\psi^\mu$$
$$\delta_\epsilon \psi^\mu = \frac{1}{2}\rho^\alpha \partial_\alpha X^\mu \epsilon. \qquad (7.24)$$

These equations follow from the transformation rules eq.(7.8). To see this we note that the zweibein is not taken out of superconformal gauge if a supersymmetry transformation with parameter ϵ is accompanied by a Lorentz transformation with parameter $l = \frac{i}{2}\bar{\epsilon}\bar{\rho}\lambda$. The Weyl degree of freedom ϕ then changes according to $\delta\phi = \frac{i}{2}\bar{\epsilon}\lambda$. Likewise, the gravitino stays in the gauge eq.(7.18) if the supersymmetry parameter satisfies $\rho^\beta \rho_\alpha D_\beta \epsilon = 0$

which is the condition $(\Pi\epsilon) = 0$ found above. If we now redefine $e^{\phi/2}\psi = \tilde{\psi}$ and $e^{-\phi/2}\epsilon = \tilde{\epsilon}$ we find, after dropping tildes, eqs.(7.24). It is also easy to show that the condition $\rho^\beta\rho_\alpha D_\beta\epsilon = 0$ reduces in superconformal gauge to $\rho^\beta\rho_\alpha\partial_\beta\tilde{\epsilon} = 0$. It is of course also straightforward to verify directly that the action eq.(7.23) is invariant under the transformations (7.24) with ϵ satisfying $\rho^\beta\rho_\alpha\partial_\beta\epsilon = 0$.

The equations of motion derived from the action eq.(7.23) are

$$\partial_\alpha\partial^\alpha X^\mu = 0,$$
$$\rho^\alpha\partial_\alpha\psi^\mu = 0. \tag{7.25}$$

As in the bosonic theory they have to be supplemented by boundary conditions. For the bosonic coordinates X^μ they are as given in eq.(2.37). For the fermionic fields ψ^μ we make variations such that $\delta\psi(\tau_0) = \delta\psi(\tau_1) = 0$. This leads to the condition that $\delta\bar{\psi}\rho^1\psi$ is periodic.

For theories with fermions the energy momentum tensor is defined as

$$T_{\alpha\beta} = -\frac{2\pi}{e}\frac{\delta S}{\delta e^\beta{}_a}e_{\alpha a}. \tag{7.26}$$

We can analogously define the supercurrent as the response to variations of the gravitino; we will denote it by T_F, indicating that it is a fermionic object related to the energy-momentum tensor T by supersymmetry. This means that $\delta S = \frac{1}{2\pi}\int d^2\sigma\, e\, i\delta\bar{\chi}^\alpha T_{F\alpha}$ with

$$T_{F\alpha} = \frac{2\pi}{e}\frac{\delta S}{i\delta\bar{\chi}^\alpha}. \tag{7.27}$$

The equations of motion for the metric and gravitino are

$$T_{\alpha\beta} = 0 \quad , \quad T_{F\alpha} = 0. \tag{7.28}$$

They are constraints on the system and generate symmetries, analogous to the bosonic case. We will have much to say about this below. After going to superconformal gauge we find

$$T_{\alpha\beta} = \frac{1}{2}\partial_\alpha X^\mu \partial_\beta X_\mu - \frac{1}{4}\partial^\gamma X^\mu \partial_\gamma X_\mu \eta_{\alpha\beta},$$
$$+ \frac{i}{4}\bar\psi^\mu \rho_\alpha \partial_\beta \psi_\mu + \frac{i}{4}\bar\psi^\mu \rho_\beta \partial_\alpha \psi_\mu = 0, \qquad (7.29)$$
$$T_{F\alpha} = \frac{1}{4}\rho^\beta \rho_\alpha \psi^\mu \partial_\beta X_\mu = 0.$$

Here we have used the equations of motion for ψ^μ to cast $T_{\alpha\beta}$ into its symmetric form. Tracelessness follows also upon using the equations of motion. Note that

$$\rho^\alpha T_{F\alpha} = 0 \qquad (7.30)$$

which is the analogue of $T^\alpha{}_\alpha = 0$. It is a consequence of super-Weyl invariance. Again, with the help of the equations of motion it is easy to show that the energy-momentum tensor and the supercurrent are conserved:

$$\partial^\alpha T_{\alpha\beta} = 0,$$
$$\partial^\alpha T_{F\alpha} = 0. \qquad (7.31)$$

These conservation laws lead, as in the bosonic theory, to an infinite number of conserved charges. This is most easily analyzed in light-cone coordinates on the world-sheet. In terms of these, eqs.(7.23) and (7.25) become:

$$S = \frac{1}{2\pi}\int d^2\sigma \left\{ \partial_+ X \cdot \partial_- X + i(\psi_+ \cdot \partial_- \psi_+ + \psi_- \cdot \partial_+ \psi_-) \right\} \qquad (7.32)$$

and

$$\partial_+ \partial_- X^\mu = 0,$$
$$\partial_- \psi^\mu_+ = \partial_+ \psi^\mu_- = 0. \qquad (7.33)$$

The conditions on the allowed reparametrizations and supersymmetry transformations take the simple form

$$\partial_+ \xi^- = \partial_- \xi^+ = 0,$$
$$\partial_+ \epsilon^- = \partial_- \epsilon^+ = 0. \qquad (7.34)$$

We have defined $\psi_A = \begin{pmatrix} \psi_+ \\ \psi_- \end{pmatrix}$ and $\epsilon^A = \begin{pmatrix} \epsilon^+ \\ \epsilon^- \end{pmatrix}$ following the conventions of the appendix. Note that for spinors $\pm$ denote their spinor components whereas for vectors they denote vector components in conformal coordinates.

As already mentioned above, the equations of motion have to be supplemented by a periodicity condition which now reads $(\psi_+\delta\psi_+ - \psi_-\delta\psi_-)(\sigma) = (\psi_+\delta\psi_+ - \psi_-\delta\psi_-)(\sigma + 2\pi)$. Its solutions are:

$$\psi_+(\sigma) = \pm\psi_+(\sigma + 2\pi)$$
$$\psi_-(\sigma) = \pm\psi_-(\sigma + 2\pi)$$

(7.35)

with the same conditions on $\delta\psi_\pm$. Anti-periodicity of $\psi_\pm$ is possible as they are fermions on the world-sheet. Periodic boundary conditions in σ are referred to as Ramond (R) boundary conditions whereas anti-periodic boundary conditions are called Neveu-Schwarz (NS) boundary conditions. This means that all quantities which are fermions on the world-sheet satisfy $\psi(\sigma + 2\pi) = e^{2\pi i\phi}\psi(\sigma)$, where $\phi = 0$ for the R-sector and $\phi = \frac{1}{2}$ for the NS-sector. The conditions for the two spinor components ψ_+ and ψ_- can be chosen independently, leading to a total of four possibilities: (R,R), (NS,NS), (NS,R) and (R,NS). Obviously, the two components of the supersymmetry parameter have to be chosen such that $\delta X^\mu = i\bar{\epsilon}\psi^\mu$ is periodic. We will show in the next chapter that string states in the R sector are space-time fermions and states in the NS sector are space-time bosons. Therefore, the two sectors (R,R) and (NS,NS) lead to space-time bosons and the remaining sectors, (NS,R) and (R,NS) to space-time fermions.

The energy-momentum tensor becomes

$$T_{++} = \frac{1}{2}\partial_+ X \cdot \partial_+ X + \frac{i}{2}\psi_+ \cdot \partial_+\psi_+$$
$$T_{--} = \frac{1}{2}\partial_- X \cdot \partial_- X + \frac{i}{2}\psi_- \cdot \partial_-\psi_-$$
$$T_{+-} = T_{-+} = 0$$

(7.36)

with

$$\partial_- T_{++} = \partial_+ T_{--} = 0. \tag{7.37}$$

Due to eq.(7.30), two of the four components of $T_{F\alpha}$ vanish identically. The only non-vanishing components are $T_{F++} \equiv T_{F+}$ and $T_{F--} \equiv T_{F-}$ where $++$ denotes the upper spinor component of the σ^+ vector component. Eq.(7.29) gives

$$\begin{aligned}
T_{F+} &= \frac{1}{2}\,\psi_+ \cdot \partial_+ X \\
T_{F-} &= \frac{1}{2}\,\psi_- \cdot \partial_- X
\end{aligned} \tag{7.38}$$

with

$$\partial_- T_{F+} = \partial_+ T_{F-} = 0.$$

From the equations of motion we learn that X^μ can again be split into left- and right-movers and that $\psi_+^\mu = \psi_+^\mu(\sigma^+)$ and $\psi_-^\mu = \psi_-^\mu(\sigma^-)$. The conservation laws tell us that T_{++} and T_{F+} are functions of σ^+ only whereas T_{--} and T_{F-} only depend on σ^-.

We have discussed in Chapter 2 how energy-momentum conservation results in an infinite number of conserved charges which generate the transformations $\sigma^\pm \to \sigma^\pm + f(\sigma^\pm)$ under which the action is invariant after going to conformal gauge. These are precisely the transformations which do not lead out of conformal gauge. This carries over to the fermionic string. But now in addition we have the conserved supercharges $\int d\sigma \epsilon^+(\sigma^+) T_{F+}(\sigma^+)$ which reflects the fact that the action and the superconformal gauge condition are invariant under supersymmetry transformations with parameters satisfying the second of eq.(7.34).

Next, let us find the algebra of T and T_F, which is the supersymmetric extension of the algebra eq.(2.70). To do this we need the basic Poisson brackets[3]

---

[3] For anticommuting variables they are defined as

$$\{\psi_A^\mu(\sigma,\tau),\Pi_B^\nu(\sigma',\tau)\}_{P.B.} = -\delta_{AB}\delta(\sigma-\sigma')\eta^{\mu\nu}. \tag{7.39}$$

The bracket of ψ with itself vanishes. For the momentum canonically conjugate to ψ_A^μ we find

$$\Pi_A^\mu = -\frac{i}{4\pi}\psi_A^\mu. \tag{7.40}$$

If, however, we use this definition of Π in eq.(7.39), we find a contradiction. Let us explain the way out of this. We notice that eq.(7.40) constitutes a (primary) constraint. We define

$$\phi_A^\mu = \Pi_A^\mu + \frac{i}{4\pi}\psi_A^\mu \tag{7.41}$$

and find

$$\{\phi_A^\mu(\sigma,\tau),\phi_B^\nu(\sigma',\tau)\}_{P.B.} = -\frac{i}{2\pi}\delta(\sigma-\sigma')\delta_{AB}\eta^{\mu\nu}. \tag{7.42}$$

In contrast to the constraints we have encountered so far (primary and secondary), the Poisson bracket of this constraint with itself does not vanish on the constrained hypersurface of phase space. Constraints of this kind are called second class constraints; the constraints we have encountered so far have all been first class. If second class constraints are presents, Poisson brackets have to be replaced by Dirac brackets. If ϕ_i are a complete set of second class constraints, we define $\{\phi_i,\phi_j\}_{P.B.} = C_{ij}$. The Dirac bracket is defined as

$$\{A,B\}_{D.B.} = \{A,B\}_{P.B.} - \{A,\phi_i\}_{P.B.}C_{ij}^{-1}\{\phi_j,B\}_{P.B.}. \tag{7.43}$$

This leads to

$$\begin{aligned}
\{\psi_+^\mu(\sigma,\tau),\psi_+^\nu(\sigma',\tau)\}_{D.B.} &= -2\pi i\delta(\sigma-\sigma')\eta^{\mu\nu},\\[2mm]
\{\psi_-^\mu(\sigma,\tau),\psi_-^\nu(\sigma',\tau)\}_{D.B.} &= -2\pi i\delta(\sigma-\sigma')\eta^{\mu\nu}.
\end{aligned} \tag{7.44}$$

$$\{A,B\}_{P.B.} = -(-1)^{\epsilon_A\epsilon_B}\left\{\frac{\partial A}{\partial\psi}\frac{\partial B}{\partial\Pi} + (-1)^{\epsilon_A\epsilon_B}\frac{\partial B}{\partial\psi}\frac{\partial A}{\partial\Pi}\right\}$$

where $\epsilon_A = 1$ if A is a commuting expression and $\epsilon = 0$ if it is anticommuting. The canonical momentum is defined as $\pi^\mu = \frac{\partial\mathcal{L}}{\partial\dot\psi^\mu}$.

Using this and the basic brackets for $X^\mu(\sigma, t)$ we find

$$\{T_{++}(\sigma), T_{++}(\sigma')\}_{D.B.} = -\left\{2T_{++}(\sigma')\partial' + \partial' T_{++}(\sigma')\right\} 2\pi\delta(\sigma - \sigma'),$$

$$\{T_{++}(\sigma), T_{F+}(\sigma')\}_{D.B.} = -\left\{\frac{3}{2}T_{F+}(\sigma')\partial' + \partial' T_{F+}(\sigma')\right\} 2\pi\delta(\sigma - \sigma'),$$

$$\{T_{F+}(\sigma), T_{F+}(\sigma')\}_{D.B.} = -\frac{i}{2}T_{++}(\sigma')\, 2\pi\delta(\sigma - \sigma').$$

$$(7.45)$$

It is also easily verified that

$$\{T_{F+}(\sigma), X^\mu(\sigma')\}_{D.B.} = -\frac{1}{2}\psi_+^\mu(\sigma')\, 2\pi\delta(\sigma - \sigma')\,,$$

$$\{T_{F+}(\sigma), \psi_+^\mu(\sigma')\}_{D.B.} = -\frac{i}{2}\partial_+ X^\mu(\sigma')\, 2\pi\delta(\sigma - \sigma'),$$

$$(7.46)$$

which is just the supersymmetry algebra. Under the transformations generated by T_{++}, ψ transforms as

$$\{T_{++}(\sigma), \psi_+(\sigma')\}_{D.B.} = -\left\{\frac{1}{2}\psi_+(\sigma')\partial' + \partial' \psi_+(\sigma')\right\} 2\pi\delta(\sigma - \sigma'). \quad (7.47)$$

This last equation tells us that the world-sheet fermions transform under conformal transformations with weight $\frac{1}{2}$.

We now proceed as in the bosonic string case and solve the equations of motion for the unconstrained system. The treatment for the bosonic coordinates is identical to the one in Chapter 2 and will not be repeated here. The fermionic fields require some care. We have to distinguishing between two choices of boundary conditions. The general solution of the two-dimensional Dirac equation for the cases of periodic (R) and antiperiodic (NS) boundary conditions is

$$\psi_+^\mu(\sigma, \tau) = \sum_{r \in \mathbf{Z}+\phi} \bar{b}_r^\mu e^{-ir(\tau+\sigma)}$$

$$\psi_-^\mu(\sigma, \tau) = \sum_{r \in \mathbf{Z}+\phi} b_r^\mu e^{-ir(\tau-\sigma)} \qquad \text{where} \qquad \begin{cases} \phi = 0 & \text{(R)} \\[2mm] \phi = \frac{1}{2} & \text{(NS)} \end{cases} \qquad (7.48)$$

and the reality of the Majorana spinors translates into the following conditions for the modes:

$$\left(b_r^\mu\right)^\dagger = b_{-r}^\mu \quad , \qquad \left(\bar{b}_r^\mu\right)^\dagger = \bar{b}_{-r}^\mu. \tag{7.49}$$

In terms of the fermionic oscillator modes the basic Dirac bracket eq.(7.44) translates to

$$\{b_r^\mu, b_s^\nu\}_{D.B.} = -i\eta^{\mu\nu}\delta_{r+s},$$

$$\{\bar{b}_r^\mu, \bar{b}_s^\nu\}_{D.B.} = -i\eta^{\mu\nu}\delta_{r+s}, \tag{7.50}$$

$$\{b_r, \bar{b}_s\}_{D.B.} = 0.$$

Next we decompose the generators of conformal and superconformal transformations into modes. In the following we will restrict ourselves to one sector, say the right-moving one. The expressions for the left-moving sector are then obtained by merely putting bars over all modes. We define

$$L_m = \frac{1}{2\pi}\int_0^{2\pi}d\sigma\, e^{-im\sigma}T_{--},$$

$$\tag{7.51}$$

$$G_r = \frac{1}{\pi}\int_0^{2\pi}d\sigma\, e^{-ir\sigma}T_{F-}.$$

From eq.(7.38) it is clear that T_{F-} satisfies the same boundary condition as the ψ^μ: periodic in the R-sector and antiperiodic in the NS-sector. Consequently, the modings are integer and half-integer respectively. In terms of oscillators we find $L_m = L_m^{(\alpha)} + L_m^{(b)}$ where

$$L_m^{(\alpha)} = \frac{1}{2}\sum_{n\in\mathbf{Z}}\alpha_{-n}\cdot\alpha_{m+n} \qquad \text{(as before)},$$

$$L_m^{(b)} = \frac{1}{2}\sum_r (r+\frac{m}{2})b_{-r}\cdot b_{m+r}, \tag{7.52}$$

$$G_r = \sum_r \alpha_{-n}\cdot b_{r+n}.$$

Note that $\sum_r b_{-r}\cdot b_{m+r} = 0$. This term has been included to make the expression for $L_m^{(b)}$ look more symmetric. It corresponds to the mode expansion of $\partial_-(\psi_-\psi_-)$ which vanishes by the Grassmann property of ψ_-. The generators L_m and G_r satisfy the following hermiticity conditions

$$L_m^\dagger = L_{-m} \quad , \qquad G_r^\dagger = G_{-r}. \tag{7.53}$$

One now verifies, using the basic brackets eqs.(2.74) and (7.50), the following (classical) algebra:

$$\{L_m, L_n\}_{D.B.} = -i(m-n)L_{m+n} \, ,$$

$$\{L_m, G_r\}_{D.B.} = -i(\tfrac{1}{2}m - r)G_{m+r} \, , \qquad (7.54)$$

$$\{G_r, G_s\}_{D.B.} = -2iL_{r+s} \, .$$

It can also be derived from eq.(7.45) and the definitions eq.(7.51). This algebra is called the super-Virasoro algebra. In the next chapter we will show how it is modified in the quantum theory. This concludes our discussion of the classical fermionic string theory.

Appendix B. Spinor algebra in two dimensions

In this appendix we summarize our notation for spinors in two dimensions and provide some identities which will prove useful in this and the following chapter. The two-dimensional Dirac matrices satisfy

$$\{\rho^\alpha, \rho^\beta\} = 2h^{\alpha\beta}. \qquad (B.1)$$

They transform under coordinate transformations and are related to the constant Dirac matrices through the zweibein: $\rho^\alpha = e^\alpha{}_a \rho^a$ from which $\{\rho^a, \rho^b\} = 2\eta^{ab}$ with $\eta^{ab} = \begin{pmatrix} -1 & 0 \\ 0 & +1 \end{pmatrix}$ follows. A convenient basis for the ρ^a is

$$\rho^0 = \begin{pmatrix} 0 & 1 \\ -1 & 0 \end{pmatrix} \quad , \quad \rho^1 = \begin{pmatrix} 0 & 1 \\ 1 & 0 \end{pmatrix} \qquad (B.2)$$

and we define $\bar{\rho} = \rho^0 \rho^1 = \begin{pmatrix} 1 & 0 \\ 0 & -1 \end{pmatrix}$ which is the analogue of γ^5 in four dimensions. We define the charge conjugation matrix as $C = \rho^0$. Then $(\rho^a)^{\mathrm{T}} = -C\rho^a C^{-1}$. A Majorana spinor satisfies $\bar{\lambda} = \lambda^\dagger \rho^0 = \lambda^{\mathrm{T}} C$. This means that Majorana spinors are real. An expression of the form $\bar{\lambda}\Gamma\psi$, where Γ is some combination of Dirac matrices, can be, using spinor indices, alternatively written as $\Lambda^A \Gamma_A{}^B \psi_B$, where $\Lambda^A = \Lambda_B \epsilon^{BA}$ with $\epsilon_{AB} = \epsilon^{AB} = \begin{pmatrix} 0 & 1 \\ -1 & 0 \end{pmatrix}$. Two-dimensional spinor indices take values $A = \pm$; i.e. $\psi^+ = -\psi_-$, $\psi^- = \psi_+$. The index structure of the Dirac matrices is $(\rho^\alpha)_A{}^B$. It is now easy to prove the following spin-flip property, valid for anti-commuting Majorana spinors:

$$\bar{\lambda}_1 \rho^{\alpha_1} \cdots \rho^{\alpha_n} \lambda_2 = (-1)^n \bar{\lambda}_2 \rho^{\alpha_n} \cdots \rho^{\alpha_1} \lambda_1. \qquad (B.3)$$

This and the following Fierz identity, again valid for anticommuting spinors, are needed to show the invariance of the action under supersymmetry transformations:

$$(\bar{\psi}\lambda)(\bar{\phi}\chi) = -\frac{1}{2}\left\{(\bar{\psi}\chi)(\bar{\phi}\lambda) + (\bar{\psi}\bar{\rho}\chi)(\bar{\phi}\bar{\rho}\lambda) + (\bar{\psi}\rho^\alpha\chi)(\bar{\phi}\rho_\alpha\lambda)\right\}. \qquad (B.4)$$

The identity

$$\rho^\alpha \rho_\beta \rho_\alpha = 0 \qquad (B.5)$$

follows trivially from the Dirac algebra in two dimensions. Another useful relation is

$$\rho^\alpha \rho^\beta = h^{\alpha\beta} - \frac{1}{e}\epsilon^{\alpha\beta}\bar{\rho} \quad , \qquad (\epsilon_{01} = 1). \qquad (B.6)$$

Let $\tilde{\chi}_a$, $\bar{\psi}$ denote a spin 3/2 fermion which satisfies $\rho \cdot \tilde{\chi} = 0$; one can show that

$$\bar{\tilde{\chi}}_\alpha \rho_\beta \tilde{\chi}_\gamma = 0 \, ,$$

$$\rho_\alpha \tilde{\chi}_\beta = \rho_\beta \tilde{\chi}_\alpha \, ,$$

$$\frac{1}{e}\epsilon^{\alpha\beta}\bar{\tilde{\psi}}^\gamma \bar{\rho}\tilde{\chi}_\gamma = h^{\alpha\beta}\bar{\tilde{\psi}} \cdot \tilde{\chi} - 2\bar{\tilde{\psi}}^\alpha \tilde{\chi}^\beta \, . \qquad (B.7)$$

These identities will be of use in the following chapter.

REFERENCES

[1] P. Ramond, *"Dual theory for free fermions"*, Phys. Rev. **D3** (1971) 2415.

[2] A. Neveu and J.H. Schwarz, *"Factorizable dual model of pions"*, Nucl. Phys. **B31** (1971) 86.

[3] F. Gliozzi, J. Scherk and D. Olive, *"Supergravity and the spinor dual model"*, Phys. Lett. **65B** (1976) 282; *"Supersymmetry, supergravity theories and the dual spinor model"*, Nucl. Phys. **B122** (1977) 253.

[4] M.B. Green and J.H. Schwarz, *"Supersymmetric dual string theory"*, Nucl. Phys. **B181** (1981) 502; *"Supersymmetrical string theories"*, Phys. Lett. **109 B** (1982) 444.

[5] L. Brink, P. Di Vecchia and P. Howe, *"A locally supersymmetric and reparametrization invariant action for the spinning string"*, Phys. Lett. **65B** (1976) 471.

[6] S. Deser and B. Zumino, *"A complete action for the spinning string"*, Phys. Lett. **65B** (1976) 369.

[7] P.S. Howe, *"Super Weyl transformations in two dimensions"*, **J. Phys. A12** (1979) 393.

Chapter 8

The Quantized Closed Fermionic String

In analogy to the bosonic string, consistent quantization of the fermionic (spinning) string implies the existence of a critical dimension ($d = 10$). The proof of the no-ghost theorem and the determination of $d = 10$ first appeared in the work of Schwarz [1], Goddard and Thorn [2] and Brower and Friedman [3]. The path integral quantization of the fermionic string was initiated by Polyakov [4].

8.1 Canonical quantization

We proceed in the same way as in the bosonic theory by making the replacement (3.1) and, in addition, by replacing the Dirac bracket for the anticommuting world-sheet fermions by an anticommutator:

$$\{\,,\,\}_{D.B.} \rightarrow \frac{1}{i}\{\,,\,\}. \tag{8.1}$$

We then get

$$\{\psi_+^\mu(\sigma,\tau), \psi_+^\nu(\sigma',\tau)\} = 2\pi\eta^{\mu\nu}\delta(\sigma - \sigma')\,,$$
$$\{\psi_-^\mu(\sigma,\tau), \psi_-^\nu(\sigma',\tau)\} = 2\pi\eta^{\mu\nu}\delta(\sigma - \sigma')\,, \tag{8.2}$$
$$\{\psi_+^\mu(\sigma,\tau), \psi_-^\nu(\sigma',\tau)\} = 0\,,$$

or, in terms of oscillators[1]

---

[1] Again, we will only write down the expressions for the right-moving sector of the closed string. The left-moving expressions are easily obtained by simply putting bars over all mode operators.

$$\{b_r^\mu, b_s^\nu\} = \eta^{\mu\nu} \delta_{r+s}. \tag{8.3}$$

It is again easy to see that oscillators with positive mode numbers are annihilation operators whereas oscillators with negative mode numbers are creation operators. We have seen in Chapter 3 that α_0^μ and $\bar\alpha_0^\mu$ correspond to the center of mass momentum of the string. We will see below how the zero-mode operators b_0^μ and $\bar b_0^\mu$ in the R-sector are to be interpreted. But we note already here that they satisfy, with suitable normalization, a Clifford algebra:

$$\{b_0^\mu, b_0^\nu\} = \eta^{\mu\nu}. \tag{8.4}$$

The level number operator is

$$N = N^{(\alpha)} + N^{(b)} \tag{8.5}$$

where

$$\begin{aligned}
N^{(\alpha)} &= \sum_{m=1}^{\infty} \alpha_{-m} \cdot \alpha_m \,, \\
N^{(b)} &= \sum_{r \in \mathbf{Z}+a > 0} r b_{-r} \cdot b_r \,.
\end{aligned} \tag{8.6}$$

The oscillator expressions of the super-Virasoro generators are again undefined without giving an operator ordering prescription. As in Chapter 3 we define them by their normal ordered expressions, i.e.

$$L_m = L_m^{(\alpha)} + L_m^{(b)} \tag{8.7}$$

with

$$L_m^{(\alpha)} = \frac{1}{2} \sum_{n \in \mathbf{Z}} : \alpha_{-n} \cdot \alpha_{m+n} :$$

$$L_m^{(b)} = \frac{1}{2} \sum_{r \in \mathbf{Z}+a} (r + \frac{m}{2}) : b_{-r} \cdot b_{m+r} : \tag{8.8}$$

and

$$G_r = \sum_{n \in \mathbf{Z}} \alpha_{-n} \cdot b_{r+n}. \tag{8.9}$$

Obviously, normal ordering is only required for L_0 and we include again an as yet undetermined normal ordering constant a in all formulas containing L_0.

The algebra satisfied by the L_m and G_r can now be determined. Great care is again required due to normal ordering. One obtains

$$[L_m, L_n] = (m - n)L_{m+n} + \frac{d}{8}m(m^2 - 2a)\delta_{m+n},$$

$$[L_m, G_r] = (\frac{m}{2} - r)G_{m+r}, \tag{8.10}$$

$$\{G_r, G_s\} = 2L_{r+s} + \frac{d}{2}(r^2 - \frac{a}{2})\delta_{r+s}.$$

This is the super-Virasoro algebra. A straightforward and save way to derive it is to use superconformal field theory. This will be presented in Chapter 12. We note that the R ($a = 0$) and NS ($a = \frac{1}{2}$) algebras agree formally except for the linear terms in the anomalies. As we have already remarked in Chapter 3, these can be changed by shifting L_0 by a constant (with a compensating change in the normal ordering constant). Indeed, if we define $L_0^{\text{R}} \to L_0^{\text{R}} + \frac{d}{16}$, it takes the form eq.(8.10) with $a = \frac{1}{2}$ in both sectors.

The algebra in the form of eq.(7.45) is modified by quantum effects as follows:

$$[T_{++}(\sigma), T_{++}(\sigma')] = \frac{i\pi d}{4}\partial_{\sigma'}^3 \delta(\sigma - \sigma')$$

$$\{T_{F+}(\sigma), T_{F+}(\sigma')\} = -\frac{\pi d}{4}\partial_{\sigma'}^2 \delta(\sigma - \sigma') \tag{8.11}$$

where we have only written the quantum corrections. Let us now examine the states of the theory. In doing so we have to distinguish between two sectors, the R and the NS sectors. The oscillator ground state in both sectors is defined by

$$\alpha_m^\mu |0\rangle = b_r^\mu |0\rangle = 0 \qquad m, r > 0 \tag{8.12}$$

(we suppress the dependence on the center of mass momentum). In the R sector we still have the b_0^μ zero modes. They do not change the mass of a given state, in particular the ground state. The mass operators for the fermionic string are given by the same expressions as in the bosonic case but with the level numbers as in eq.(8.5). It is then easy to check that $[b_0^\mu, m^2] = 0$, i.e. the states $|0\rangle$ and $b_0^\mu |0\rangle$ are degenerate in mass. But α_n^μ, b_r^μ for $n, r < 0$ increase $\alpha' m^2$ by $2n$ and $2r$ units respectively. This means that in the NS sector there is a unique ground state which must therefore be spin zero. In the R sector the ground state is degenerate. Since the b_0^μ are the generators of a Clifford algebra (c.f. eq.(8.4)) we conclude that the R ground state is a spinor of $SO(d-1,1)$. This is why states in the R sector are space-time fermions whereas states in the NS sector are space-time bosons. The oscillators, all being space-time vectors, cannot change bosons into fermions or vice versa. Whether a state belongs to the R or the NS sector depends on the ground state it is built on. We will come back to this important point in Chapter 12. We will write the R ground state as $|a\rangle$ where a is a $SO(d-1,1)$ spinor index; then $b_0^\mu |a\rangle = \frac{1}{\sqrt{2}} (\Gamma^\mu)^a{}_b |b\rangle$ where Γ^μ is a Dirac matrix in d dimensions, satisfying $\{\Gamma^\mu, \Gamma^\nu\} = 2\eta^{\mu\nu}$.

We now have to implement the constraints on the states of the theory. Due to the anomalies in the super-Virasoro algebra it is again impossible to impose $L_m |\text{phys}\rangle = G_r |\text{phys}\rangle = 0$ for all m and r. The most we can do is to demand that

$$
\begin{aligned}
G_r |\text{phys}\rangle &= 0 & r &> 0 & & \\
L_m |\text{phys}\rangle &= 0 & m &> 0 & \text{(NS)} & \quad (8.13a) \\
(L_0 - a) |\text{phys}\rangle &= 0 & & & &
\end{aligned}
$$

in the NS sector, and

$$
\begin{aligned}
G_r |\text{phys}\rangle &= 0 & r &\geq 0 & & \\
L_m |\text{phys}\rangle &= 0 & m &> 0 & \text{(R)} & \quad (8.13b) \\
L_0 |\text{phys}\rangle &= 0 & & & &
\end{aligned}
$$

in the R sector. Note that we do not have included a normal ordering constant in the last equation. There are several reasons for this. From the super-Virasoro algebra we find that $G_0^2 = L_0$, i.e. if we have $(L_0 - \mu^2)|\text{phys}\rangle = 0$ we also need $(G_0 - \mu)|\text{phys}\rangle = 0$. However, G_0 has no normal ordering ambiguity and the normal ordering constants arising from the bosonic and the fermionic oscillators cancel in L_0 in the Ramond sector. Also, G_0 is anti-commuting whereas the normal ordering constant is a commuting c-number. When we discuss the spectrum we will find that setting $\mu = 0$ is indeed correct. There is of course a second set of conditions for the left movers and we also have to demand that

$$(L_0 - \bar{L}_0)|\text{phys}\rangle = 0 \tag{8.14}$$

which again expresses the fact that no point on a closed string is distinct.

So far the quantization has been canonical and covariant. We know, however, that due to the negative eigenvalue of $\eta^{\mu\nu}$ there are negative norm states (ghosts). As in the bosonic theory one can prove a no-ghost theorem which states that the negative norm states decouple in the critical dimension d for a particular value of the normal ordering constant a. It turns out that for the fermionic string $d = 10$ and $a = 1/2$. As it was the conformal symmetry in the bosonic case, in the fermionic theory the superconformal symmetry is just big enough to allow for the ghost decoupling. We will not prove the no-ghost theorem here but instead follow our treatment of the bosonic theory and discuss the non-covariant light-cone quantization which provides a solution of the constraints. At the end of this chapter we will discuss the covariant path integral quantization. Both approaches will also lead to the above values for the critical dimension and the normal ordering constant.

8.2 Light cone quantization

In the bosonic theory the light cone gauge was obtained by the choice

$$X^+ = \alpha' p^+ \tau \tag{8.15}$$

(cf. eq.(3.25)) which fixed the gauge completely. This choice is again possible in the fermionic theory and also completely eliminates the reparametrization invariance. But now we still have local supersymmetry transformations. In going to super-conformal gauge we have fixed it partially leaving only transformations satisfying $\partial_+\epsilon^- = \partial_-\epsilon^+ = 0$. This freedom can now be used to transform ψ^+ away; i.e. in addition to eq.(8.15) the light-cone gauge condition in the fermionic theory is

$$\psi^+ = 0 \tag{8.16}$$

or, equivalently, $b_r^+ = 0$, $\forall r$. (Here and below the superscript denotes the light-cone component; i.e. $\psi^\pm = \frac{1}{\sqrt{2}}(\psi^0 \pm \psi^{d-1}$.) Now there is no gauge freedom left and we can solve the constraints. Eq.(3.26) is replaced by $(\alpha' = 2)$

$$\partial_\pm X^- = \frac{1}{2p^+}\left[(\partial_\pm X^i)^2 + i\,\psi_\pm^i \partial_\pm \psi_\pm^i\right] \tag{8.17}$$

and

$$\psi_\pm^- = \frac{1}{p^+}\psi_\pm^i \partial_\pm X^i \tag{8.18}$$

which leaves only the transverse components X^i and ψ^i as independent degrees of freedom. In terms of oscillators we get

$$\alpha_m^- = \frac{1}{2p^+}\left\{\sum_n :\alpha_n^i \alpha_{m-n}^i: + \sum_r (\frac{m}{2} - r) :b_r^i b_{m-r}^i: -2a\delta_m\right\} \tag{8.19}$$

and

$$b_r^- = \frac{1}{p^+}\sum_q \alpha_{r-q}^i b_q^i. \tag{8.20}$$

These expressions are valid for the right-moving part of the closed string and have to be supplemented by the corresponding expressions for the left-moving part. We have included a normal ordering constant which will not be the same for the two sectors (NS and R). The arguments given above suggests that it vanishes in the R sector. We will verify this shortly. The mass operator is now

$$m^2 = m_R^2 + m_L^2 \tag{8.21}$$

with

$$\alpha' m_R^2 = 2\left\{ \sum_{n>0} \alpha^i_{-n}\alpha^i_n + \sum_{r>0} r b^i_{-r} b^i_r - a \right\} \tag{8.22}$$

with a similar expression for m_L^2. Condition (8.14) translates to

$$m_L^2 = m_R^2 \tag{8.23}$$

for physical states.

The light cone action is simply

$$S^{\text{l.c.}} = \frac{1}{8\pi} \int \mathrm{d}^2\sigma \left((\dot{X}^i)^2 - (X'^i)^2 - 2i\bar{\psi}^i \rho^\alpha \partial_\alpha \psi^i \right) \tag{8.24}$$

and the Hamiltonian is

$$\begin{aligned}
H &= L_0 - \bar{L}_0 - 2a \\
&= (p^i)^2 + \sum_{n>0} (\alpha^i_{-n}\alpha^i_n + \bar{\alpha}^i_{-n}\bar{\alpha}^i_n) + \sum_{r>0} r(b^i_{-r} b^i_r + \bar{b}^i_{-r} \bar{b}^i_r) - 2a.
\end{aligned} \tag{8.25}$$

Let us now look at the spectrum of the fermionic string where we have to distinguish between Ramond and Neveu-Schwarz sectors. Let us first discuss the right-moving part of the closed fermionic string spectrum. This is in fact, up to a mass rescaling by a factor of two, identical to the spectrum of the open fermionic string.

(i) NS-sector: The ground state is the oscillator vacuum $|0\rangle$ with $\alpha' m^2 = -a$. The first excited state is $b^i_{-1/2}|0\rangle$ with $\alpha' m^2 = \frac{1}{2} - a$. This is a vector of $SO(d-2)$ and, following the argument of Chapter 3, must be massless, leading to a value $a = \frac{1}{2}$ for the normal ordering constant. In the fermionic theory the normal ordering constant is formally

Table 8.1 Open fermionic string spectrum

$\alpha'(\text{mass})^2$	states and their $SO(8)$ representation contents	little group	$(-1)^F$	representation contents with respect to the little group
NS-sector				
$-\frac{1}{2}$	$\lvert 0 \rangle$ (1)	$SO(9)$	-1	(1)
0	$b^i_{-1/2}\lvert 0 \rangle$ $(8)_v$	$SO(8)$	$+1$	$(8)_v$
$+\frac{1}{2}$	$\alpha^i_{-1}\lvert 0 \rangle \quad b^i_{-1/2}b^j_{-1/2}\lvert 0 \rangle$ $(8)_v \qquad (28)$	$SO(9)$	-1	(36)
$+1$	$b^i_{-1/2}b^j_{-1/2}b^k_{-1/2}\lvert 0 \rangle$ $(56)_v$ $\alpha^i_{-1}b^j_{-1/2}\lvert 0 \rangle \qquad b^i_{-3/2}\lvert 0 \rangle$ $(1)+(28)+(35)_v \qquad (8)_v$	$SO(9)$	$+1$ $+1$	$(84)+(44)$
R-sector				
0	$\lvert a \rangle$ $(8)_s$ $\lvert \bar{a} \rangle$ $(8)_c$	$SO(8)$	$+1$ -1	$(8)_s$ $(8)_c$
$+1$	$\alpha^i_{-1}\lvert a \rangle \qquad b^i_{-1}\lvert \bar{a} \rangle$ $(8)_c+(56)_c \quad (8)_s+(56)_s$ $\alpha^i_{-1}\lvert \bar{a} \rangle \qquad b^i_{-1}\lvert a \rangle$ $(8)_s+(56)_s \quad (8)_c+(56)_c$	$SO(9)$	$+1$ -1	(128) (128)

$a = -\frac{d-2}{2}(\sum_{n=0}^{\infty} n - \sum_{r=1/2}^{\infty} r)$ which, using ζ-function regularization gives $a = \frac{d-2}{2}(\frac{1}{12} + \frac{1}{24}) = \frac{d-2}{16}$ from which we derive the value $d = 10$ for the critical dimension.[2] At the next excitation level we have the states $\alpha^i_{-1}|0\rangle$ and $b^i_{-1/2} b^j_{-1/2}|0\rangle$ with $\alpha' m^2 = \frac{1}{2}$, comprising $8 + 28$ bosonic states. It can again be shown that these and all other massive light-cone states, which are tensors of $SO(8)$, combine uniquely to tensors of $SO(9)$, the little group for massive states in ten dimensions.

(ii) R-sector: We already know that the R ground state is a spinor of $SO(9,1)$. A Dirac spinor in ten space-time dimensions has 2^5 independent complex or 64 real components. On shell this reduces to 32 components since the Dirac equation $\gamma^\mu \partial_\mu \psi = 0$ relates half of the components to the other half which satisfies the Klein-Gordon equation. We can now still impose a Weyl or a Majorana condition, each of which reduces the number of independent components further by a factor of two. In ten dimensional space-time it is however possible to impose both simultaneously[3] leaving 8 independent on-shell components. They can also be viewed as the components of a Majorana-Weyl spinor of $SO(8)$, the corresponding little group for massless states. It is easy to see that the Ramond ground state is indeed massless. We can now choose the ground state to have either one of two possible chiralities, which we will denote by $|a\rangle$ and $|\bar{a}\rangle$ respectively. The first excitation level consists of states $\alpha^i_{-1}|a\rangle$ and $b^i_{-1}|a\rangle$ plus their chiral partners with $\alpha' m^2 = 1$. Again, for $d = 10$, all the massive light-cone states can be uniquely assembled into representations of $SO(9)$. In table 8.1 this is demonstrated for the first few mass levels.

---

[2] The general formula is $\sum_{n \geq 0}(n + a) = \zeta(-1, a) = -\frac{1}{12}(6a^2 - 6a + 1)$.

[3] The general statement is that we can impose Majorana and Weyl conditions simultaneously on spinors of $SO(p,q)$ if and only if $p - q = 0 \bmod 8$. For Minkowski space-times ($q = 1$) this is the case for $d = 2 + 8n$ and for Euclidean spaces ($q = 0$) for $d = 8n$.

However, it can be shown that the fermionic string theory with all the states in both the R and NS sectors is inconsistent. We have to make a truncation of the spectrum, called the GSO projection [5], which leaves us with a tachyon free space-time supersymmetric theory. We will prove both statements, the necessity of the truncation and the space-time supersymmetry of the resulting spectrum, in Chapter 9 where the first assertion follows from the requirement of modular invariance and the second from the vanishing of the one-loop partition function.

Here we will turn the argument around and motivate the GSO projection by requiring a space-time supersymmetric spectrum. By inspection of table 8.1 we see that at the massless level this can be achieved by projecting out one of the two possible chiralities of the R ground state. This leaves us with the on-shell degrees of freedom of $N = 1$, $d = 10$ Super-Yang-Mills theory: a massless spinor and a massless vector [6]. Obviously, we also have to get rid of the tachyon. Let us define a quantum number G which is the eigenvalue of the operator $G = (-1)^F$ where F is the world-sheet fermion number. If we assign the NS vacuum $(-1)^F|0\rangle = -|0\rangle$, i.e. $G = -1$, we can write in the NS sector $F = \sum_{r>0} b^i_{-r} b^i_r - 1$. If we then require that all states satisfy $G = 1$, we remove all states with half-integer $\alpha' m^2$ (for which there are no space-time fermions) and some of the other states. In particular the tachyon disappears. A general state in the NS sector, $\alpha^{i_1}_{-n_1} \cdots \alpha^{i_N}_{-n_N} b^{j_1}_{-r_1} \cdots b^{j_M}_{-r_M}|0\rangle$ has $G = (-1)^M$ and all states with M even are projected out. In the R sector the equivalent of G is a generalized chirality operator $\Gamma = (-1)^F = b^1_0 \cdots b^8_0 (-1)^{\sum_{n>0} b^i_{-n} b^i_n}$ where $b^1_0 \cdots b^8_0$ is the chirality operator in the 8 transverse dimensions and $\sum_{n>0} b^i_{-n} b^i_n$ the world-sheet fermion number operator. It is easy to see that $\{\Gamma, \psi^\mu\} = 0$ and the eigenvalues of the R ground states are ± 1, depending on their chirality, if we define $\Gamma|a\rangle = \prod_{i=1}^8 b^i_0|a\rangle = +1$ and $\Gamma|\bar{a}\rangle = -1$. Then a general state in the R-sector $\alpha^{i_1}_{-n_1} \cdots \alpha^{i_N}_{n_N} b^{j_1}_{-m_1} \cdots b^{j_M}_{-m_M}|a\rangle$ has $\Gamma = (-1)^M (-1)^{\sum_i \delta_{m_i,0}}$

and $\alpha^{i_1}_{-n_1} \cdots \alpha^{i_N}_{n_N} b^{j_1}_{-m_1} \cdots b^{j_M}_{-m_M} |\bar{a}\rangle$ has $\Gamma = -(-1)^M (-1)^{\sum_i \delta_{m_i,0}}$. The GSO projection then corresponds to demanding that all states have either $\Gamma = 1$ or $\Gamma = -1$. We see from table 8.1 that making the GSO projection we arrive at a supersymmetric spectrum (at least up to the level displayed there).

To obtain the full closed string spectrum we have to tensor the left- and right-moving states together obeying the constraint eq.(8.14). We have to distinguish between four sectors, two of which ((NS,NS) and (R,R)) lead to space-time bosons and two ((NS,R) and (R,NS)) to space-time fermions. An additional complication arises because we can choose between two possible chiralities for the left and right R ground state. Since in each sector we have to satisfy the constraint $L_0 - \bar{L}_0 = 0$, or, equivalently $m_R^2 = m_L^2$, the closed string states are tensor products of open string states at the same mass level. The possible states up to the massless level are shown in table 8.2. There are too many states at the massive level to display there but it is straightforward to work out the continuation of the table. Again, we have to make the GSO projection. One way to perform it is for the right- and left-movers separately. For the NS states we require $(-1)^F = +1$ and $(-1)^{\bar{F}} = +1$ and for the R states $\Gamma = +1$ or $\Gamma = -1$ and likewise for $\bar{\Gamma}$. This leads to several possibilities. For instance, the theory with $\Gamma = \bar{\Gamma} = +1$ has no tachyon and the following massless states:

$$\text{Bosons} : \left[(1) + (28) + (35)_v\right] + \left[(1) + (28) + (35)_s\right]$$

$$\text{Fermions} : \left[(8)_c + (56)_c\right] + \left[(8)_c + (56)_c\right]$$

i.e. we get a total of 128 bosonic and fermionic states, indicating a supersymmetric spectrum. The projection as given above defines the type IIB string theory whose massless spectrum is that of type IIB supergravity in ten dimensions. The $(35)_v$ represents the on-shell degrees of freedom of a

Table 8.2 Closed fermionic string spectrum

$\alpha'(\text{mass})^2$	states and their $SO(8)$ representation contents	little group	$(-1)^F$	$(-1)^{\bar F}$	representation contents with respect to the little group
	(NS,NS)-sector				
-2	$\|0\rangle_L \times \|0\rangle_R$ (1) (1)	$SO(9)$	-1	-1	(1)
0	$b^i_{-1/2}\|0\rangle_L \times b^j_{-1/2}\|0\rangle_R$ $(8)_v$ $(8)_v$	$SO(8)$	$+1$	$+1$	$(1) + (28) + (35)_v$
	(R,R)-sector				
0	$\|a\rangle_L \times \|b\rangle_R$ $(8)_s$ $(8)_s$	$SO(8)$	$+1$	$+1$	$(1) + (28) + (35)_s$
	$\|\bar a\rangle_L \times \|\bar b\rangle_R$ $(8)_c$ $(8)_c$		-1	-1	$(1) + (28) + (35)_c$
	$\|\bar a\rangle_L \times \|b\rangle_R$ $(8)_c$ $(8)_s$		-1	$+1$	$(8)_v + (56)_v$
	$\|a\rangle_L \times \|\bar b\rangle_R$ $(8)_s$ $(8)_c$		$+1$	-1	$(8)_v + (56)_v$
	(R,NS)-sector				
0	$\|a\rangle_L \times b^i_{-1/2}\|0\rangle_R$ $(8)_s$ $(8)_v$	$SO(8)$	$+1$	$+1$	$(8)_c + (56)_c$
	$\|\bar a\rangle_L \times b^i_{-1/2}\|0\rangle_R$ $(8)_c$ $(8)_v$		-1	$+1$	$(8)_s + (56)_s$
	(NS,R)-sector				
0	$b^i_{-1/2}\|0\rangle_L \times \|a\rangle_R$ $(8)_v$ $(8)_s$	$SO(8)$	$+1$	$+1$	$(8)_c + (56)_c$
	$b^i_{-1/2}\|0\rangle_L \times \|\bar a\rangle_R$ $(8)_v$ $(8)_c$		$+1$	-1	$(8)_s + (56)_s$

graviton, the two (28)'s represent two antisymmetric tensor fields and the $(35)_s$ a rank four selfdual antisymmetric tensor. I addition there are two real scalars. The fermionic degrees of freedom correspond to two gravitinos (the $(56)_c$) with spin 3/2 and two spin 1/2 fermions. The presence of two gravitinos means that this theory has $N = 2$ supersymmetry. Since both gravitinos are of the same handedness, it is a chiral theory.

The choice $\Gamma = -\bar{\Gamma} = 1$ leads to the following massless spectrum:

$$\text{Bosons} : \left[(1) + (28) + (35)_v\right] + \left[(8)_v + (56)_v\right]$$

$$\text{Fermions} : \left[(8)_c + (56)_c\right] + \left[(8)_s + (56)_s\right]$$

representing the degrees of freedom of a graviton $((35)_v)$, an antisymmetric rank three tensor $((56)_v)$, an antisymmetric rank two tensor $((28))$, one vector $(8)_v)$ and one real scalar, the dilaton. The fermions can be interpreted as two gravitinos of spin 3/2 and to dilatinos of spin 1/2, one each for each handedness. Again, we have $N = 2$ supersymmetry but non-chiral. This theory is called type IIA.

8.3 Path integral quantization

Let us now turn to the path integral quantization of Polyakov. In Chapter 3 we have seen that to obtain the ghost action we had to re-express the Faddeev-Popov determinant as an integral over anti-commuting ghost fields. The ghost action was than simply $S_{\text{gh}} \sim \int b^{\alpha\beta} \frac{\delta h_{\alpha\beta}}{\delta \xi^\gamma} c^\gamma$ where $b_{\alpha\beta}$ was symmetric and traceless, the tracelessness following from the Weyl-invariance of the theory (at least in the critical dimension). In other words, the ghost Lagrangian followed immediately from the traceless symmetric variation of the metric by replacing the gauge parameter ξ^α by a ghost field c^α of opposite statistic and introducing the antighost $b_{\alpha\beta}$ with the same tensor structure as the gauge field $h_{\alpha\beta}$, but opposite statistics. We will now apply the same procedure to the fermionic string.

The transformations of the zweibein and the gravitino under reparametrizations and supersymmetry transformations are

$$e_{\alpha a}\delta e_\beta{}^a = \frac{1}{2}(P\xi)_{\alpha\beta} - \frac{i}{2}(\bar{\chi}_\beta\rho_\alpha\epsilon)$$

$$\delta\chi_\alpha = 2(\Pi\epsilon)_\alpha - \frac{3}{4e}\bar{\rho}\chi_\alpha(\epsilon^{\beta\gamma}\nabla_\beta\xi_\gamma) + \frac{1}{4}\chi_\alpha\nabla_\beta\xi^\beta + (\nabla_\beta\chi_\alpha)\xi^\beta \qquad (8.26)$$

where we have made compensating Weyl, Lorentz and super-Weyl transformations to eliminate the trace and antisymmetric part in the first line and to get $\rho\cdot\delta\chi = 0$ in the second. The parameters of these transformations were

$$\Lambda = -\frac{1}{2}\nabla^\alpha\xi_\alpha,$$

$$l = -\frac{1}{2e}\epsilon^{\alpha\beta}\nabla_\alpha\xi_\beta, \qquad (8.27)$$

$$\eta = -\rho^\alpha\nabla_\alpha\epsilon - \frac{1}{4}\rho^\alpha\chi^\beta(P\xi)_{\alpha\beta}.$$

In the critical dimension these are good symmetries and we can chose χ to be ρ-traceless, which is what we have done. Then the torsion piece in the connection vanishes and all covariant derivatives will be without torsion. The ghost Lagrangian is then

$$\mathcal{L}^{\text{ghost}} = -\frac{i}{2\pi}\left\{b^{\alpha\beta}\left[e_{\alpha a}\frac{\delta e_\beta{}^a}{\delta\xi^\gamma}c^\gamma - \frac{i}{2}e_{\alpha a}\frac{\delta e_\beta{}^a}{\delta\epsilon}\gamma\right] + i\bar{\beta}^\alpha\left[\frac{\delta\chi_\alpha}{\delta\xi^\beta}c^\beta - \frac{i}{2}\frac{\delta\chi_\alpha}{\delta\epsilon}\gamma\right]\right\}$$

$$= -\frac{i}{2\pi}\left\{b^{\alpha\beta}\nabla_\alpha c_\beta + \bar{\beta}^\alpha\nabla_\alpha\gamma + i\bar{\chi}_\gamma\left[\frac{3}{2}\beta^\gamma\nabla\cdot c - \frac{3}{2}\beta^\alpha\nabla_\alpha c^\gamma\right.\right.$$

$$\left.\left. - (\nabla\cdot\beta)c^\gamma + (\nabla_\beta\beta^\gamma)c^\beta - \frac{i}{4}b^{\gamma\delta}\rho_\delta\gamma\right]\right\}.$$

$$(8.28)$$

β_α and γ are commuting spin 3/2 and spin 1/2 ghosts, with $\rho\cdot\beta = 0$. $b_{\alpha\beta}$ and c^γ are as in Chapter 5. The factors of i have been included to make γ real and β imaginary. In deriving eq.(8.28) we have redefined $\epsilon \to \epsilon - \frac{1}{2}\xi^\gamma\chi_\gamma$ and made use of the identities given in the appendix of the previous chapter. Note that the term in brackets will be absent in superconformal gauge.[4]

---

[4] We could also have introduced ghosts for Weyl, Lorentz and Super-Weyl transformations, but they would have been integrated out, giving constraints on $b_{\alpha\beta}$ and β_α,

From the ghost action we can now derive the ghost energy-momentum tensor and the ghost supercurrent. Using the equations of motion and a gravitino in superconformal gauge, we get

$$
\begin{aligned}
T_{\alpha\beta} = \ & i\big\{ b_{\alpha\gamma}\nabla_\beta c^\gamma + b_{\beta\gamma}\nabla_\beta c^\gamma - c^\gamma\nabla_\gamma b_{\alpha\beta} \\
& + \frac{3}{4}(\bar\beta_\alpha\nabla_\beta + \bar\beta_\beta\nabla_\alpha)\gamma + \frac{1}{4}(\nabla_\alpha\bar\beta_\beta + \nabla_\beta\bar\beta_\alpha)\gamma \big\} ,
\end{aligned}
\tag{8.29}
$$

$$
T_{F\gamma} = -i\Big\{ \frac{3}{2}\beta^\alpha\nabla_\gamma c_\alpha + (\nabla_\alpha\beta_\gamma)c^\alpha - \frac{i}{4}b_{\gamma\delta}\,\rho^\delta\gamma \Big\} .
$$

We could now proceed as in Chapter 3, expand the ghost fields in modes, find their contribution to the super Virasoro operators and show that the conformal anomaly vanishes in the critical dimension. We will however not do this here but rather postpone the discussion until Chapter 12, where we will be able to derive the same results in a much easier way using superconformal field theory.

---

resulting in their being symmetric-traceless and ρ-traceless respectively. Also, we would find that only the helicity $\pm 3/2$ components of the gravitino couple, reflecting super-Weyl invariance [7].

REFERENCES

[1] J.H. Schwarz, *"Physical states and pomeron poles in the dual pion model"*, Nucl. Phys. **B46** (1972) 61.

[2] P. Goddard and C.B. Thorn, *"Compatibility of the dual Pomeron with unitarity and the absence of ghosts in the dual resonance model"*, Phys. Lett. **40B** (1972) 235.

[3] R.C. Brower and K.A. Friedman, *"Spectrum-generating algebra and no-ghost theorem for the Neveu-Schwarz model"*, Phys. Rev. **D7** (1973) 535.

[4] A.M. Polyakov, *"Quantum geometry of fermionic strings"*, Phys. Lett. **103B** (1981) 211.

[5] F. Gliozzi, J. Scherk and D. Olive, *"Supergravity and the spinor dual model"*, Phys. Lett. **65B** (1976) 282; *"Supersymmetry, supergravity theories and the dual spinor model"*, Nucl. Phys. **B122** (1977) 253.

[6] L. Brink, J.H. Schwarz and J. Scherk, *"Supersymmetric Yang-Mills theories"*, Nucl. Phys. **B121** (1977) 77.

[7] D.Z. Freedman and N.P. Warner, *"Locally supersymmetric string Jacobian"*, Phys. Rev. **D34** (1986) 3084.

Chapter 9

Spin Structures and Superstring Partition Function

In the first part of this chapter we compute the one-loop partition function of the closed fermionic string. We will do this in light cone gauge. The possibility to assign to the world-sheet fermions periodic or anti-periodic boundary conditions leads to the concept of spin structures which we will introduce. The requirement of modular invariance is then shown to result in the GSO projection. In the last part we generalize some of the results of Chapter 6 to the case of fermions.

Spin structures in string theory and their relation to modular invariance were first discussed by Seiberg and Witten [1] and Alvarez-Gaumé, Ginsparg, Moore and Vafa [2]. Reference [3] gives a more detailed insight into the subject.

Let us begin by explaining what spin structures on a genus g Riemann surface Σ_g are. As we know from Chapter 6, there are two non-contractible loops associated with each of the g holes. All other non-contractible loops can be generated by deforming and joining elements of this basis. When we have spinors defined over Σ_g we can assign to them either periodic or anti-periodic boundary conditions around each of the $2g$ loops. Each of these 2^{2g} possible assignments is called a spin structure on Σ_g. An important distinction is that of even and odd spin structures which is connected to the zero mode structure of the chiral Dirac operator which, on Σ_g, is simply D_z and $D_{\bar{z}}$ for the two chiralities. We will treat these operators in more detail below. We call a spin structure even if the number of zero modes of the chiral Dirac operator is even and we call it odd otherwise. Let us study the situation for the torus and then generalize to arbitrary genus.

We can put a flat metric on the torus for which the chiral Dirac operator is simply ∂_z. It is then clear that the only global zero mode is the constant spinor. Obviously, only $(+,+)$ boundary conditions allow for a constant spinor where the two entries refer to the boundary conditions along the two non-contractible loops. This means that there are three even and one odd spin structure on the torus. For the generalization to arbitrary genus and the properties under modular transformations the following facts, which we will state without proof, are important:

(i) for a given spin structure, the number of chiral Dirac zero modes is a topological invariant modulo two;

(ii) the number of chiral Dirac zero modes is additive modulo two when we glue together two Riemann surfaces.

The second fact together with our result for the torus can be used to find the number of even and odd spin structures for arbitrary Riemann surfaces. It is not hard to see that there are $\sum_{m\,\text{odd}} \binom{g}{m} 3^{g-m} = 2^{g-1}(2^g - 1)$ odd and $\sum_{m\,\text{even}} \binom{g}{m} 3^{g-m} = 2^{g-1}(2^g + 1)$ even spin structures. Since the number of zero modes of the Dirac operator mod 2 is a modular invariant (this follows from (i)), the two classes of spin structures transform separately under modular transformations. In fact it can be shown that they transform irreducibly. This means that in the computation of the partition function or any correlation function we have to sum over all boundary conditions leading to even or odd spin structures. The relative phases between the different contributions are then determined by modular invariance.

Let us illustrate the above and work out the details for the vacuum amplitude on the torus, the one-loop partition function. As already discussed in Chapter 6 we parametrize the torus by two coordinates $\xi^1, \xi^2 \in [0,1]$ and define complex coordinates $z = \xi^1 + \tau\xi^2$ and $\bar{z} = \xi^1 + \bar{\tau}\xi^2$ in terms of which the metric is $\mathrm{d}s^2 = |\mathrm{d}z|^2$. τ is the Teichmüller parameter distinguishing different complex structures. Recall that modular transformations are those changes of τ which lead to identical complex structures. They are generated by $S : \tau \to -1/\tau$ and $T : \tau \to \tau + 1$. Under a general modular transformation $\tau \to \frac{a\tau+b}{c\tau+d}$ the metric changes to $\mathrm{d}s^2 \to \frac{1}{|c\tau+d|^2}|\mathrm{d}\xi'^1 + \tau\mathrm{d}\xi'^2|^2$ where $(\xi'^1, \xi'^2) = (d\xi^1 + b\xi^2, c\xi^1 + a\xi^2)$. We have the following possible boundary conditions for fermions, leading to four spin structures:

$$\psi(\xi^1 + 1, \xi^2) = \pm\psi(\xi^1, \xi^2)$$
$$\psi(\xi^1, \xi^2 + 1) = \pm\psi(\xi^1, \xi^2). \tag{9.1}$$

Periodic boundary conditions in ξ^1 correspond to the Ramond sector and antiperiodic boundary conditions to the Neveu-Schwarz sector. Under an S transformation with modular matrix $\begin{pmatrix} 0 & +1 \\ -1 & 0 \end{pmatrix} : (\xi^1, \xi^2) \to (\xi^2, -\xi^1)$. This means that the fermions transform as $\psi(\xi^1, \xi^2) \to \psi'(\xi^1, \xi^2) \propto \psi(\xi^2, -\xi^1)$ from which we easily derive the following action of S on the boundary conditions or spin structures:

$$S: \qquad (++) \to (++), \qquad \begin{array}{l} (--) \to (--) \\ (+-) \to (-+) \\ (-+) \to (+-) \end{array} \tag{9.2}$$

In the same way we find that under $\tau \to \tau + 1$ the fermions transform as $\psi(\xi^1, \xi^2) \to \psi'(\xi^1, \xi^2) \propto \psi(\xi^1 + \xi^2, \xi^2)$ which leads to the following action of T:

$$T: \qquad (++) \to (++), \qquad \begin{array}{l} (--) \to (-+) \\ (+-) \to (+-) \\ (-+) \to (--) \end{array} \tag{9.3}$$

This demonstrates our general statement above that even and odd spin structures transform irreducibly under modular transformations. In string theory one of the basic principles is invariance under diffeomorphisms of the world-sheet, also global ones. Since, as we have seen, they do change the spin structure, we have, to get modular invariant expressions, to sum over all different spin structures in each class (even and odd). At the one loop level the (++) spin structure is invariant by itself, being the only odd one and so is its contribution to the partition function. The other three must all be included in a modular invariant way. This means in particular that we must include both the R and the NS sectors.

It is now important to note that due to the world-sheet supersymmetry algebra, world-sheet fermions ψ^μ as well as the gravitino χ_α and consequently also the superconformal ghosts β, γ, all have the same spin structure. Left- and right-movers however can have different spin structures. We then denote the contribution to the partition function of the right-moving fermions with spin structure (++) by $A^{(++)}(\tau)$ and likewise for the other three cases and the left-movers. In light-cone gauge we get the following expressions which are trivial generalizations of the corresponding expression for the bosonic string:

$$
\begin{aligned}
A^{(++)}(\tau) &= \eta_{(++)} \operatorname{Tr} e^{2\pi i \tau H_R} (-1)^F \\
A^{(+-)}(\tau) &= \eta_{(+-)} \operatorname{Tr} e^{2\pi i \tau H_R} \\
A^{(--)}(\tau) &= \eta_{(--)} \operatorname{Tr} e^{2\pi i \tau H_{NS}} \\
A^{(-+)}(\tau) &= \eta_{(-+)} \operatorname{Tr} e^{2\pi i \tau H_{NS}} (-1)^F
\end{aligned}
\tag{9.4}
$$

where the η are phases to be determined by modular invariance. Let us comment on the $(-1)^F$ factors. For anticommuting variable the trace automatically implies that the fermions satsify anti-periodic boundary conditions along ξ^2. If we want to have periodic boundary conditions, we have to insert the operator $(-1)^F$ [4]. The light-cone Hamiltonians in the two sectors are (cf. Chapter 8):

$$H_R = \sum_{m=1}^{\infty} m b^i_{-m} b^i_m + \frac{1}{3}$$

$$H_{\mathrm{NS}} = \sum_{r=\frac{1}{2}}^{\infty} r b^i_{-r} b^i_r - \frac{1}{6}$$

$$\text{(9.5)}$$

The normal ordering constants follow most easily by subtracting the bosonic contribution $-\frac{d-2}{24}$ from the total normal ordering constant in each sector, namely 0 (R) and $-\frac{1}{2}$ (NS). It is now easy to evaluate the different contributions to the partition function. For instance, for $A^{(--)}(\tau)$ we get ($q = e^{2\pi i \tau}$):

$$
\begin{aligned}
A^{(--)}(\tau) &= \eta_{(--)} \operatorname{Tr} q^{H_{\mathrm{NS}}} \\[2mm]
&= \eta_{(--)} q^{-1/6} \operatorname{Tr} q^{\sum_{r=1/2}^{\infty} r b^i_{-r} b^i_r} \\[2mm]
&= \eta_{(--)} q^{-1/6} \prod_r \Big(\sum_{N_r} q^{r N_r} \Big)^8 \\[2mm]
&= \eta_{(--)} q^{-1/6} \Big(\prod_{n=1}^{\infty} (1 + q^{n-1/2}) \Big)^8 .
\end{aligned}
$$

$$\text{(9.6)}$$

The calculation is completely analogous to the bosonic case only that the occupation numbers are now restricted by the Pauli principle to $N_r = 0$ and 1. (This is just the grand partition function for an ideal Fermi gas with energy levels $E_r = r$.) We can now write

$$
\begin{aligned}
q^{-1/6} \prod_{n=1}^{\infty} \big(1 + q^{n-1/2} \big)^8 &\\[2mm]
&= \Big\{ q^{-1/24} \prod_{n=1}^{\infty} (1 - q^n)^{-1} \Big\}^4 \Big\{ \prod_{n=1}^{\infty} (1 - q^n)^4 (1 + q^{n-1/2})^8 \Big\} \\[2mm]
&= \frac{\theta_3^4(0|\tau)}{\eta^4(\tau)}
\end{aligned}
$$

$$\text{(9.7)}$$

where $\eta(\tau)$ is the previously encountered eta function and θ_3 one of the four Jacobi theta functions. In general, we can define the theta functions

$$\theta[^{\theta}_{\phi}](0|\tau) = \eta(\tau)e^{2\pi i\theta\phi}q^{\frac{\theta^2}{2}-\frac{1}{24}}\prod_{n=1}^{\infty}\left(1+q^{n+\theta-1/2}e^{2\pi i\phi}\right)\left(1+q^{n-\theta-1/2}e^{-2\pi i\phi}\right)$$

$$= \sum_{n=-\infty}^{+\infty}\exp[i\pi(n+\theta)^2\tau + 2\pi i(n+\theta)\phi].$$

$$\tag{9.8}$$

Through the one-loop partition function the θ-functions for arbitrary θ and ϕ are in correspondence to the generalized fermion boundary conditions[1] as:

$$\begin{aligned}
\psi(\xi^1+1,\xi^2) &= -e^{-2\pi i\theta}\psi(\xi^1,\xi^2)\,,\\
\psi(\xi^1,\xi^2+1) &= -e^{-2\pi i\phi}\psi(\xi^1,\xi^2)\,.
\end{aligned}$$

$$\tag{9.9}$$

The different spin structures then correspond to

$$
\begin{array}{llll}
(++) & \theta = \phi = 1/2 & \theta[^{1/2}_{1/2}] = \theta_1 & \\
(+-) & \theta = 1/2, \phi = 0 & \theta[^{1/2}_{0}] = \theta_2 & \\
(--) & \theta = \phi = 0 & \theta[^{0}_{0}] = \theta_3 & \\
(-+) & \theta = 0, \phi = 1/2 & \theta[^{0}_{1/2}] = \theta_4 &
\end{array}
$$

$$\tag{9.10}$$

The Jacobi theta functions and their generalizations to higher genus Riemann surfaces (the Riemann theta functions) play an important role in string theory and conformal field theory. They satisfy many amazing identities. At one loop, some of the most important ones are

$$\theta_2^4(0|\tau) - \theta_3^4(0|\tau) + \theta_4^4(0|\tau) = 0$$

$$\theta_2(0|\tau)\theta_3(0|\tau)\theta_4(0|\tau) = 2\eta^3(\tau) \qquad \text{(Jacobi triple product identity)}$$

$$\theta_1'(0|\tau) = 2\pi\eta^3(\tau)$$

$$\tag{9.11}$$

where the prime denotes differentiation with respect to the first argument (cf. the appendix to this chapter). Also, it is easy to see that $\theta_1(0|\tau) = 0$. In the same way that we have derived the partition function for the $(--)$ spin structure, we easily show

---

[1] This will be derived in the appendix to this chapter.

$$A^{(--)}(\tau) = \eta_{(--)} \frac{\theta_3^4(0|\tau)}{\eta^4(\tau)} \,,$$

$$A^{(-+)}(\tau) = \eta_{(-+)} \frac{\theta_4^4(0|\tau)}{\eta^4(\tau)} \,,$$

$$A^{(+-)}(\tau) = \eta_{(+-)} \frac{\theta_2^4(0|\tau)}{\eta^4(\tau)} \,,$$

$$A^{(++)}(\tau) = \eta_{(++)} \frac{\theta_1^4(0|\tau)}{\eta^4(\tau)} \,. \tag{9.12}$$

Obviously, $A^{(++)}(\tau) = 0$, i.e. the partition function for the odd spin structure vanishes. This is not surprising since we know that the Dirac operator has a zero mode and $A^{(++)} \sim \int D\psi e^{-\psi \slashed{D} \psi} = 0$.

We have stated above that odd and even spin structures transform irreducibly among one another under modular transformations. This should be reflected by the transformation properties of the θ-functions. Indeed, from the series expansion it is not hard to show that

$$\begin{aligned}
\theta_1'(0|\tau+1) &= e^{i\pi/4}\theta_1'(0|\tau)\,, \\
\theta_2(0|\tau+1) &= e^{i\pi/4}\theta_2(0|\tau)\,, \\
\theta_3(0|\tau+1) &= \theta_4(0|\tau)\,, \\
\theta_4(0|\tau+1) &= \theta_3(0|\tau)\,, \\
\eta(1+\tau) &= e^{i\pi/12}\eta(\tau)\,,
\end{aligned} \tag{9.13}$$

which reflects the transformation properties of the spin structures under T (cf. eq.(9.3)). Under S they transform as

$$\begin{aligned}
\theta_1'(0|-1/\tau) &= (-i\tau)^{3/2}\theta_1'(0|\tau)\,, \\
\theta_2(0|-1/\tau) &= (-i\tau)^{1/2}\theta_4(0|\tau)\,, \\
\theta_3(0|-1/\tau) &= (-i\tau)^{1/2}\theta_3(0|\tau)\,, \\
\theta_4(0|-1/\tau) &= (-i\tau)^{1/2}\theta_2(0|\tau)\,, \\
\eta(-1/\tau) &= (-i\tau)^{1/2}\eta(\tau)\,,
\end{aligned} \tag{9.14}$$

which corresponds to eq.(9.2).

Let us now determine the phases η. We will first require that the spin structure sum is modular invariant separately both in the left- and right-moving sectors. Since only relative phases are relevant we will arbitrarily set $\eta_{(--)} = +1$, i.e. $A^{(--)}(\tau) = \dfrac{\theta_3^4(0|\tau)}{\eta^4(\tau)}$. Using the transformation rules of the theta and eta functions we easily find

$$A^{(--)}(\tau+1) = A^{(-+)}(\tau) = \frac{\theta_4^4(0|\tau)}{\eta^4(\tau)} e^{-i\pi/3}. \tag{9.15}$$

The contribution from the eight transverse bosonic degrees of freedom is $\sim \dfrac{1}{\eta^8(\tau)}$ which contributes an extra factor of $e^{-2\pi i/3}$ so that we get for the phase $\eta_{(-+)} = -1$. Similarly we show that $\eta_{(+-)} = -1$. Clearly, $\eta_{(++)}$ cannot be determined from modular invariance; we will show below that it has to be ± 1. With these phases the contribution of the right-moving world-sheet fermions to the superstring partition function is

$$A(\tau) = \text{Tr}\, e^{2\pi i \tau H_{\text{NS}}} \tfrac{1}{2}\left(1 + (-1)^{F+1}\right) - \text{Tr}\, e^{2\pi i \tau H_{\text{R}}} \tfrac{1}{2}\left(1 - \eta_{(++)}(-1)^F\right)$$

$$= \frac{1}{2}\frac{1}{\eta(\tau)^4}\left\{\theta_3^4(0|\tau) - \theta_4^4(0|\tau) - \theta_2^4(0|\tau) + \eta_{(++)}\theta_1^4(0|\tau)\right\}$$

$$\tag{9.16}$$

with a similar expression for the left-movers. The relative sign between the two sectors reflects the fact that states in the NS sector are bosons whereas states in the R sector are fermions. The $\tfrac{1}{2}(1 + (-1)^F)$ in the NS sector is just the GSO projection. In the R sector it is $(-1)^F = \pm 1$ according to $\eta_{(++)} = \pm 1$ which agrees with Chapter 8. Due to the first identity in eq.(9.11) and the vanishing of θ_1, the partition function vanishes. This reflects a supersymmetric spectrum: the contributions from space-time bosons and fermions cancel.

It is worthwhile mentioning that there is one other modular invariant combination of boundary conditions: it consists of summing over the same boundary conditions for the left- and right-movers. It follows that the left- and right-moving sectors are not separately modular invariant due to the non-trivial connection between their boundary conditions. This leads to the following partition function:

$$A(\tau) = \mathrm{Tr}\, e^{2\pi i \tau H_{\mathrm{NS}} - 2\pi i \bar{\tau} \bar{H}_{\mathrm{NS}}} \tfrac{1}{2}\big(1 + (-1)^{F+\bar{F}}\big)$$

$$+ \mathrm{Tr}\, e^{2\pi i \tau H_{\mathrm{R}} - 2\pi i \bar{\tau} \bar{H}_{\mathrm{R}}} \tfrac{1}{2}\big(1 - \eta_{++}(-1)^{F+\bar{F}}\big). \tag{9.17}$$

If we include the contribution from the bosons we get

$$\chi(\tau, \bar{\tau}) = \frac{1}{2} \frac{1}{(\mathrm{Im}\tau)^4} \frac{|\theta_2(0|\tau)|^8 + |\theta_3(0|\tau)|^8 + |\theta_4(0|\tau)|^8}{|\eta(\tau)|^{24}}. \tag{9.18}$$

Modular invariance of this expression is easily checked. This theory has only space-time bosons and contains a tachyon. The GSO projection in the NS sector is $(-1)^{F+\bar{F}} = 1$ which does allow the tachyon in table 8.2.

Let us now give the argument why the phase $\eta_{(++)}$ can only be ± 1. Clearly, for the partition function to have the interpretation as a sum over states we can only allow $\eta_{(++)} = 0$ or ± 1. If we look at the partition function at two loops it will be expressible in terms of the appropriate Riemann theta functions, ten of which correspond to even and six to odd spin structures. In the limit where the genus two surface degenerates to two tori, the genus two theta functions become simply products of Jacobi theta functions. Especially $\theta_1(\tau_1)\theta_1(\tau_2)$, where $\tau_{1,2}$ are the Teichmüller parameters of the two resulting tori, is the degeneration limit of an even theta function at genus two which has to be part of the partition function since the even theta functions transform irreducibly under global diffeomorphisms (the Dehn twists) of the genus two surface; that means that $\eta_{(++)} = 0$ is excluded.

Let us close this chapter with the extension of some results of Chapter 6 to the case of fermions. We know from Chapter 7 that the covariant derivative acts on a spin 1/2 world-sheet fermion as $D_\alpha \psi = \partial_\alpha \psi - \frac{1}{2}\omega_\alpha \bar{\rho}\psi$ where ω_α is the spin connection. In conformal gauge, with zweibein $e_\alpha{}^a = e^{\sigma/2}\delta_\alpha^a$ the spin connection is $\omega_\alpha = \frac{1}{2}\epsilon_\alpha{}^\beta \partial_\beta \sigma$. In local complex coordinates the covariant derivatives act on the two helicity components as $D_z \psi_\pm =$

$(\partial_z \mp \frac{1}{4}\partial_z \sigma)\psi_\pm$. Now recall our discussion in Chapter 7, where we had to rescale the fermions to arrive at the action in conformal gauge. Under conformal transformations they transform with weight 1/2 (this follows from the invariance of the action or from eq.(7.47)) and the factor by which we had to rescale them was exactly the square root of the zweibein. But multiplying the fermions with the square root of the zweibein converts their tangent space index to a world index. Let us denote the transformed spinors by $e^{\sigma/4}\psi_\pm = \tilde\psi_\pm$. Since the zweibein is covariantly constant, the covariant derivative acts on the $\tilde\psi_\pm$ as $D_z\tilde\psi_+ = (\partial_z - \frac{1}{2}\partial_z\sigma)\tilde\psi_+$ and $D_z\tilde\psi_- = \partial_z\tilde\psi_-$. Extending the notation of Chapter 6 to tensors of half-integer rank, we find that $\tilde\psi_+ \in \mathcal{T}^{(1/2)}$ and $(h^{z\bar z})^{1/2}\tilde\psi_- \in \mathcal{T}^{(-1/2)}$. We can then extend the definition of covariant derivatives eqs.(6.9,10) and to general half-integer n. The scalar product eq.(6.4) also generalizes. Of particular interest is the Riemann-Roch theorem which holds without modification. We can then complete table 6.1 for half-integer n in the same way as we did for $n \in \mathbf{Z}$ in Chapter 6. The only subtlety is at genus one. For integer $n > 0$ there is always a constant zero mode of $\nabla_z^{(n)}$. For $n \in \mathbf{Z} + \frac{1}{2}$ this is only true for the odd spin structure. The results are collected in table 9.1.

Table 9.1:

g	dim $\ker\nabla_z^{(n)}$	dim $\ker\nabla^z_{(n+1)}$	
0	$2n+1$	0	
1	1	1	odd spin structure
1	0	0	even spin structure
>1	1 for $n=0$	g	
	0 for $n>0$	$(2n+1)(g-1)$	

We find that there are two zero modes of $\nabla_z^{(1/2)}$ at $g = 0$ corresponding to conformal Killing spinors or zero modes of the superconformal ghost γ. The zero modes of $\nabla^z_{(3/2)}$ indicate the presence of super-moduli. We will however not discuss them here. The ghost zero modes for arbitrary (integer and half-integer) n will be discussed in Chapter 13.

Appendix C.

In this appendix we want to generalize the computation of the fermionic partition function for general boundary conditions

$$\psi(\sigma + 2\pi) = -e^{-2\pi i\theta}\psi(\sigma) \qquad (C.1)$$

which requires the world-sheet fermions to be complex. The action for one chirality (say the right-movers) is

$$S = \frac{i}{\pi} \int d^2\sigma\, \psi^\dagger \partial_+ \psi \qquad (C.2)$$

with energy-momentum tensor

$$T = \frac{i}{2}(\psi^\dagger \partial_- \psi + \psi \partial_- \psi^\dagger)\,. \qquad (C.3)$$

The anti-commutation relations are

$$\{\psi^\dagger(\sigma), \psi(\sigma')\} = 2\pi\delta(\sigma - \sigma')\,. \qquad (C.4)$$

The mode expansion is

$$\psi(\sigma^+) = \sum_{n \in \mathbf{Z}} b_{n+\theta-\frac{1}{2}}\, e^{-i(n+\theta-\frac{1}{2})(\tau+\sigma)}$$

$$\psi^\dagger(\sigma^+) = \sum_{n \in \mathbf{Z}} b^\dagger_{n+\theta-\frac{1}{2}}\, e^{i(n+\theta-\frac{1}{2})(\tau+\sigma)} \qquad (C.5)$$

and we find

$$\{b^\dagger_{m+\theta-\frac{1}{2}}, b_{n+\theta-\frac{1}{2}}\} = \delta_{m,n}\,. \qquad (C.6)$$

The b_η and $b^\dagger_{-\eta}$ are annihilation operators for $\eta > 0$. The Hamiltonian is

$$
\begin{aligned}
H &= \frac{1}{2\pi} \int_0^{2\pi} T(\sigma)\mathrm{d}\sigma \\
&= \sum_{n \in \mathbf{Z}} (n + \theta - \frac{1}{2}) : b^\dagger_{n+\theta-\frac{1}{2}} b_{n+\theta-\frac{1}{2}} : + \Big(\frac{\theta^2}{2} - \frac{1}{24}\Big) \\
&= \sum_{n=1}^{\infty} \Big\{\Big(n + \theta - \frac{1}{2}\Big) N_{n+\theta-\frac{1}{2}} + \Big(n - \theta - \frac{1}{2}\Big) \bar{N}_{n+\theta-\frac{1}{2}}\Big\} + \Big(\frac{\theta^2}{2} - \frac{1}{24}\Big),
\end{aligned}
$$

$$\qquad (C.7)$$

where we have introduced the mode number operators ($\eta > 0$)

$$N_\eta = b_\eta^\dagger b_\eta \, ,$$
$$\bar{N}_\eta = b_{-\eta} b_{-\eta}^\dagger \, . \qquad (C.8)$$

The normal ordering constant follows from the general formula given in footnote 2 of Chapter 8. (We have taken $|\theta| \leq \frac{1}{2}$.) The partition function is then ($N = \sum_{n>0} N_{n+\theta-\frac{1}{2}}$)

$$A(\theta, \phi) = \mathrm{Tr}\, e^{2\pi i \phi(N - \bar{N})} q^H$$
$$= e^{2\pi i \theta \phi} q^{\frac{\theta^2}{2} - \frac{1}{24}} \prod_n \left(1 + q^{n+\theta-\frac{1}{2}} e^{2\pi i \phi}\right)\left(1 + q^{n-\theta-\frac{1}{2}} e^{-2\pi i \phi}\right) \qquad (C.9)$$
$$= \frac{\theta[^\theta_\phi](0|\tau)}{\eta(\tau)} = \det(\partial_{\theta,\phi}).$$

We have inserted the operator $e^{2\pi i \phi(N - \bar{N})}$ to enforce the boundary conditions in the ξ^2-direction on the torus (cf. eq.(9.9)). We have also included an extra phase for convenience. This result is valid (up to a phase) for all θ and ϕ.

Let us finally give some more information on theta functions. They are functions of two variables $\theta[^\theta_\phi](z|\tau)$ and have a series expansion ($\mathrm{Im}\tau > 0$)

$$\theta[^\theta_\phi](z|\tau) = \sum_{n \in \mathbf{Z}} \exp[i\pi(n+\theta)^2 \tau + 2\pi i(n+\theta)(z+\phi)]. \qquad (C.10)$$

Clearly

$$\theta[^\theta_\phi](z|\tau) = \theta[^{\ \theta}_{\phi+z}](0|\tau). \qquad (C.11)$$

The first argument, or the shift in the ξ^2 boundary condition if we use eq.(C.11), is important when we couple the fermions to an external field. However it will not enter in our applications. An alternative representation of the theta-functions as an infinite product was given in eq.(9.8). It is not hard to show the transformation properties under S and T transformations, the generators of the modular group:

$$\theta[{}^{\theta}_{\phi}](0|\tau+1) = e^{-i\pi(\theta^2-\theta)}[{}^{\theta}_{\theta+\phi-\frac{1}{2}}](0|\tau)$$

$$\theta[{}^{\theta}_{\phi}](0|-\frac{1}{\tau}) = \sqrt{-i\tau}\,e^{2\pi i\theta\phi\tau}\theta[{}^{-\phi}_{\theta}](0|\tau) \qquad |\arg\sqrt{-i\tau}| < \frac{\pi}{2}. \qquad (C.12)$$

The first equation follows directly from the sum representation. To show the second one uses Poisson resummation (cf. Chapter 10).

REFERENCES

[1] N. Seiberg and E. Witten, *"Spin structures in string theory"*, *Nucl. Phys.* **B276** (1986) 272.

[2] L. Alvarez-Gaumé, P. Ginsparg, G. Moore and C. Vafa, *"An $O(16) \times O(16)$ heterotic string"*, *Phys. Lett.* **171 B** (1986) 155.

[3] L. Alvarez-Gaumé, G. Moore and C. Vafa, *"Theta functions, modular invariance and strings"*, *Comm. Math. Phys.* **106** (1986) 1.

[4] L. Alvarez-Gaumé, *"Supersymmetry and index theory"*, preprint HUTP-84/A075 (1984), in proceedings of 1984 NATO School on Supersymmetry, Bonn.

Toroidal Compactification of the Closed Bosonic String – 10-Dimensional Heterotic String

So far we have described two kinds of closed oriented string theories. First, the closed bosonic string theory was formulated consistently in 26 space-time dimensions. The spectrum of physical states contains a negative $(mass)^2$ scalar tachyon and, at the next level, with $(mass)^2 = 0$, a symmetric traceless tensor (the graviton), an antisymmetric tensor field and a scalar dilaton. These states are accompanied by an infinite tower of massive excitations. As such, the closed bosonic string has many serious, phenomenological drawbacks: flat 26-dimensional space-time, the appearance of a tachyon and the non-existence of space-time fermions.

Some of these difficulties could be overcome by the 10-dimensional fermionic string theories. Modular invariance forces to project onto a space-time supersymmetric spectrum which excludes the scalar tachyon. The lowest states are again the massless graviton, antisymmetric tensor field and dilaton which are now accompanied by their superpartners, namely two gravitinos and two dilatinos. Therefore the theory has $N = 2$ space-time supersymmetry in 10 dimensions.

In conclusion, both the bosonic and fermionic closed string theories describe a higher dimensional theory of pure (super) gravity at the massless level. However, one would like to include also non-Abelian gauge interactions, massless scalars and fermions and formulate the theory in four space-time dimensions. Both of these goals can be reached simultaneously when "compactifying" some of the string coordinates on an internal compact space. However, the notion of compactification should not be taken too literally – we will eventually learn that, in the field theory sense, the "string compactifications" are, in general, no compactifications at all. With the most general ansatz of constructing four-dimensional string theories, the concept of critical dimension is replaced by the requirement that the central charge of the Virasoro algebra is zero. This can be realized by introducing only four "usual" string coordinates corresponding to four-dimensional Minkowski space-time and in addition a two-dimensional (super) conformal field theory which has to satisfy the consistency constraints of unitarity, locality, conformal invariance, modular invariance etc. Simple realizations of these internal conformal field theories are two-dimensional bosons living on a torus or free two-dimensional fermions. Moreover, many constructions may turn out to be quantum mechanically equivalent, e.g. via the two-dimensional equivalence between bosons and fermions. This will be discussed in Chapter 11.

In this chapter we restrict ourselves to discussing the toroidal compactification of the closed bosonic string and the construction of the supersymmetric 10-dimensional heterotic string theory. In Chapter 14 we will discuss theories in four dimensions.

Compactification of closed strings on a torus was first discussed by Cremmer and Scherk [1] and by Green, Schwarz and Brink [2]. References [3,4] initiated further developments. Additional literature can be found in [5]. The heterotic string was invented by Gross, Harvey, Martinec and Rohm [6].

10.1 Toroidal compactification of the closed bosonic string

For illustrative purposes we first consider the simplest case of one coordinate compactified on a circle of radius R. It means that for one spatial coordinate, i.e. X^{25}, we require periodicity such that points on the real axis are identified according to

$$X^{25} \sim X^{25} + 2\pi RL \quad , \quad L \in \mathbf{Z}. \qquad (10.1)$$

Thus, X^{25} describes the one-dimensional circle S^1. In other words, S^1 is obtained by dividing the real line by the integers times $2\pi R$

$$S^1 = \mathbf{R}/2\pi RL \quad , \quad L \in \mathbf{Z}, \qquad (10.2)$$

which also defines the equivalence relation eq.(10.1). Now, the coordinate $X^{25}(\sigma, \tau)$, $0 \le \sigma \le 2\pi$, maps the closed string onto the spatial circle $0 \le X^{25} \le 2\pi R$. Therefore we have to reformulate the periodicity condition a closed string has to obey in the following way:

$$X^{25}(\sigma + 2\pi, \tau) = X^{25}(\sigma, \tau) + 2\pi RL. \qquad (10.3)$$

The second, new term describes string states which are only closed on the circle however not on the real axis. These states correspond to so-called winding states; they are characterized by the winding number L that counts how many times the string wraps around the circle. This phenomenon has no counterpart in the theory of point particles. The winding states are topologically stable solitons; the winding number cannot be changed without breaking the string. Such solitons always exist if the internal manifold contains non-contractible loops.

We get the following mode expansion for $X^{25}(\sigma, \tau)$ which respects eq.(10.3)

$$X^{25}(\sigma, \tau) = x^{25} + 2p^{25}\tau + LR\sigma + i\sum_{n \neq 0} \frac{1}{n}\left(\alpha_n^{25} e^{-in(\tau - \sigma)} + \bar{\alpha}_n^{25} e^{-in(\tau + \sigma)}\right).$$

$$(10.4)$$

x^{25} and p^{25} obey the usual commutation relation

$$[x^{25}, p^{25}] = i. \qquad (10.5)$$

p^{25} generates translations of x^{25}. Singlevaluedness of the wave function $\exp(ip^{25}x^{25})$ restricts the allowed internal momenta to discrete values:

$$p^{25} = \frac{M}{R} \quad , \quad M \in \mathbf{Z}. \tag{10.6}$$

We split $X^{25}(\sigma, \tau)$ into left and right movers.

$$X_L^{25}(\tau + \sigma) = \frac{1}{2}x^{25} + (p^{25} + \frac{1}{2}LR)(\tau + \sigma) + i \sum_{n \neq 0} \frac{1}{n}\bar{\alpha}_n^{25}e^{-in(\tau+\sigma)},$$

$$X_R^{25}(\tau - \sigma) = \frac{1}{2}x^{25} + (p^{25} - \frac{1}{2}LR)(\tau - \sigma) + i \sum_{n \neq 0} \frac{1}{n}\alpha_n^{25}e^{-in(\tau-\sigma)}. \tag{10.7}$$

The mass operator gets contributions from the soliton states (remember that $\alpha' = 2$):

$$m_L^2 = \frac{1}{2}\Big(\frac{M}{R} + \frac{1}{2}LR\Big)^2 + N_L - 1,$$

$$m_R^2 = \frac{1}{2}\Big(\frac{M}{R} - \frac{1}{2}LR\Big)^2 + N_R - 1, \tag{10.8}$$

$$m^2 = m_L^2 + m_R^2 = \frac{M^2}{R^2} + \frac{1}{4}L^2R^2 + N_L + N_R - 2,$$

where $m^2 = -\sum_{\mu=0}^{24} p_\mu p^\mu$. The factor $\frac{M^2}{R^2}$ is the contribution from the momenta in the compact dimension and the term $\frac{1}{4}L^2R^2$ is the energy required to wrap the string around the circle L times.

Physical string states have to satisfy the reparametrization constraint eq.(3.18)

$$m_L^2 = m_R^2 \quad \leftrightarrow \quad N_R - N_L = ML. \tag{10.9}$$

Let us examine the spectrum of the effectively 25-dimensional string theory. First consider states with no winding number and internal momentum excitations.

(i) The lowest energy state is again the scalar tachyon with $m^2 = -2$ (we suppress the space-time momentum)

$$|\text{tachyon}\rangle = |0\rangle. \tag{10.10}$$

(ii) At the massless level with $N_L = N_R = 1$ there are now the 25-dimensional graviton, antisymmetric tensor and dilaton.

$$|G^{\mu\nu}\rangle = \alpha^\mu_{-1}\bar{\alpha}^\nu_{-1}|0\rangle \qquad \mu,\nu = 0,\dots,24 \qquad (10.11)$$

with the oscillators in the 25 uncompactified space-time directions.

(iii) In addition to these states which were already present in the uncompactified theory there are also new states which arise from the compactification. We can replace one space-time oscillator by an internal oscillator to get two vector states:

$$|V_1^\mu\rangle = \alpha^\mu_{-1}\bar{\alpha}^{25}_{-1}|0\rangle \,,$$
$$|V_2^\mu\rangle = \alpha^{25}_{-1}\bar{\alpha}^\mu_{-1}|0\rangle \,. \qquad (10.12)$$

These massless vectors originate from the Kaluza-Klein compactification of the bosonic string on the circle – they are just part of the originally 26-dimensional graviton and antisymmetric tensor field. They give rise to a $U(1)_L \times U(1)_R$ gauge symmetry which corresponds to the left and right isometries of the circle. Of course, the appearance of these two gauge bosons is expected in any field theoretical compactification.

(iv) Finally, acting with two internal oscillators on the vacuum we obtain a massless scalar field which is also a compactified degree of freedom of the 26-dimensional metric; its vacuum expectation value corresponds to the radius R of the circle:

$$|\phi\rangle = \alpha^{25}_{-1}\bar{\alpha}^{25}_{-1}|0\rangle \,. \qquad (10.13)$$

Let us now turn to the more interesting case, namely states with non-trivial internal momentum and winding number. Therefore we act now with the oscillators on the soliton vacua $|M,L\rangle$. We will concentrate on the first winding sector, $M = \pm L = \pm 1$.

Choosing $M = L = \pm 1$ we derive from eq.(10.8) that

$$m_L^2 = \frac{1}{2R^2} + \frac{1}{8}R^2 + N_L - \frac{1}{2},$$

$$m_R^2 = \frac{1}{2R^2} + \frac{1}{8}R^2 + N_R - \frac{3}{2}, \tag{10.14}$$

$$m^2 = \frac{1}{R^2} + \frac{1}{4}R^2 + N_L + N_R - 2.$$

The level matching constraint eq.(10.9) is satisfied if $N_L = 0, N_R = 1$. Thus we have two vector states of the form

$$|V_a^\mu\rangle = \alpha_{-1}^\mu \,|\pm 1, \pm 1\rangle, \qquad a = 1, 2, \quad \mu = 0, \ldots, 24 \tag{10.15}$$

and also two scalars

$$|\phi_a\rangle = \alpha_{-1}^{25}|\pm 1, \pm 1\rangle \tag{10.16}$$

with mass which depends on the radius of the circle

$$m^2(R) = \frac{1}{R^2} + \frac{1}{4}R^2 - 1. \tag{10.17}$$

Analogously we can set $M = -L = \pm 1$. Then eq.(10.9) is satisfied if $N_L = 1$, $N_R = 0$ and we obtain again two vectors and two scalars

$$|V_a^{\prime\mu}\rangle = \bar{\alpha}_{-1}^\mu|\pm 1, \mp 1\rangle, \qquad a = 1, 2, \quad \mu = 0, \ldots, 24$$

$$|\phi_a'\rangle = \bar{\alpha}_{-1}^{25}|\pm 1, \mp 1\rangle \tag{10.18}$$

with mass also given by eq.(10.17). It is easy to see that $m^2(R) \geq 0$ with equality holding for $R = \sqrt{2} = \sqrt{\alpha'}$. This means that for this particular radius, which is determined by the string tension, we get extra massless states, of which the massless vectors are of particular interest. This phenomenon is of utmost importance in the theory of the compactified bosonic string. The four additional massless vectors form, together with the massless vectors of eq.(10.12), the adjoint representation of $SU(2)_L \times SU(2)_R$. The oscillator excitations eq.(10.12) with zero winding number correspond to the $U(1)_L \times U(1)_R$ Cartan subalgebra generators of $SU(2)_L \times SU(2)_R$, the soliton states of eqs.(10.15) and (10.18) to the (non-commuting) roots (this will be put on more rigorous grounds in the next chapter). $(M + L, M - L)$ are the $U(1)_L \times U(1)_R$ quantum numbers. It is easy to convince oneself that these additional massless vectors are the only ones possible for any choice of M, L and R. We then also have extra tachyons $M = 0$, $L = \pm 1$ and $M = \pm 1$, $L = 0$.

We have seen that for a special value of the radius of the compact circle one gets an enhancement of the gauge symmetry due to the soliton states. This can never occur in any point particle compactification. Furthermore, for $R = \sqrt{2}$ we get also eight additional massless scalars

$$\alpha_{-1}^{25} | \pm 1, \pm 1\rangle, \quad \bar{\alpha}_{-1}^{25} | \pm 1, \mp 1\rangle, \quad | \pm 2, 0\rangle, \quad |0, \pm 2\rangle \qquad (10.19)$$

which, together with $\alpha_{-1}^{25} \bar{\alpha}_{-1}^{25} |0, 0\rangle$ (cf. eq.(10.13)) form the $(\underline{3}, \underline{3})$ representation of $SU(2)_L \times SU(2)_R$. However, for arbitrary values of the radius, both the four non-commuting gauge bosons of $SU(2)_L \times SU(2)_R$ and the four scalars with internal oscillator excitations are massive – the gauge symmetry is broken to $U(1)_L \times U(1)_R$. Therefore this phenomenon can be interpreted as a stringy Higgs effect. For arbitrary radii these four massive scalars build the longitudinal components of the four massive vector particles. Two of the remaining scalars in eq.(10.19) become massive and the other two become tachyons.

The $U(1)_L \times U(1)_R$ gauge bosons on the other hand stay massless for all values of R. This also the case for the single scalar of eq.(10.13). In a low energy effective field theory this neutral scalar will have a completely flat potential which corresponds to the freedom of choosing the radius of the circle as a free parameter. In summary, the spectrum of the bosonic string compactified on S^1 is characterized by a single parameter, also called modulus, namely the radius of the circle which is the vacuum expectation value of the scalar field eq.(10.13). However at this point one must be quite careful. Inspection of the mass formula eq.(10.8) shows that the spectrum is invariant under transformations $R \to \frac{2}{R}$ if one simultaneously also interchanges the winding numbers L and the momenta M. This transformation is called a duality transformation where the point $R = \sqrt{2}$ is a fixed point under this transformation. It means that the compactified string theory looks the same regardless whether we consider it at large or small radius of the internal circle.[1] Therefore the spectrum of the compactified bosonic string is already completely characterized by the values $R \geq \sqrt{2}$, or, equivalently, $R \leq \sqrt{2}$; i.e. the so-called moduli space of this theory is not the whole real axis but only one of the above intervals. The invariance under duality indicates that in string theory one cannot probe distances smaller than the string size.

We now want to generalize this mechanism to the case where we compactify D bosonic coordinates on a D-dimensional torus T^D. The resulting theory is therefore $(26-D)$-dimensional. The torus is defined by identifying points in the D-dimensional internal space as follows (compact dimensions are labeled with capital letters):

$$X^I \sim X^I + \sqrt{2}\pi \sum_{i=1}^{D} n_i R_i e_i^I = X^I + 2\pi L^I, \qquad n_i \in \mathbf{Z} \tag{10.20}$$

with

$$L^I = \sqrt{\frac{1}{2}} \sum_{i=1}^{D} n_i R_i e_i^I. \tag{10.21}$$

The $e_i = \{e_i^I\}$ $(i = 1 \ldots D)$ are D linear independent vectors normalized to $(e_i)^2 = 2$. The $\mathbf{L} = \{L^I\}$ can be thought of as lattice vectors of a D-dimensional lattice Λ^D: $\mathbf{L} \in \Lambda^D$. This lattice has as basis the D vectors $\sqrt{\frac{1}{2}} R_i e_i$. Therefore the torus on which we compactify is obtained by dividing $\mathbf{R}^D$ by $2\pi \Lambda^D$:

$$T^D = \mathbf{R}^D / 2\pi \Lambda^D. \tag{10.22}$$

---

[1] Strictly speaking one has to make sure that the S-matrix elements transform properly. One can show that this is indeed the case.

Center of mass position and momentum satisfy canonical commutation relations

$$[x^I, p^J] = i\delta^{IJ}, \qquad (10.23)$$

i.e. p^I generates translations of x^I and single valuedness of $e^{ix^I p^I}$ requires that $L^I p^I \in \mathbf{Z}$, i.e. the allowed momenta have to lie on the lattice which is dual to Λ^D, denoted by $(\Lambda^D)^*$:[2]

$$p^I = \sqrt{2} \sum_{i=1}^{D} \frac{m_i}{R_i} e_i^{*I} \qquad (10.24)$$

where the e_i^{*I} are dual to the e_i^I, i.e.

$$\sum_{I=1}^{D} e_i^I e_j^{*I} = \delta_{ij}. \qquad (10.25)$$

Their normalization is $(e_i^*)^2 = \frac{1}{2}$. From eq.(10.25) it follows that

$$\sum_{i=1}^{D} e_i^I e_i^{*J} = \delta^{IJ}. \qquad (10.26)$$

The basis vectors of $(\Lambda^D)^*$ are $\frac{\sqrt{2}}{R_i} e_i^*$.

The condition a closed string in the compact directions has to satisfy now looks like:

$$X^I(\sigma + 2\pi, \tau) = X^I(\sigma, \tau) + 2\pi L^I. \qquad (10.27)$$

Thus the L^I play the role of winding numbers. The mode expansion becomes:

$$X_L^I(\tau + \sigma) = \frac{1}{2} x^I + (p^I + \frac{1}{2} L^I)(\tau + \sigma) + i \sum_{n \neq 0} \frac{1}{n} \bar{\alpha}_n^I \, e^{-in(\tau + \sigma)},$$

$$X_R^I(\tau - \sigma) = \frac{1}{2} x^I + (p^I - \frac{1}{2} L^I)(\tau - \sigma) + i \sum_{n \neq 0} \frac{1}{n} \alpha_n^I \, e^{-in(\tau - \sigma)}. \qquad (10.28)$$

The mass formula is $(m^2 = -\sum_{\mu=0}^{25-D} p_\mu p^\mu)$

---

[2] Some basic facts about lattices are collected at the beginning of section 11.2.

$$m_L^2 = \frac{1}{2}\sum_{I=1}^{D}(p^I + \tfrac{1}{2}L^I)^2 + N_L - 1 = \frac{1}{2}p_L^2 + N_L - 1\,,$$

$$m_R^2 = \frac{1}{2}\sum_{I=1}^{D}(p^I - \tfrac{1}{2}L^I)^2 + N_R - 1 = \frac{1}{2}p_R^2 + N_R - 1\,,$$

$$m^2 = m_L^2 + m_R^2 = N_R + N_L - 2 + \sum_{I=1}^{D}\left(p^I p^I + \tfrac{1}{4}L^I L^I\right) \tag{10.29}$$

$$= N_L + N_R - 2 + \sum_{i,j=1}^{D}\left(m_i g_{ij}^* m_j + \tfrac{1}{4}n_i g_{ij} n_j\right)$$

with $\boldsymbol{p}_{L,R} = \boldsymbol{p} \pm \tfrac{1}{2}\boldsymbol{L}$. g_{ij} and g_{ij}^* are the metrics of Λ^D and $(\Lambda^D)^*$:

$$g_{ij} = \frac{1}{2}\sum_{I=1}^{D} R_i e_i^I R_j e_j^I\,,$$

$$g_{ij}^* = 2\sum_{I=1}^{D}\frac{1}{R_i}e_i^{*I}\frac{1}{R_j}e_j^{*I}\,. \tag{10.30}$$

With eqs.(10.25) and (10.26) it follows that $g_{ij}^* = (g^{-1})_{ij}$. The volumes of the unit cells are $\mathrm{vol}(\Lambda^D) = \sqrt{\det g}$ and $\mathrm{vol}((\Lambda^D)^*) = \sqrt{\det g^*} = \frac{1}{\sqrt{\det g}}$.

The constraint eq.(10.9) generalizes to

$$N_R - N_L = \boldsymbol{p}\cdot\boldsymbol{L} = \sum_{i=1}^{D} m_i n_i\,. \tag{10.31}$$

Using this information one can easily show that the $2D$-dimensional vectors $\boldsymbol{P} = (\boldsymbol{p}_L, \boldsymbol{p}_R)$ build an even self-dual lattice[3] $\Gamma_{D,D}$ if we choose the signature of the metric of this lattice to be of the form $((+1)^D,(-1)^D)$; i.e. $\boldsymbol{P}\cdot\boldsymbol{P}' = \sum_I (p_L^I p_L^{I\prime} - p_R^I p_R^{I\prime})$. Therefore $\Gamma_{D,D}$ is called a Lorentzian lattice. Self-duality of $\Gamma_{D,D}$ follows from the definition of the mutual scalar product between two different vectors $\boldsymbol{P}$ and $\boldsymbol{P}'$: $\boldsymbol{P}\cdot\boldsymbol{P}' = \sum_{i=1}^{D}(m_i n_i' + n_i m_i') \in \mathbf{Z}$. Note however that, given the vectors $(\boldsymbol{p}_L, \boldsymbol{p}_R) \in \Gamma_{D,D}$, the set of vectors $\boldsymbol{p}_L$, $\boldsymbol{p}_R$ do

---

[3] Our notation is such that we denote the left- and right-moving momentum lattices by $\Gamma_{L,R}$ and the winding vector lattice by Λ.

not, for a general metric on the torus, form separate lattices Γ_L, Γ_R in spite of closing under addition. For example, consider the two-dimensional even Lorentzian self-dual lattice $\Gamma_{1,1}$ consisting of points $(\frac{M}{R} + \frac{1}{2}LR, \frac{M}{R} - \frac{1}{2}LR)$. The left (or right) components alone, even though they close under addition, do not form a one-dimensional lattice. For general real values of R we cannot write all possible p_L as the integer multiple of one basis vector. A torus compactification where the left and right momenta p_L, p_R build separately Euclidean lattices Γ_L and Γ_R is called rational. The notation 'rational' is used since Γ_L, Γ_R can be decomposed into a finite number of cosets (see Chapter 11). In this case the possible $U(1)$ charges are rational numbers and the corresponding conformal field theory is rational. For example, the lattice $\Gamma_{1,1}$ is only rational if R^2 is a rational number.

As before, the sector without any winding and internal momentum contains a $(26 - D)$-dimensional tachyon, a massless graviton, antisymmetric tensor and dilaton. Furthermore there exist $2D$ massless vectors of the form

$$\begin{aligned}
|V_1^{\mu I}\rangle &= \alpha^\mu_{-1}\bar{\alpha}^I_{-1}|0\rangle\,, \\
|V_2^{\mu I}\rangle &= \bar{\alpha}^\mu_{-1}\alpha^I_{-1}|0\rangle
\end{aligned}$$

$$(10.32)$$

which are the gauge bosons of $[U(1)]^D_L \times [U(1)]^D_R$ reflecting the isometry group of the torus $[U(1)]^D$. Finally, there are D^2 massless scalars

$$|\phi^{IJ}\rangle = \alpha^I_{-1}\bar{\alpha}^J_{-1}|0\rangle\,. \qquad (10.33)$$

These scalar fields correspond to the moduli of D-dimensional torus compactifications of the bosonic string. Of the D^2 scalars $\frac{D}{2}(D+1)$ are the internal graviton components; their vacuum expectation values give the constant background parameters g_{ij} which describe the shape of the D-dimensional torus T^D. The remaining $\frac{D}{2}(D-1)$ scalars are the internal components of the antisymmetric tensor field B_{ij} which may also acquire constant vacuum expectation values. This kind of background fields will also influence the string spectrum and therefore enter the mass formula eq.(10.29). The

$B^{IJ} = \frac{\sqrt{2}}{R_i} e_i^{*I} B_{ij} e_j^{*J} \frac{\sqrt{2}}{R_j}$ background fields are coupled to the free bosonic fields X^I via an additional term in the bosonic string action:

$$S = \frac{1}{8\pi} \int \mathrm{d}^2\sigma \, \epsilon^{\alpha\beta} B_{IJ} \partial_\alpha X^I(\sigma,\tau) \partial_\beta X^J(\sigma,\tau) \qquad (10.34)$$

where we have assumed that the antisymmetric tensor has nonvanishing components only in the compactified directions and that it is constant. This term induces a change in the internal canonical momenta $\Pi^I = \frac{\partial S}{\partial(\partial_\tau X^I)} = \frac{1}{2\pi}(p^I + \frac{1}{2} B_{IJ} L^J + \text{oscillators})$. This does however only affect the center of mass momenta. They are now given by $\pi^I = p^I + \frac{1}{2} B_{IJ} L^J$. These vectors now generate translations and lie on the lattice Λ_D^* which is dual to Λ_D. The center of mass momentum which enters the mass formula is still given by p^I. p_L and p_R can be expressed in terms of the π^I instead of the p^I:

$$\begin{aligned}
p_{L,R}^I &= \pi^I \pm \frac{1}{2}(\delta^{IJ} \mp B^{IJ}) L_J \\
&= \sqrt{2}\frac{m_i}{R_i} e_i^{*I} \pm \frac{1}{2}\sqrt{\frac{1}{2}} n_i R_i e_i^I - \sqrt{\frac{1}{2}\frac{1}{R_i}} B_{ij} n_j e_i^{*I}.
\end{aligned} \qquad (10.35)$$

It is again instructive to calculate the inner product of two vectors $\boldsymbol{P} = (\boldsymbol{p_L}, \boldsymbol{p_R})$, $\boldsymbol{P}' = (\boldsymbol{p_L'}, \boldsymbol{p_R'}) \in \Gamma_{D,D}$: $\boldsymbol{P} \cdot \boldsymbol{P}' = \sum_{i=1}^D (m_i n_i' + n_i m_i')$. We see that the inner product does not depend on the choice of background parameters g_{ij} and B_{ij}. Using this background independence we can take for example $g_{ij} = \delta_{ij}$ ($e_i^I = \sqrt{2}\delta_i^I$, $R_i = R = 1$) and $B_{ij} = 0$. Then $\Gamma_{D,D}$ is manifestly an even self-dual Lorentzian lattice and we conclude that $\Gamma_{D,D}$ is self-dual for any value of the background fields. We have seen that the torus compactification of the bosonic string is described by D^2 real parameters. The D^2 dimensional parameter space of the background fields is therefore called moduli space of the torus compactification. Different values of the D^2 parameters correspond to different choices of the Lorentzian, self-dual lattices $\Gamma_{D,D}$. This fact is very useful to obtain more information about the geometrical structure of the moduli space. It is known that all possible Lorentzian self-dual lattices can be obtained from each other by $SO(D,D)$

Lorentz rotations of some reference lattice Γ_0 which can always be chosen to correspond to $B_{ij} = 0$ and $g_{ij} = \delta_{ij}$. However not every Lorentz rotation leads to a different string theory since the string spectrum is invariant under separate rotations $SO(D)_L$, $SO(D)_R$ of the vectors $\boldsymbol{p}_L$ and $\boldsymbol{p}_R$ (see the mass formula eq.(10.29)). Therefore distinct compactified string theories, i.e. different points in the moduli space correspond to points in the coset manifold $\frac{SO(D,D)}{SO(D)\times SO(D)}$ which is of dimension D^2. We conclude that the geometrical structure of the moduli space is given by this manifold. However the string spectrum is again invariant under generalized, discrete duality transformations involving the background fields g_{ij} and B_{ij}. Therefore the global structure of the moduli space is quite complicate in the sense that those points in the above coset which are connected by the duality transformations have to be identified.

Let us now turn to the soliton states and assume that $B_{ij} = 0$. By the same arguments as before we might expect additional massless state for special values of the radii R_i. We are again particularly interested in massless vectors. Inspection of eq.(10.29) shows that we need either $p_L^2 = 2$, $N_L = 0$ with $\boldsymbol{p}_R = 0$, $N_R = 1$ or L and R interchanged. Together with eq.(10.31) this means that the only possibilities to get massless vectors are $m_i = \pm n_i = \pm 1$, $m_j = n_j = 0$ for $i \neq j$ and $g_{ij} = 2\delta_{ij}$. This corresponds to $R_i = \sqrt{2}$, $\forall\, i = 1, \ldots, D$, $e_i^I = \sqrt{2}\delta_i^I$ and the gauge group is $[SU(2)]_L^D \times [SU(2)]_R^D$. In this case $\boldsymbol{p}_{L,R}$ build the weight lattice of $[SU(2)]^D$. This is a trivial extension of the case considered before in the sense that the bosonic string is compactified on D orthogonal circles with radii $R = \sqrt{2}$. We do however want to get more general and in particular larger gauge groups such that $\Gamma_{L,R}$ contain the root lattice of some gauge group $G_{L,R}$. This is only possible if one considers also a non-trivial antisymmetric tensor field background B_{ij}.

As the simplest non-trivial example with non-vanishing B_{ij} consider the toroidal compactification of two dimensions. Choose for the two-dimensional lattice Λ_2 the root lattice of $SU(3)$ with basis vectors $e_1 = (\sqrt{2}, 0)$ and $e_2 = (\frac{1}{\sqrt{2}}, \sqrt{\frac{3}{2}})$. We further set $R_1 = R_2 = R$. This fixes two metric background fields, namely the ratio of the two radii and the relative angle between e_1 and e_2. Only the overall scale R is left as a free parameter. The antisymmetric tensor field background is given by $B_{ij} = B\epsilon_{ij}$. For generic R and B the gauge group is $[U(1)]_L^2 \times [U(1)]_R^2$. However at the critical point $R = \sqrt{2}$, $B = \frac{1}{2}$ the bosonic string has an enlarged gauge symmetry $[SU(3)]_L \times [SU(3)]_R$. In this case the lattice $\Gamma_{2,2}$ contains lattice vectors $\boldsymbol{P} = (\boldsymbol{p}_L, 0)$ and $\boldsymbol{P}' = (0, \boldsymbol{p}_R)$ with $\boldsymbol{p}_L$, $\boldsymbol{p}_R$ being the six root vectors of $SU(3)$. These states correspond to the non-Abelian gauge bosons of $[SU(3)]_L \times [SU(3)]_R$. In fact, one can easily verify that the lattice $\Gamma_{2,2}$ is the weight lattice of $[SU(3)]_L \times [SU(3)]_R$ specified by the three allowed conjugacy classes $(0,0)$, $(1,1)$ and $(2,2)$ where 0, 1 and 2 are the three conjugacy classes of $SU(3)$ (see next chapter). We are therefore dealing with a rational lattice for this choice of background fields.

Instead of discussing more examples with non-vanishing antisymmetric tensor field background let us consider the problem from a different point of view which also provides the key for the construction of the heterotic string. Consider again the mode expansion eq.(10.28). So far, only the oscillators $\alpha_n^I, \bar{\alpha}_n^I$ were treated as independent variables, however, the center of mass coordinate x^I and the momenta p^I were not. This is in fact necessary if one wants to maintain the interpretation that the $X^I(\sigma, \tau)$ are coordinates in some D-dimensional manifold. In this case the left- and right-moving modes must have common center of mass and common momentum. However, for general two-dimensional world-sheet bosons this is not necessary; we are free to regard X_L^I, X_R^I as completely independent two-dimensional fields with expansions

$$X_L^I(\tau + \sigma) = x_L^I + p_L^I(\tau + \sigma) + i \sum_{n \neq 0}^{\infty} \frac{1}{n} \bar{\alpha}_n^I e^{-in(\tau+\sigma)},$$

$$X_R^I(\tau - \sigma) = x_R^I + p_R^I(\tau - \sigma) + i \sum_{n \neq 0}^{\infty} \frac{1}{n} \alpha_n^I e^{-in(\tau-\sigma)}.$$

$$(10.36)$$

It means that one gives up the naive picture of compactifying the string on an internal manifold. The proper way of understanding this is to regard the resulting theory as a string theory in $(26 - D)$ space-time dimensions; X_L^I, X_R^I are boson fields which are needed as internal degrees of freedom to cancel the conformal anomaly. Since for closed strings the fields have to satisfy $X_{L,R}^I(\sigma + 2\pi) \simeq X_{L,R}^I(\sigma)$ where the identification is up to a vector of a lattice $\Lambda_{L,R}$, we find that treating X_L^I and X_R^I as independent necessitates that they are compactified, however not necessarily on the same torus. The periodicity requirement entails that p_L^I and p_R^I have to be interpreted as winding vectors, i.e. $p_{L,R} \in \Lambda_{L,R}$. But $p_{L,R}^I$ also generates translations of $x_{L,R}^I$. The commutation relations are

$$[x_{L,R}^I, p_{L,R}^J] = i\delta^{IJ},$$

$$[x_{L,R}^I, p_{R,L}^J] = 0.$$

$$(10.37)$$

The second commutator follows from our assumption that left- and right-movers are independent. There is actually a subtlety here. Eq.(10.37) are not the canonical commutation relations. For a purely left-moving boson with the normalization of x^I and p^I as in eq.(10.36), the canonical momentum would be $\Pi_L^I = \frac{1}{4\pi}\partial_\tau X^I$ from which we would get the canonical commutation relations $[x_L^I, p_L^J] = 2i\delta^{IJ}$. However, requiring the X^I to be purely left-moving constitutes a constraint: $\phi_L^I = (\partial_\tau - \partial_\sigma)X_L^I = 0$. From $\{\phi_L^I(\sigma,\tau), \phi_L^J(\sigma',\tau)\}_{P.B.} = -8\pi\partial_\sigma\delta(\sigma - \sigma')$, we conclude that it is second class. As described in Chapter 7, we have to replace the Poisson bracket by a Dirac bracket. This leads to eq.(10.37). (With these commutators we also get that $[x^I, p^J] = i$ where $x^I = x_L^I + x_R^I$ and $2p^I = p_L^I + p_R^I$.

Cf. also the discussion in Section 3.1.) Singlevaluedness of $e^{i p_{L,R}^I X_{0L,R}^I}$ requires that $p_{L,R} \in \Lambda_{L,R}^*$; i.e. we find that $p_{L,R} \in \Lambda_{L,R} \cap \Lambda_{L,R}^* := \Gamma_{L,R}$. The 2D-dimensional vectors $p = (p_L, p_R)$ again build a Lorentzian lattice $\Gamma_{D,D} = \Gamma_L \otimes \Gamma_R$, and modular invariance forces this lattice to be even, self-dual (cf. the discussion in the next section).

Let us now discuss the spectrum of this theory. The mass formula and reparametrization constraint are

$$
\begin{aligned}
m_{L,R}^2 &= \frac{1}{2} p_{L,R}^2 + N_{L,R} - 1 \,, \\
N_L - N_R &= \frac{1}{2} (p_R^2 - p_L^2) \,.
\end{aligned}
\tag{10.38}
$$

Clearly, we still have the $U(1)$ gauge bosons of eq.(10.32). Additional massless $(26 - D)$-dimensional vectors are obtained if there exist lattice vectors $p = (p_L, p_R) \in \Gamma_{D,D}$ with the property $p_L^2 = 2, p_R = 0$ or $p_R^2 = 2, p_L = 0$. The corresponding massless vectors have the form

$$
\begin{aligned}
|V_L^\mu\rangle &= \alpha_{-1}^\mu | p_L^2 = 2, p_R = 0 \rangle \,, \\
|V_R^\mu\rangle &= \bar{\alpha}_{-1}^\mu | p_L = 0, p_R^2 = 2 \rangle \,.
\end{aligned}
\tag{10.39}
$$

So if $\Gamma_{L,R}$ contains $l_{L,R}$ vectors $p_{L,R}$ with $p_{L(R)}^2 = 2$ we get $l_{L,R}$ massless vectors $|V_{L,R}^\mu\rangle$. These vectors correspond to the non-commuting generators of a non-abelian Lie group $G_{L,R}$. The $p_{L,R}^2 = 2$ vectors must therefore be roots of $G_{L,R}$ and $G_{L,R}$ must be simply laced.[4] This means that $\Gamma_{L,R}$ must contain the root lattice of a simply laced group $G_{L,R}$. Then the massless vectors of eq.(10.39) build, together with the states eq.(10.32), the gauge bosons of the non-Abelian gauge group $G_L \times G_R$ with dim $(G_{L(R)}) = l_{L(R)} + D$ and rank$(G_L) =$ rank$(G_R) = D$. The oscillator excitations eq.(10.32)

---

[4]A Lie group G is simply laced if all its roots α_i have the same length which can be normalized to $\alpha_i^2 = 2 \,\forall\, i = 1, \ldots, \dim G$. Dots in the Dynkin diagram (corresponding to simple roots) are then either disconnected or connected by a single line. This leaves only $D_n \sim SO(2n)$, $A_n \sim SU(n+1)$ and $E_{6,7,8}$ or products thereof.

correspond to the $[U(1)]_L^D \times [U(1)]_R^D$ Cartan subalgebra of $G_L \times G_R$. Note that G_L and G_R are in general different.

In conclusion, toroidal compactification of the bosonic string may be viewed in two ways: as compactification of independent left- and right-movers on different tori or as compactification on the same torus in the presence of background B_{ij} and g_{ij} fields.

10.2 The heterotic string

The important observation is now that since we have treated the left- and right-moving compactified coordinates as completely independent, we can drop either one of them. This is the starting point of the heterotic string construction which we will discuss in the following.

The heterotic string is a hybrid construction a left-moving 26-dimensional bosonic string together with a right-moving 10-dimensional superstring. By the arguments given before it is a string theory in 10-dimensions. We deal with the following two-dimensional fields: As left moving coordinates we have 10 uncompactified bosonic fields $X_L^\mu(\tau + \sigma)\,(\mu = 0,\ldots,9)$ and, in addition, 16 internal bosons $X_L^I(\tau + \sigma)\,(I = 1,\ldots,16)$ which live on a 16-dimensional torus. The right moving degrees of freedom consist of 10 uncompactified bosons $X_R^\mu(\tau - \sigma)\,(\mu = 0,\ldots,9)$, and their two-dimensional fermionic superpartners $\psi_R^\mu(\tau - \sigma)$. Finally, we have left- and right-moving reparametrization ghosts b,c and only right moving superconformal ghosts β,γ. $X_L^\mu(\tau + \sigma)$ and $X_R^\mu(\tau - \sigma)$ have common center of mass coordinates and common space-time momentum which can take on continuous values. On the other hand, the momenta of the additional bosons $X_L^I(\tau + \sigma)$ take only discrete values; they are vectors of a 16-dimensional lattice Γ^{16}:

$$p_L \in \Gamma^{16}, \qquad p_L^I = p_i e_i^I, \quad I = 1\ldots 16, \quad p_i \in \mathbf{Z}. \tag{10.40}$$

e_i^I are the basis vectors of Γ^{16} whose metric is given by

$$g_{ij} = \sum_{I=1}^{16} e_i^I e_j^I. \tag{10.41}$$

Γ^{16} cannot be any lattice that contains the root lattice of some simply laced rank 16 group. One-loop modular invariance puts severe restrictions on it. We will limit ourselves to consider only the vacuum amplitude, i.e. the partition function. In the Hamiltonian formalism the partition function is given by

$$\chi(\bar{\tau},\tau) = \mathrm{Tr}\, \bar{q}^{H_L} q^{H_R}, \qquad q = e^{2\pi i \tau}. \tag{10.42}$$

H_L and H_R are the left- and right-moving Hamiltonians in the light cone gauge:

$$\begin{aligned}
H_L &= \frac{1}{2} p_i^2 + N_L + \frac{1}{2} p_L^2 - 1 \\
H_R &= \frac{1}{2} p_i^2 + N_R + H^{\mathrm{NS(R)}}.
\end{aligned} \tag{10.43}$$

p_i ($i = 1 \ldots 8$) are the transverse space-time momenta, p_L the 16 internal left-moving momenta; N_L contains the 8 external as well as the 16 internal left-moving bosonic oscillators, where N_R gets contribution only from the 8 right-moving external bosonic oscillators. Finally, $H^{\mathrm{NS}}, H^{\mathrm{R}}$ are the Neveu-Schwarz and Ramond Hamiltonians of the fermionic string and the right-moving normal ordering constant is included there. This leads to the following partition function:

$$\chi(\bar{\tau},\tau) \sim (\mathrm{Im}\,\tau)^{-4} \frac{1}{[\eta(\bar{\tau})]^{24}} \left(\sum_{p_L \in \Gamma^{16}} q^{\frac{1}{2} p_L^2} \right)$$

$$\times \frac{1}{[\eta(\tau)]^{12}} \Big[[\theta_3(\tau)]^4 - [\theta_4(\tau)]^4 - [\theta_2(\tau)]^4 \Big). \tag{10.44}$$

$\eta(\bar{\tau})^{-24} \eta(\tau)^{-8}$ is the bosonic oscillator contribution, the $(\mathrm{Im}\,\tau)^{-4}$ factor arises from the zero modes of the uncompactified transverse coordinates and $\eta(\tau)^{-4} \times (\theta\text{-functions})$ comes from the world-sheet fermions (cf. Chapter 9). The novel and most interesting part of this partition function is the lattice (or soliton) sum (from now on we suppress the bar over τ)

$$P(\tau) = \sum_{\boldsymbol{p}_L \in \Gamma^{16}} q^{\frac{1}{2}\boldsymbol{p}_L^2} . \tag{10.45}$$

The summation is over all lattice vectors of Γ^{16}. From the known modular transformation properties of $\mathrm{Im}\,\tau$ and $\eta(\tau)$ under S and T we conclude that in order for $\chi(\bar{\tau}, \tau)$ to be modular invariant, $P(\tau)$ must be invariant under the T transformation,

$$P(\tau + 1) = P(\tau) \tag{10.46}$$

and must transform under S like

$$P(-\frac{1}{\tau}) = \tau^8 P(\tau) \tag{10.47}$$

So let us first check eq.(10.46):

$$P(\tau + 1) = \sum_{\boldsymbol{p}_L \in \Gamma^{16}} q^{\frac{1}{2}\boldsymbol{p}_L^2} e^{\pi i \boldsymbol{p}_L^2} . \tag{10.48}$$

Invariance under T clearly demands that $\boldsymbol{p}_L^2 \in 2\mathbf{Z}$, $\forall\, \boldsymbol{p}_L \in \Gamma^{16}$ which means that Γ^{16} must be an even lattice. Note that this already follows from eq.(10.38) with $\boldsymbol{p}_R = 0$. Since $\boldsymbol{p}_L^2 = \sum_I p_L^I p_L^I = \sum_{ij} p_i g_{ij} p_j = \sum_i p_i^2 g_{ii} + 2\sum_{i<j} p_i g_{ij} p_j$ $(p_{i,j} \in \mathbf{Z})$ we find that for an even lattice the diagonal elements of the metric g_{ij} must be even integers: $g_{ii} \in 2\mathbf{Z}$, $\forall\, i = 1, \ldots, 16$.

The more subtle part is the transformation of $P(\tau)$ under S. To study it we recall the Poisson resummation formula. Consider the function

$$F(\boldsymbol{x}) = \sum_{\boldsymbol{p} \in \Lambda} e^{-\pi \alpha (\boldsymbol{p}+\boldsymbol{x})^2 + 2\pi i \boldsymbol{y} \cdot (\boldsymbol{p}+\boldsymbol{x})} \tag{10.49}$$

where the sum runs over the points of an n-dimensional lattice and $\mathrm{Re}\,\alpha > 0$. $\boldsymbol{x}$ and $\boldsymbol{y}$ are arbitrary vectors. Since

$$F(\boldsymbol{x} + \boldsymbol{p}) = F(\boldsymbol{x}) \tag{10.50}$$

for $\boldsymbol{p} \in \Lambda$, we can expand it in a Fourier series

$$F(\boldsymbol{x}) = \sum_{\boldsymbol{q} \in \Lambda^*} e^{2\pi i \boldsymbol{x} \cdot \boldsymbol{q}} F^*(\boldsymbol{q}) \tag{10.51}$$

where

$$F^*(q) = \frac{1}{\text{vol}(\Lambda)} \int_{\substack{\text{unit} \\ \text{cell}}} d^n x \, e^{-2\pi i q \cdot x} F(x). \qquad (10.52)$$

Inserting eq.(10.49), combining the sum over Λ and the integral over the unit cell to an integral over all of $\mathbf{R}^n$, we get, after doing the Gaussian integral,

$$F(x) = \frac{1}{\text{vol}(\Lambda) \, \alpha^{n/2}} \sum_{q \in \Lambda^*} e^{-2\pi i q \cdot x} e^{-\frac{\pi}{\alpha}(y+q)^2}. \qquad (10.53)$$

This is the key formula to derive the implications of modular invariance for $P(\tau)$. Applying it to $P(-1/\tau)$ we get

$$P(-\frac{1}{\tau}) = \frac{\tau^8}{\sqrt{\det g}} \sum_{p \in (\Gamma^{16})^*} q^{\frac{1}{2}p^2}. \qquad (10.54)$$

Therefore, in order to satisfy eq.(10.47), Γ^{16} must be a self-dual lattice, i.e.

$$\Gamma^{16} = (\Gamma^{16})^*. \qquad (10.55)$$

Then $\det g = 1$ and $\text{vol}(\Gamma) = \text{vol}(\Gamma^*) = 1$.

In summary, modular invariance of the partition function implies that the internal 16-dimensional momentum lattice Γ^{16} must be an Euclidean even self-dual lattice.

These lattices are very rare. We will study them move carefully in the next chapter. The result is that in 16 dimensions there are only two Euclidean even self-dual lattices namely the direct product lattice $\Gamma_{E_8} \otimes \Gamma_{E_8}$, where Γ_{E_8} is the root lattice of E_8, and $\Gamma_{D_{16}}$ which is the weight lattice of Spin $(32)/\mathbf{Z}_2$ which contains the root lattice of $SO(32)$. The metric g_{ij} of Γ_{E_8} is the Cartan matrix of E_8:

$$g_{ij} = \begin{pmatrix} 2 & -1 & & & & & & \\ -1 & 2 & -1 & & & & & \\ & -1 & 2 & -1 & & & & \\ & & -1 & 2 & -1 & & & \\ & & & -1 & 2 & -1 & & -1 \\ & & & & -1 & 2 & -1 & \\ & & & & & -1 & 2 & \\ & & & & -1 & & & 2 \end{pmatrix} \tag{10.56}$$

One can check that $\det g_{ij} = 1$. The construction of the weight lattice of Spin $(32)/\mathbf{Z}_2$ will be discussed in the next chapter. Both, the root lattice of $E_8 \times E_8$ and the weight lattice of Spin $(32)/\mathbf{Z}_2$ contain 480 vectors of $(\text{length})^2 = 2$ which are the roots of $E_8 \times E_8$ and $SO(32)$ respectively. Therefore, according to our previous discussion the dimension 496 gauge group of the heterotic string is either $E_8 \times E_8$ or $SO(32)$. This is required by modular invariance.

Let us investigate the spectrum of the heterotic string more carefully. First consider left-moving excitations. As usual, there is the tachyonic vacuum of the bosonic string. At the massless level we have oscillator excitations $\bar{\alpha}^i_{-1}|0\rangle$, $\bar{\alpha}^I_{-1}|0\rangle$. The former transform like 10-dimensional space-time vectors whereas the internal oscillator excitations correspond to the left-moving part of the Abelian $U(1)^{16}$ gauge boson. They build the Cartan subalgebra of $E_8 \times E_8$ or $SO(32)$. We also have the states in the soliton sector with non-trivial internal momenta p_L. The states $|p_L^2 = 2\rangle$, $N_L = 0$ are massless, p_L is a $(\text{length})^2 = 2$ root vector of $E_8 \times E_8$ or $SO(32)$ and generate the non-Abelian gauge bosons of these groups.

The right-moving excitations are those of the 10-dimensional superstring – the spectrum is space-time supersymmetric. The NS tachyon $|0\rangle_{\text{NS}}$ is projected out by the GSO projection which was enforced by modular invariance. Therefore the lowest states are the 10-dimensional vector $b^i_{-1/2}|0\rangle_{\text{NS}}$ and the 10-dimensional spinor $|S^\alpha\rangle$ (previously denoted by $|a\rangle$).

Finally we take the tensor product of the left- and right-moving sectors to obtain the spectrum of the heterotic string. It is clear that there is no tachyon since the left-moving tachyonic vacuum does not satisfy the left-right level matching constraint. Due to the right-moving supersymmetry the spectrum is $N = 1$ supersymmetric in 10 dimensions. We have four kinds of massless states:

(i) The states corresponding to the ten-dimensional graviton, antisymmetric tensor field and dilaton

$$\bar{\alpha}^i_{-1}|0\rangle \otimes b^j_{-\frac{1}{2}}|0\rangle_{\text{NS}}, \qquad (10.57)$$

(ii) and their supersymmetric partners, the gravitino and dilatino

$$\bar{\alpha}^i_{-1}|0\rangle \otimes |S^\alpha\rangle_{\text{R}}. \qquad (10.58)$$

(iii) In addition we have the gauge bosons of $E_8 \times E_8$ or $SO(32)$

$$\bar{\alpha}^I_{-1}|0\rangle \otimes b^i_{-\frac{1}{2}}|0\rangle_{\text{NS}}\,,$$
$$|p_L^2 = 2\rangle \otimes b^i_{-\frac{1}{2}}|0\rangle_{\text{NS}} \qquad (10.59)$$

where in the first line we have the gauge bosons corresponding to the Cartan-Weyl subalgebra and in the second line the gauge bosons corresponding to the root vectors.

(iv) Finally there are the 496 supersymmetric partners of the gauge bosons, the gauginos

$$\bar{\alpha}^I_{-1}|0\rangle \otimes |S^\alpha\rangle_{\text{R}}\,,$$
$$|p_L^2 = 2\rangle \otimes |S^\alpha\rangle_{\text{R}}\,. \qquad (10.60)$$

It is straightforward to work out the massive spectrum but we will not do it here. However, it is useful to remember that the number of (massive) states coming from the soliton sector is encoded in the lattice partition function of the root and weight lattices of $E_8 \times E_8$ and $\text{Spin}(32)/\mathbf{Z}_2$ respectively. To see this, let us calculate the partition functions

$$P = \sum_\lambda e^{i\pi\tau\lambda^2} = \sum_\lambda q^{\frac{1}{2}\lambda^2} \tag{10.61}$$

where the sum extends over all vectors in the E_8 and $\mathrm{Spin}(32)/\mathbf{Z}_2$ weight lattices respectively. The 240 roots of E_8 are given by[5]

$$\alpha = \begin{cases} 0,\ldots \pm 1, 0, \ldots, \pm 1, 0 \ldots) \\ (\pm\frac{1}{2}, \ldots, \pm\frac{1}{2}) \end{cases} \qquad \text{even number of ``}-\text{'' signs} \tag{10.62}$$

and the weight vectors are of the form

$$\lambda = \begin{cases} (n_1, \ldots, n_8) \\ (n_1 + \frac{1}{2}, \ldots, n_8 + \frac{1}{2}) \end{cases} \qquad \sum_{i=1}^{8} n_i = \text{even integer.} \tag{10.63}$$

To implement the condition on the n_i we insert a factor $\frac{1}{2}(1 + e^{i\pi\sum n_i})$. We then get

$$\begin{aligned}
P_{E_8} &= \frac{1}{2}\Bigg\{ \prod_{i=1}^{8} \sum_{n_i \in \mathbf{Z}} e^{i\pi n_i^2 \tau} + \prod_{i=1}^{8} \sum_{n_i \in \mathbf{Z}} e^{i\pi n_i^2 \tau} e^{i\pi n_i} \\
&\qquad + \prod_{i=1}^{8} \sum_{n_i \in \mathbf{Z}} e^{i\pi(n_i+\frac{1}{2})^2\tau} + \prod_{i=1}^{8} \sum_{n_i \in \mathbf{Z}} e^{i\pi(n_i+\frac{1}{2})^2\tau} e^{i\pi(n_i+\frac{1}{2})} \Bigg\} \\
&= \frac{1}{2}\big\{ \theta_3^8(\tau) + \theta_4^8(\tau) + \theta_2^8(\tau) \big\}
\end{aligned} \tag{10.64}$$

where we have used the definitions of the theta-functions given in Chapter 9. The contribution of the last term is $\theta_1^8(\tau) = 0$. Expanding P_{E_8} in powers of q we find

$$P_{E_8}(\tau) = 1 + 240q + 2160q^2 + 6720q^3 + \ldots \tag{10.65}$$

It shows that the E_8 root lattice has 240 points of $(\text{length})^2 = 2$ corresponding to the roots, 2160 points of $(\text{length})^2 = 4$ etc.

For the $\mathrm{Spin}(32)/\mathbf{Z}_2$ case one derives in a similar way, using results from Chapter 11, that

---

[5] Properties of root and weight lattices will be discussed in detail in the next chapter.

$$P_{\text{Spin}(32)/\mathbf{Z}_2} = \frac{1}{2}\left[\theta_3^{16}(\tau) + \theta_4^{16}(\tau) + \theta_2^{16}(\tau)\right]. \tag{10.66}$$

With the help of the identity $\theta_3^4 = \theta_2^4 + \theta_4^4$ (cf. Chapter 9) we can show that

$$\begin{aligned}
P_{\text{Spin}(32)/\mathbf{Z}_2} &= \left[P_{E_8}\right]^2 \\
&= 1 + 480q + 61920q^2 + 1050240q^3 + \ldots
\end{aligned} \tag{10.67}$$

It follows that the $E_8 \times E_8$ and the $SO(32)$ heterotic string theories have the same number of states at every mass level which are however differently organized under the internal gauge symmetries. So, even though the partition functions are identical, the theories are nevertheless different. The differences show up in correlation functions.

REFERENCES

[1] E. Cremmer and J. Scherk, *"Dual models in four dimensions with internal symmetries"*, Nucl. Phys. **B103** (1976) 399.

[2] M.B. Green, J.H. Schwarz and L. Brink, *"$N = 4$ Yang-Mills and $N = 8$ supergravity as the limit of string theories"*, Nucl. Phys. **B198** (1982) 474.

[3] K.S. Narain, *" New heterotic string theories in uncompactified dimensions $d < 10$"*, Phys. Lett. **169B** (1986) 41.

[4] K.S. Narain, M.H. Sarmadi and E. Witten, *" A note on toroidal compactification of heterotic string theory"*, Nucl. Phys. **B278** (1987) 369.

[5] W. Lerche, A.N. Schellekens and N. Warner, *"Lattices and Strings"*, Phys. Rep. **177** (1989) 1.

[6] D.J. Gross, J. Harvey, E. Martinec and R. Rohm, *"Heterotic string"*, Phys. Rev. Lett. **54** (1985) 502; *"Heterotic string theory (I). The free heterotic string"*, Nucl. Phys. **B256** (1985) 253; *"Heterotic string theory (II). The interacting heterotic string"*, Nucl. Phys. **B267** (1986) 75.

Chapter 11

Conformal Field Theory II: Lattices and Kac-Moody Algebras

In the previous chapter we have learned that massless vector bosons may arise from toroidal compactification of bosonic string coordinates, a feature which is not expected by any field theoretical argument. However, we omitted to prove that these massless vectors are really gauge bosons of a non-abelian gauge group G transforming in the adjoint representation. The necessary mathematical tool to do this is the theory of infinite dimensional (current) algebras, the so-called affine Kac-Moody algebras. They are the subject of this chapter. These algebras were first discussed by Kac [1] and Moody [2]. A collection of reprinted papers and more references on this subject can be found in [3].

11.1 Kac-Moody algebras

A Kac-Moody algebra is the infinitesimal version of a certain infinite dimensional Lie group $\mathcal{G}$, namely the group of mappings of the circle S^1 into a finite dimensional compact connected Lie group G. $\mathcal{G}$ is the so-called loop group of G.

Represent S^1 as the unit circle in the complex plane

$$S^1 = \{z \in \mathbf{C} : |z| = 1\} \tag{11.1}$$

and denote a map from S^1 into G by $z \to \gamma(z) \in G$. The group operation on $\mathcal{G}$ is defined by pointwise multiplication; i.e. given two maps $\gamma_1, \gamma_2 \in \mathcal{G}$, the product of γ_1 and γ_2 is $\gamma_1 \cdot \gamma_2 \in \mathcal{G}$ where

$$\gamma_1 \cdot \gamma_2(z) = \gamma_1(z)\gamma_2(z). \tag{11.2}$$

The infinite dimensional algebra $\hat{g}_0$ of $\mathcal{G}$ can be obtained from the finite dimensional Lie algebra g of G,

$$[T^a, T^b] = if^{abc}T^c \tag{11.3}$$

where f^{abc} are the structure constants of g by writing

$$\gamma(z) = \exp[-i \sum_{a=1}^{\dim g} T^a \Theta_a(z)]. \tag{11.4}$$

$\Theta_a(z)$ are dim g functions defined on the unit circle. Expanding these functions into modes

$$\Theta_a(z) = \sum_{n=-\infty}^{\infty} \Theta_a^{-n} z^n \tag{11.5}$$

we can introduce generators T_n^a,

$$T_n^a = T^a z^n \tag{11.6}$$

such that

$$\gamma(z) = \exp[-i \sum_{n,a} T_{-n}^a \Theta_a^n]. \tag{11.7}$$

We see that the Θ_a^n's are an infinite set of parameters for $\mathcal{G}$ and the T_n^a an infinite number of generators of $\mathcal{G}$ satisfying the following algebra

$$[T_m^a, T_n^b] = if^{abc}T_{m+n}^c \tag{11.8}$$

which follows from eqs.(11.3) and (11.6). This is the (untwisted) affine Kac-Moody algebra $\hat{g}_0$, also called loop algebra. (We will not consider twisted Kac-Moody algebras.) Note that the T_0^a generate a subalgebra isomorphic to g. It corresponds to $\Theta_a(z) = \text{const.}$.

If the T^a are Hermitian generators of G,

$$T^{a\dagger} = T^a, \tag{11.9}$$

then the Kac-Moody generators satisfy ($z^* = z^{-1}$ for $|z| = 1$)

$$T_n^{a\dagger} = T_{-n}^a.$$ (11.10)

A representation of $\hat{g}_0$ satisfying this hermiticity condition is called unitary.

In Chapter 2 we have seen that for the closed bosonic string (say) right movers are functions of $(\tau - \sigma)$ only and periodicity allows an expansion in Fourier modes $e^{in(\tau - \sigma)}$. This means that the fields are defined on S^1. The Virasoro algebra $\hat{v}_0$ generates reparametrizations of S^1. The generators can be represented as

$$L_n = -z^{n+1}\partial_z$$ (11.11)

where $z = e^{i(\tau - \sigma)} \in S^1$. With the $\hat{g}_0$ being the algebra of the group of maps $S^1 \to G$ and $\hat{v}_0$ the algebra of $\mathrm{Diff}(S^1)$ it is clear that they are not unrelated. In particular, to every Kac-Moody algebra there is an associated Virasoro algebra. Using the explicit form for the generators eqs.(11.6) and (11.11) we easily find

$$[L_m, T_n^a] = -nT_{n+m}^a ;$$ (11.12)

i.e. $\hat{v}_0$ and $\hat{g}_0$ form a semidirect sum $\mathcal{A}_0 = \hat{v}_0 \Subset \hat{g}_0$.

So far we have constructed the Kac-Moody algebra from the classical Lie algebra eq.(11.3). When going to the quantum theory we have however to be careful. We saw in Chapter 3 that the Virasoro algebra gets a central extension parametrized by the central charge c. This possibility also arises in the case of the Kac-Moody algebra. Allowing a general central extension, it has the form

$$[T_m^a, T_n^b] = if^{abc}T_{m+n}^c + d_{mnj}^{ab}k^j ,$$

$$[T_m^a, k^j] = [k^i, k^j] = 0$$ (11.13)

where k^i, $(i = 1, \ldots, M)$ are central elements. Then for G compact and simple one can show that up to redefinitions of the generators by terms linear in the the k^i the only possible choice for $d_{mni}^{ab}k^i$ consistent with the

Jacobi identities is $km\delta^{ab}\delta_{m+n}$. We can then define the untwisted affine Kac-Moody algebra $\hat{g}$ by the following commutation relations:

$$[T_m^a, T_n^b] = if^{abc}T_{m+n}^c + km\delta^{ab}\delta_{m+n}. \qquad (11.14)$$

The central element k is called the level of the Kac-Moody algebra. It is a real constant in each representation. In fact, when considering irreducible unitary representations of the Kac-Moody algebra, the values of k are not arbitrary but constrained to $k \geq 0$; for $G \neq U(1)$ $k \in \mathbf{Z}_+$. This will be shown below. From eq.(11.14) we see that the Lie algebra g does not allow a non-trivial central extension. If G is compact but not simple we get a different k for each $U(1)$ and each simple factor.

The Kac-Moody algebra eq.(11.14) is closely related to a two-dimensional current algebra. Consider the conserved chiral currents $J^a(z)$ which carry the adjoint representation index of some Lie group. Since $\partial_{\bar{z}}J^a(z) = 0$ we have, as in the case of the Virasoro algebra, an infinite number of conserved charges

$$J_n^a = \oint \frac{dz}{2\pi i} z^n J^a(z) \qquad (11.15)$$

which satisfy the same algebra as the corresponding generators, namely an affine Kac-Moody algebra:

$$[J_m^a, J_n^b] = if^{abc}J_{m+n}^c + mk\delta^{ab}\delta_{m+n} \qquad (11.16)$$

and also

$$[L_m, J_n^a] = -nJ_{m+n}^a \qquad (11.17)$$

reflecting the semidirect sum structure $\mathcal{A} = \hat{v} \in \hat{g}$. Inverting eq.(11.15) we get

$$J^a(z) = \sum_n z^{-n-1} J_n^a. \qquad (11.18)$$

Using the techniques of Chapter 4 we easily find that eq.(11.16) is equivalent to the following current operator algebra:

$$J^a(z)J^b(w) = \frac{k\delta^{ab}}{(z-w)^2} + if^{abc}\frac{J^c(w)}{(z-w)} + \ldots \tag{11.19}$$

In this equation, the central charge appears as a so-called Schwinger term. In conformal field theory language eq.(11.17) means that the currents $J^a(z)$ are primary fields with weight $h=1$ of the Virasoro algebra (cf. eq.(4.35)). Indeed, eq.(11.17) is equivalent to the operator product

$$T(z)J^a(w) = \frac{J^a(w)}{(z-w)^2} + \frac{\partial_w J^a(w)}{z-w} + \ldots \tag{11.20}$$

We can now define primary fields of $\mathcal{A}$ by

$$T(z)\phi_i(w) = \frac{h\phi_i(w)}{(z-w)^2} + \frac{\partial\phi_i(w)}{z-w} + \ldots$$

$$J^a(z)\phi_i(w) = \frac{(T^a)_i^{\ j}\phi_j(w)}{z-w} + \ldots \tag{11.21}$$

where $(T^a)_i^{\ j}$ are representation matrices of g. Comparison with eq.(11.19) gives that $J^a(z)$ is not primary with respect to the combined algebra $\mathcal{A}$. Indeed, in the notation of Chapter 4 we have $J^a(z) = \hat{J}^a_{-1}I(z)$.

There is an explicit construction of the Virasoro algebra in terms of the Kac-Moody generators. This is the so-called Sugawara construction [4]. For simplicity we will only consider the case when g is simple. Define the energy momentum tensor as

$$T(z) = \frac{1}{2k+C_2}\sum_a \,:J^a(z)J^a(z):\, \tag{11.22a}$$

or

$$L_n = \frac{1}{2k+C_2}\sum_m \,:J^a_{m+n}J^a_{-m}:\, \tag{11.22b}$$

where C_2 is the quadratic Casimir of the adjoint representation defined by

$$f^{acd}f^{bcd} = C_2\delta^{ab}. \tag{11.23}$$

Normal ordering in eq.(11.22) is with respect to the modes of $J^a(z)$. It can be shown that $T(z)$ satisfies a Virasoro algebra with central charge

$$c = \frac{2k \dim G}{C_2 + 2k}. \tag{11.24}$$

Let us now specialize to simply laced groups G of rank n. Then C_2 is given by the formula

$$C_2 = 2\left(\frac{\dim G}{n} - 1\right). \tag{11.25}$$

It implies that for level one Kac-Moody algebras, i.e. $k = 1$, the central charge of the corresponding Virasoro algebra is an integer, namely the rank n of G. This suggests that the level one Kac-Moody currents of the simply laced group G can be constructed from n free bosonic fields. This construction, known as the Frenkel-Kac-Segal construction, uses n bosonic fields "compactified" on the root lattice of the corresponding group and proves the appearance of non-Abelian gauge symmetries in the heterotic string theory introduced in Chapter 10. To present it, let us therefore recall some basic facts about lattices and Lie algebras.

11.2 Lattices and Lie algebras

A discussion of the theory of roots and weights in connection with Kac-Moody algebras is given in [5]. Reference [6] also provides some material presented in this section.

A lattice Λ is defined as a set of points in a n-dimensional real vector space V:

$$\Lambda = \{\sum_{i=1}^{n} n_i e_i \,|\, n_i \in \mathbf{Z}\}. \tag{11.26}$$

The e_i $(i = 1 \ldots n)$ are n basis vectors of V.[1] We will only be interested in the cases where V is $\mathbf{R}^n$ with Euclidean inner product or $\mathbf{R}^{p,q}$ $(p + q = n)$ with Lorentzian inner product; i.e. for v, w being two lattice vectors we have

---

[1] The notation differs from that of the previous chapter. We have redefined $\frac{1}{\sqrt{2}} R_i e_i \rightarrow e_i$.

$v \cdot w = \sum_{I=1}^{n} v^I w^I$ for the Euclidean case and $v \cdot w = \sum_{I=1}^{p} v^I w^I - \sum_{I=p+1}^{n} v^I w^I$ for the Lorentzian case. The matrix $g_{ij} = e_i \cdot e_j$ is the metric on Λ; it contains all information about the angles between the basis vectors and about their lengths. The volume of the unit cell which contains exactly one lattice point, vol(Λ), is also determined by g_{ij}:

$$\text{vol}(\Lambda) = \sqrt{|\det g|}. \tag{11.27}$$

The dual lattice Λ^* is defined as

$$\Lambda^* = \{w \in V, w \cdot v \in Z, \forall v \in \Lambda\}. \tag{11.28}$$

The basis vectors e_i^* of Λ^* satisfy

$$e_i^* \cdot e_j = \delta_{ij} \tag{11.29}$$

and the metric on Λ^* is $g_{ij}^* = e_i^* \cdot e_j^*$ which is the inverse of g_{ij}. The volume of the unit cell of Λ^* is then given by

$$\text{vol}(\Lambda^*) = (\text{vol}(\Lambda))^{-1}. \tag{11.30}$$

A lattice is called unimodular if it has one point per unit volume, i.e. if vol(Λ) = 1; then also vol (Λ^*) = 1. It is integral if $v \cdot w \in Z, \forall\, v, w \in \Lambda$. Clearly Λ is integral if and only if $\Lambda \subset \Lambda^*$. Furthermore an integral lattice is called even if all lattice vectors have even $(\text{length})^2$; it is called odd otherwise. Finally, Λ is self-dual if $\Lambda = \Lambda^*$. A necessary and sufficient condition for Λ being self-dual is to be unimodular and integral.

If Λ_s is a sublattice of equal dimension of Λ, we can decompose Λ into cosets with respect to Λ_s. To do so, choose a set of vectors m_i ($i = 2, \ldots, N_s$), such that

$$m_i \in \Lambda, \quad m_i \notin \Lambda_s\,,$$
$$m_i - m_j \notin \Lambda_s \text{ if } i \neq j\,. \tag{11.31}$$

Then the lattice Λ can be written as sum over cosets

$$\Lambda = \Lambda_s \oplus (m_2 + \Lambda_s) \oplus \ldots \oplus (m_{N_s} + \Lambda_s). \tag{11.32}$$

This notation means that every vector in Λ can be written as $m_i + v_s$, $v_s \in \Lambda_s$, $i = 1, \ldots, N_s$, if we define $m_1 = 0$. The vectors m_i are called coset representatives. The volume of Λ can be expressed as

$$\mathrm{vol}(\Lambda) = \frac{1}{N_s}\mathrm{vol}(\Lambda_s). \tag{11.33}$$

The lattices we are most interested in are the so-called Lie algebra lattices. To discuss them we need some basic facts about Lie algebras, which we will now review. We will especially concentrate on the properties of their root and weight lattices. This is most conveniently done in the so-called Cartan-Weyl basis. Choose a maximal set of hermitian commuting generators H^i $(i = 1, \ldots, n)$

$$[H^i, H^j] = 0 \tag{11.34}$$

where the dimension n of this subalgebra is called the rank of G. The H^i generate the Cartan subalgebra. Given a choice of a Cartan subalgebra we can diagonalize the remaining generators in the sense that they have definite eigenvalues with respect to the H^i:

$$[H^i, E^\alpha] = \alpha_i E^\alpha. \tag{11.35}$$

The real non-zero n-dimensional vector α is called a root and E^α a step operator corresponding to α. Note that from eq.(11.35) it follows that the E^α are necessarily non-hermitian. Indeed, we find that

$$E^{-\alpha} = (E^\alpha)^\dagger, \tag{11.36}$$

i.e. if α is a root then so is $-\alpha$. A root is called positive if its first non-zero component is positive. The E^α with α positive are called raising operators and lowering operators otherwise. If all roots have the same length, the group is called simply laced. We can then normalize the roots to $\alpha^2 = 2$. In the following we will only consider simply laced groups. Some of the expressions given below will have to be modified for the general case.

To complete the Lie algebra we need to determine the commutation relation between the step operators E^α, E^β. The commutators are constrained by the Jacobi identities and the result can be summarized as follows:

$$[E^\alpha, E^\beta] = \begin{cases} \epsilon(\alpha, \beta) E^{\alpha+\beta} & \text{if } \alpha + \beta \text{ is a root} \\ \alpha \cdot \boldsymbol{H} & \text{if } \alpha = -\beta \\ 0 & \text{otherwise.} \end{cases} \qquad (11.37)$$

The constants $\epsilon(\alpha, \beta)$, antisymmetric in α and β, can be arranged to be ± 1. With each root we can associate a $SU(2)$ subalgebra generated by E^α, $E^{-\alpha}$ and $\alpha \cdot \boldsymbol{H}$. If we identify them with J_+, J_- and $2J_3$, we recognize the angular momentum algebra. It is well known that for unitary representations the eigenvalues of $2J_3$ or $\alpha \cdot \boldsymbol{H}$ have to be integer.

Taking arbitrary integer linear combinations of root vectors one generates a n-dimensional Euclidean lattice, called root lattice Λ_R. Since the number of root pairs $\pm\alpha$ in general exceeds the rank n of G, it is convenient to select a set of roots α_i $(i = 1 \ldots n)$ which serve as a basis for Λ_R. These are the so-called simple roots. They are those positive roots which cannot be written as sum of two positive roots. The Cartan matrix, defined by

$$g_{ij} = \alpha_i \cdot \alpha_j \qquad (11.38)$$

is an integer $n \times n$ matrix; its diagonal elements are 2 and its off diagonal elements are -1 or 0; i.e. the root lattice of any simply laced Lie algebra is an integral, even lattice. Therefore it is contained in its dual lattice Λ_R^*. g_{ij} is the metric on Λ_R. From a given Cartan matrix one can construct a basis of simple roots and from that all roots.

Let us look at the classification of simply laced Lie algebras. The first class is the D_n $(n \geq 1)$ series[2] corresponding to the orthogonal groups $SO(2n)$ with rank n. The root vectors have the following form:

---

[2]Note that we also include the case $n = 1$ with $D_1 \sim U(1)$.

$$\alpha_i = (\pm 1, \pm 1, 0, \dots, 0) + \text{all permutations.} \tag{11.39}$$

Counting all combinations of distributing two "± 1" entries, one easily verifies that there are $2(n^2 - n)$ root vectors. They build, together with the n Cartan subalgebra generators H_i, the $2n^2 - n$ generators of D_n in the Cartan-Weyl basis. The n simple roots of D_n are

$$(1, -1, 0^{n-2}), (0, 1, -1, 0^{n-3}), \dots, (0^{n-2}, 1, -1), (0^{n-2}, 1, 1) \tag{11.40}$$

where 0^i denotes the i-dimensional null vector.

The next class of simply laced Lie algebras is the A_n series ($n \geq 1$) corresponding to $SU(n + 1)$ with rank n and dimension $n^2 + 2n$. Let us take as an example A_2. The six roots of $SU(3)$ have the form

$$\begin{aligned}
\alpha_1 &= (\sqrt{\tfrac{1}{2}}, \sqrt{\tfrac{3}{2}}), & \alpha_2 &= (\sqrt{\tfrac{1}{2}}, -\sqrt{\tfrac{3}{2}}), & \alpha_3 &= (\sqrt{2}, 0), \\
\alpha_4 &= -(\sqrt{\tfrac{1}{2}}, \sqrt{\tfrac{3}{2}}), & \alpha_5 &= -(\sqrt{\tfrac{1}{2}}, -\sqrt{\tfrac{3}{2}}), & \alpha_6 &= -(\sqrt{2}, 0).
\end{aligned} \tag{11.41}$$

The two simple roots are α_1 and α_2.

Besides the D_n and A_n series there are the exceptional simply laced Lie algebras E_6, E_7 and E_8 of dimensions 78, 133 and 248 and rank 6,7 and 8 respectively. The 72 roots of E_6 (in a suitably chosen basis) look like

$$(\pm 1, \pm 1, 0^3; 0) + 36 \text{ permutations}$$

$$(\pm\tfrac{1}{2}, \pm\tfrac{1}{2}, \pm\tfrac{1}{2}, \pm\tfrac{1}{2}, \pm\tfrac{1}{2}; \tfrac{\sqrt{3}}{2}) \quad \text{even number of minus signs} \tag{11.42}$$

$$(\pm\tfrac{1}{2}, \pm\tfrac{1}{2}, \pm\tfrac{1}{2}, \pm\tfrac{1}{2}, \pm\tfrac{1}{2}; -\tfrac{\sqrt{3}}{2}) \quad \text{odd number of minus signs}$$

and the root vectors of E_7 have the form

$$(\pm 1, \pm 1, 0^4; 0) + 56 \text{ permutations}$$

$$(0^6; \pm\sqrt{2}) \tag{11.43}$$

$$(\pm\frac{1}{2}, \pm\frac{1}{2}, \pm\frac{1}{2}, \pm\frac{1}{2}, \pm\frac{1}{2}, \pm\frac{1}{2}; \pm\frac{\sqrt{2}}{2}) \qquad \begin{array}{l} \text{even number of minus signs} \\ \text{in first six components} \end{array}$$

The roots of E_8 have a particularly simple form; they are given by the 112 root vector of D_8 (see eq.(11.39)) and in addition to these the following 128 eight-dimensional vectors:

$$\alpha_i = (\pm\frac{1}{2}, \pm\frac{1}{2}, \ldots, \pm\frac{1}{2}) \tag{11.44}$$

where the number of minus-signs is restricted to be even.

So far we have only considered root lattices Λ_R which are constructed from the adjoint representation of G. However, any Lie group G has infinitely many irreducible representations which are characterized by their so-called weight vectors. Consider a finite dimensional irreducible representation of G. The states which transform in a specific representation are denoted by $|m_I, D\rangle$ where D is the dimension of the representation and I runs from 1 to D. These states are eigenstates of the Cartan subalgebra generators:

$$H^i|m_I, D\rangle = m_I^i|m_I, D\rangle. \tag{11.45}$$

The n-dimensional vector m_I is called weight vector of $|m_I, D\rangle$. The weight vectors characterize the representation. We can reach all states in a given representation by acting with lowering operators on the so-called highest weight state. Thus the weights in a given representation differ by vectors in the root lattice. For simply laced groups $\alpha_i \cdot m \in \mathbf{Z}$ for all roots and weights. Also, if $\beta \cdot m \in \mathbf{Z}$, $\forall m$, then $\beta \in \Lambda_R$.

Irreducible representations fall into different conjugacy classes. Two different representations are said to be in the same conjugacy class if the difference between their weight vectors is a vector of the root lattice. Of course, all weights of a given representation belong to the same conjugacy class. Therefore, a particular conjugacy class k is specified by choosing a representative weight vector m_k, e.g. the heighest weight of the lowest dimensional representation contained in this conjugacy class.[3] Then all vectors within the same conjugacy class are constructed by adding the whole root lattice to this representative. Since the root lattice is an integral lattice and also the scalar product between the roots and m_k are integer, it follows that the mutual inner product between all vectors of two conjugacy classes depends only on the scalar product of their representative vectors, i.e. on the conjugacy classes, modulo integers.

All weights of all conjugacy classes including the (0) conjugacy class (the root lattice itself) form the weight lattice Λ_w. Our results above clearly imply

$$\Lambda_R \supset \Lambda_w, \qquad \Lambda_R = \Lambda_w^* \tag{11.46}$$

which entails that

$$\mathrm{vol}(\Lambda_R) = (\mathrm{vol}(\Lambda_w))^{-1}. \tag{11.47}$$

The decomposition of the weight lattice into conjugacy classes is then simply a coset decomposition of Λ_w with respect to Λ_R and we can write

$$\Lambda_w = \Lambda_R \oplus (\Lambda_R + m_2) \oplus \ldots \oplus (\Lambda_R + m_{N_c}) \tag{11.48}$$

where the m_k $(k = 2, \ldots, N_c)$ are the representative vectors of each non-trivial conjugacy class and N_c is the number of conjugacy classes which is

---

[3] A highest weight state $|m_0, D\rangle$ satisfies $E^\alpha |m_0, D\rangle = 0$ $\forall$ positive α. It means that $\alpha + m_0$ is not a weight vector for any positive root α. The other states in the same representation are obtained by acting with lowering operators on the highest weight state. Any irreducible representation of g has a unique highest weight state – the other weights have the property that $m_0 - m$ is a sum of positive roots. The highest weight of the adjoint representation is called highest root ψ with $\psi^2 = 2$.

finite for finite dimensional Lie algebras. The cosets form an abelian dis-
rete group under addition, isomorphic to the center of G. The direct sum
decomposition eq.(11.48) attributes to each root N_c weights (including the
root itself); it follows that (cf. eq.(11.33))

$$\text{vol}(\Lambda_w) = \frac{1}{N_c}\text{vol}(\Lambda_R) \tag{11.49}$$

and with eq.(11.47)

$$\text{vol}(\Lambda_w) = \frac{1}{\sqrt{N_c}},$$
$$\text{vol}(\Lambda_R) = \sqrt{N_c}. \tag{11.50}$$

One can also consider so-called Lie algebra lattices, which are the direct
sum of only a subset of all possible conjugacy classes. The choice of possible
conjugacy classes is restricted; it must be closed under addition of all lattice
vectors which in particular means that the root lattice is always present.
The possible subsets of conjugacy classes are in one-to-one correspondence
with the subgroups of the center of G.

Since the volume of the unit cell of Λ_R is $\sqrt{N_c}$, the Lie algebra lattice is
unimodular if it contains $\sqrt{N_c}$ conjugacy classes (cf. eq.(11.33) with $\Lambda_s =
\Lambda_R$). This means that N_c must be the square of an integer. Furthermore,
the Lie algebra lattice will be self-dual if all mutual scalar products between
the different conjugacy classes are integer.

Let us illustrate this by considering specific simply laced Lie algebras.
The D_n algebras have four inequivalent conjugacy classes. The already
discussed (0) conjugacy class, the root lattice, contains vectors of the form

$$(k_1 \ldots k_n), \quad k_i \in \mathbf{Z}, \quad \sum_{i=1}^{n} k_i = 0 \bmod 2. \tag{11.51}$$

Next, the vector conjugacy class, denoted by (V) contains as smallest representation the vector representation of dimension $2n$. Its weight vectors
are

$$m = (\pm 1, 0, \ldots, 0) + \text{all permutations.} \tag{11.52}$$

A representative vector of the (V) conjugacy class can be choosen to be $(1, 0^{n-1})$ which implies that all vectors of the V conjugacy class have the form

$$(k_1 \ldots k_n), \quad k_i \in \mathbf{Z}, \quad \sum_{i=1}^{n} = 1 \bmod 2. \tag{11.53}$$

It also follows that the (length)^2 of any vector in the V conjugacy class is 1 mod 2. The spinor conjugacy class, denoted by S, has as smallest representation the spinor representation of dimension 2^{n-1}. The corresponding weight vectors are

$$m = (\pm\frac{1}{2}, \pm\frac{1}{2}, \ldots, \pm\frac{1}{2}), \quad \text{even number of " } - \text{ " signs.} \tag{11.54}$$

Thus, a representative vector of the S conjugacy class can be choosen to be $((\frac{1}{2})^n)$. Finally, the C conjugacy class possesses as lowest dimensional representation the anti-spinor representation with weights

$$m = (\pm\frac{1}{2}, \pm\frac{1}{2}, \ldots, \pm\frac{1}{2}), \quad \text{odd number of " } - \text{ " signs.} \tag{11.55}$$

Its representative vector is $\left(-\frac{1}{2}, (\frac{1}{2})^{n-1}\right)$. The (length)^2 of all vectors in the S and C conjugacy classes is $\frac{n}{4}$ mod 2.

The center of D_n is Z_4 for n odd and $Z_2 \times Z_2$ for n even. It has the same number of elements as there are conjugacy classes, namely four. The addition rules of the conjugacy classes (which correspond to the tensor products of the representations) are determined by the addition rules of the different representative vectors and are summarized in table 11.1. The mutual scalar products (defined modulo 1) between the four conjugacy classes are shown in table 11.2.

From the above discussion it is clear that the D_n Lie algebra lattices are unimodular if they contain in addition to the root lattice one further conjugacy class. Inspection of tables 11.1 and 11.2 shows that the (0) and (V) conjugacy classes of D_n together form an odd self-dual lattice for any value

Table 11.1: Addition rules for D_n conjugacy classes

n even

	(0)	(V)	(S)	(C)
(0)	(0)	(V)	(S)	(C)
(V)	(V)	(0)	(C)	(S)
(S)	(S)	(C)	(0)	(V)
(C)	(C)	(S)	(V)	(0)

n odd

	(0)	(V)	(S)	(C)
(0)	(0)	(V)	(S)	(C)
(V)	(V)	(0)	(C)	(S)
(S)	(S)	(C)	(V)	(0)
(C)	(C)	(S)	(0)	(V)

Table 11.2: Multiplication rules (mod 1) for D_n conjugacy classes

	(0)	(V)	(S)	(C)
(0)	0	0	0	0
(V)	0	0	$\frac{1}{2}$	$\frac{1}{2}$
(S)	0	$\frac{1}{2}$	$n/4$	$(n-2)/4$
(C)	0	$\frac{1}{2}$	$(n-2)/4$	$n/4$

of n. This lattice is identical to the n-dimensional cubic lattice $\mathbf{Z}^n$. For even n, the lattice with (0) and either (S) or (C) conjugacy class is also a unimodular Lie algebra lattice. It is the weight lattice of $\mathrm{Spin}(2n)/\mathbf{Z}_2$.[4] Furthermore, we obtain an odd self-dual lattice of this type for $n = 4$ mod 8 and an even self-dual lattice if $n = 0$ mod 8.

The A_n weight lattice consists of $n + 1$ conjugacy classes denoted by (p) $(p = 0, \ldots, n)$ where the (0) conjugacy class corresponds again to the root lattice. The center of A_n is Z_{n+1}. The smallest representations in each conjugacy class are the symmetric rank p tensors. The addition rules of the conjugacy classes are very simple

---

[4] $\mathrm{Spin}(2n)$, n even, is simply connected and has center $\mathbf{Z}_2 \times \mathbf{Z}_2$. If we divide by the diagonal $\mathbf{Z}_2$ we get $SO(2n)$ with only (0) and (V) conjugacy classes. If we divide by one of the $\mathbf{Z}_2$ we are left with the (0) and one of the spinor conjugacy classes.

$$(p) + (q) = (p + q) \tag{11.56}$$

where $(p + q)$ is defined modulo $n + 1$. The mutual scalar products are

$$(p) \cdot (q) = \frac{p(n + 1 - q)}{n + 1} \bmod 1, \quad p \leq q. \tag{11.57}$$

Using this one can verify that the Lie algebra lattices $A_{k^2 - 1}$ with conjugacy classes $(0), (k), (2k), \ldots \big((k-1)k\big)$ are odd self-dual for k even and even self-dual for k odd.

The E_6 weight lattice contains three conjugacy classes $(0), (1)$ and $(\bar{1})$ corresponding to the singlet, the $\underline{27}$ and the $\underline{\overline{27}}$ representations of E_6. The addition rules of these conjugacy classes are the same as for A_2, the $(\text{length})^2$ of the weights of the $\underline{27}$, $\underline{\overline{27}}$ is $4/3$ and the mutual scalar product between (1) and $(\bar{1})$ is $2/3$ mod 1.

E_7 has two conjugacy classes, (0) and (1) where the minimal representation of the (1) is the $\underline{56}$ with weights of $(\text{length})^2 = \frac{3}{2}$.

Finally E_8 has only one, namely the (0) conjugacy class. Therefore, the weight lattice of E_8 is identical to its root lattice which implies that it is even self-dual. Recall that the root vectors of E_8 are of the form

$$\alpha = \begin{cases} (\pm 1, \pm 1, 0^6) & + \quad \text{permutations} \\ (\pm \frac{1}{2}, \pm \frac{1}{2}, \ldots, \pm \frac{1}{2}) & \quad \text{even number of ``} - \text{'' signs.} \end{cases} \tag{11.58}$$

We recognize that the E_8 root lattice is identical to the D_8 lattice with (0) and (S) conjugacy classes and one can also show that (up to rotations) it is the A_8 Lie algebra lattice with $(0), (3)$ and (6) conjugacy classes. On the other hand, if a Lie algebra lattice contains besides the roots additional weight vectors of $(\text{length})^2 = 2$, then these lattice vectors are roots of a larger Lie algebra with fewer conjugacy classes. E.g. the adjoint of E_8 decomposes into $SO(16)$ as

$$\underline{248} = \underline{120} + \underline{128} \tag{11.59}$$

where the $\underline{128}$ is the spinor representation and the $\underline{120}$ the adjoint representation of $SO(16)$.

Now consider direct products of Lie algebra lattices like $D_n \otimes D_m$, $A_n \otimes A_m$, $D_n \otimes A_m$ etc. These contain of course the root system of the corresponding semi-simple Lie algebras. Furthermore, the Lie algebra lattice is specified by the so-called glue vectors which generate upon addition all other conjugacy classes. Take e.g. $D_2 \otimes D_3$ with glue vector (V,S). Then we obtain, according table 11.1, a Lie algebra lattice with the following conjugacy classes: (0,0), (V,S), (0,V), (V,C). Since $D_2 \otimes D_3$ contains 16 conjugacy classes, the above specified Lie algebra lattice is unimodular. It is however not self-dual. A different example is $E_6 \otimes A_2$ with glue vector (1,1). The conjugacy classes are now (0,0), (1,1) and $(\bar{1}, 2)$. We now get additional (length)$^2 = 2$ vectors, and in fact this lattice is again the even self-dual E_8 root lattice which can be also seen by decomposition of E_8 to $E_6 \times SU(3)$:

$$\underline{248} = (\underline{78}, \underline{1}) + (\underline{1}, \underline{8}) + (\underline{27}, \underline{3}) + (\overline{\underline{27}}, \overline{\underline{3}}) \, . \tag{11.60}$$

Let us now come to the classification of Euclidean even self-dual lattices. They only exist in dimensions which are a multiple of 8. In eight dimensions the only Euclidean even self-dual lattice is the root lattice of E_8. In 16 dimensions there are two even self-dual lattices, namely the root lattice of $E_8 \otimes E_8$ and the Lie algebra lattice of D_{16} with (0) and (S) conjugacy classes, called Spin(32)/$\mathbf{Z}_2$. These two lattices arise in the construction of the 10-dimensional heterotic string theory. They are the only ones satisfying the constraint of one-loop modular invariance, as discussed at the end of Chapter 10.

In 24 dimensions there are 24 even self-dual lattices called Niemeier lattices. 23 of them are Lie algebra lattices of semi-simple Lie groups. The following table, taken from ref. [7], summarizes them together with the relevant glue vectors.

Table 11.3: The 23 Euclidean self-dual semi-simple Lie algebra lattices in 24 dimensions. (Conjugacy classes in square brackets should be cyclically permuted)

Lie algebra	glue vector
D_{24}	(S)
$D_{16}E_8$	(S, 0)
E_8^3	(0, 0, 0)
A_{24}	(5)
D_{12}^2	(S, V), (V, S)
$A_{17}E_7$	(3, 1)
$D_{10}E_7^2$	(S, 1, 0), (C, 0, 1)
$A_{15}D_9$	(2, S)
D_8^3	(S, V, V), (V, S, V), (V, V, S)
A_{12}^2	(1, 5)
$A_{11}D_7E_6$	(1, S, 1)
E_6^4	$(1, 0, 1, \bar{1}), (1, \bar{1}, 0, 1), (1, 1, \bar{1}, 0)$
$A_9^2D_6$	(2, 4, 0), (5, 0, S), (0, 5, C)
D_6^4	even permutations of (0, S, V, C)
A_8^3	(1, 1, 4), (4, 1, 1), (1, 4, 1)
$A_7^2D_5^2$	(1, 1, S, V), (1, 7, V, S)
A_6^4	(1, 2, 1, 6), (1, 6, 2, 1), (1, 1, 6, 2)
$A_5^4D_4$	(2, [0, 2, 4], 0), (3, 3, 0, 0, S), (3, 0, 3, 0, V), (3, 0, 0, 3, C)
D_4^6	(S, S, S, S, S, S), (0, [0, V, C, C, V])
A_4^6	(1, [0, 1, 4, 4, 1])
A_3^8	(2, [2, 0, 0, 1, 0, 1, 1])
A_2^{12}	(2, [1, 1, 2, 1, 1, 1, 2, 2, 2, 1, 2])
A_1^{24}	(1, [0, 0, 0, 0, 0, 1, 0, 1, 0, 0, 1, 1, 0, 0, 1, 1, 0, 1, 0, 1, 1, 1, 1])

The 24th even self-dual lattice is the so-called Leech lattice. It contains no vectors of $(\text{length})^2 = 2$. Its shortest vectors have $(\text{length})^2 = 4$.

Above 24 dimensions the number of even self-dual lattices increases rapidly and most of them are not known explicitly.

So far we have considered only Euclidean Lie algebra lattices. However there exis two types of Lorentzian "Lie algebra" lattices. First consider $\Gamma_{n,m} = \Gamma_{n_L} \otimes \Gamma_{m_R}$ in $\mathbf{R}^{n_L,m_R}$ where Γ_{n_L} and Γ_{m_R} are semi-simple Lie algebra lattices of dimensions n_L and n_R respectively. Again $\Gamma_{n,m}$ is completely specified by the knowledge of the relevant glue vectors. The second type of Lorentzian Lie algebra lattice is obtained if the sign of the signature of the metric changes within a given Lie algebra lattice. $D_{n,m}$ with (0) and (S) conjugacy classes is a suitable example. Actually, Lorentzian lattices of the second type are always also of the first type. For example, $D_{n,m}$ with 0 and S conjugacy classes is identical to $D_{n_L} \otimes D_{m_R}$ with (0,0), (V,V), (S,S) and (C,C) conjugacy classes.

Lorentzian even self-dual lattices $\Gamma_{n,m}$ exist for $n - m = 0 \bmod 8$. They are unique up to Lorentz transformations. For $n = m + 8p$ they are Lorentz transformations of $(E_8)^p \otimes D_{m,m}$ where $D_{m,m}$ is defined by the (0) and (S) conjugacy classes.

11.3 Frenkel-Kac-Segal construction

Let us now return to our original problem namely to provide an operator construction of the level one Kac-Moody algebra from free chiral boson fields moving on a n-dimensional torus. This construction is due to Frenkel and Kac [8] and Segal [9]. An easily accessible discussion can be found in [5].

The level k Kac-Moody algebra in the Cartan-Weyl basis for a simply laced group G reads:

$$[H_n^i, H_m^j] = m\delta^{ij}\delta_{m+n}$$

$$[H_m^i, E_n^\alpha] = \alpha^i E_{m+n}^\alpha$$

$$[E_m^\alpha, E_n^\beta] = \begin{cases} \epsilon(\alpha,\beta) E_{m+n}^{\alpha+\beta} & \text{if } \alpha \cdot \beta = -1 \\ \alpha \cdot H_{m+n} + km\delta_{m+n} & \text{if } \alpha \cdot \beta = -2 \\ 0 & \text{if } \alpha \cdot \beta \geq 0 \end{cases} \tag{11.61}$$

with hermiticity properties

$$H_n^{i\dagger} = H_{-n}^i \quad , \qquad E_n^{\alpha\dagger} = E_{-n}^{-\alpha}. \tag{11.62}$$

Note that $\alpha \cdot \beta = -1$ implies that $\alpha + \beta$ is a root and $\alpha \cdot \beta = -2$ that $\alpha + \beta = 0$. Let us now try to construct conformal fields $H^i(z)$ and $E^\alpha(z)$ from free bosons $X^i(z)$, $(i = 1, \ldots, \text{rank } G)$ with mode expansion

$$X^i(z) = q^i - ip^i \ln z + i \sum_{n \neq 0} \frac{1}{n} \alpha_n^i z^{-n} \tag{11.63}$$

and two-point function

$$\langle X^i(z) X^j(w) \rangle = -\delta^{ij} \ln(z - w). \tag{11.64}$$

(Cf. Chapter 4.) The moments of $H^i(z)$ and $E^\alpha(z)$ are the Kac-Moody generators H_m^i, E_m^α (cf. eq.(11.15)):

$$\begin{aligned}
E_m^\alpha &= \oint_{C_0} \frac{dz}{2\pi i} z^m E^\alpha(z), \\
H_m^i &= \oint_{C_0} \frac{dz}{2\pi i} z^m H^i(z).
\end{aligned} \tag{11.65}$$

Consider the conformal field

$$\tilde{E}^\alpha(z) =: e^{i\alpha \cdot \boldsymbol{X}(z)}: \tag{11.66}$$

where α is a root vector of G with $(\text{length})^2 = 2$. This implies that $\tilde{E}^\alpha(z)$ has conformal dimension $h = 1$. We will now show that $\tilde{E}^\alpha(z)$ is already almost the desired field $E^\alpha(z)$. The operator product expansion of $\tilde{E}^\alpha(z)$ and $\tilde{E}^\alpha(w)$ has the form (see eq.(4.89b):

$$\tilde{E}^\alpha(z)\tilde{E}^\beta(w) = (z - w)^{\alpha \cdot \beta} \tilde{E}^{\alpha+\beta}(w) + $$
$$+ (z - w)^{\alpha \cdot \beta + 1} \alpha \cdot i\partial \boldsymbol{X}(w)\tilde{E}^{\alpha+\beta}(w) + \ldots . \tag{11.67}$$

Since $\alpha \cdot \beta$ is an integer one derives that

$$\tilde{E}^{\alpha}(z)\tilde{E}^{\beta}(w) = (-1)^{\alpha\cdot\beta}\tilde{E}^{\beta}(w)\tilde{E}^{\alpha}(z).\tag{11.68}$$

This implies that with the contour integration trick of Chapter 4 we can only calculate $\tilde{E}^{\alpha}_n \tilde{E}^{\beta}_m - (-1)^{\alpha\cdot\beta}\tilde{E}^{\beta}_m \tilde{E}^{\alpha}_n$. We get

$$\tilde{E}^{\alpha}_m\tilde{E}^{\beta}_n - (-1)^{\alpha\cdot\beta}\tilde{E}^{\beta}_n\tilde{E}^{\alpha}_m$$

$$= \left\{ \oint_{C_0}\frac{dz}{2\pi i}\oint_{\substack{C_0 \\ |z|>|w|}}\frac{dw}{2\pi i} - \oint_{C_0}\frac{dz}{2\pi i}\oint_{\substack{C_0 \\ |w|>|z|}}\frac{dw}{2\pi i} \right\}z^m\tilde{E}^{\alpha}(z)w^n\tilde{E}^{\beta}(w)$$

$$= \oint_{C_0}\frac{dw}{2\pi i}\oint_{C_w}\frac{dz}{2\pi i}(z-w)^{\alpha\cdot\beta}z^m w^n\tilde{E}^{\alpha+\beta}(w)$$

$$\times\left[1 + i(z-w)\alpha\cdot\partial\boldsymbol{X}(w) + \dots\right]\tag{11.69}$$

$$= \begin{cases} \tilde{E}^{\alpha+\beta}_{m+n} & \alpha\cdot\beta = -1 \\ m\delta_{m+n} + i\alpha\cdot\oint_{C_0}\frac{dw}{2\pi i}w^{m+n}\partial\boldsymbol{X}(w) & \alpha\cdot\beta = -2 \\ 0 & \text{otherwise} \end{cases}$$

Comparing this with eq.(11.61) we recognize that, the unwanted factor $(-1)^{\alpha\cdot\beta}$ and the structure constants $\epsilon(\alpha,\beta)$ aside, we have reached our goal if we identify $H^i(z)$ with the derivative of the free bosons $X^i(z)$:

$$H^i(z) = i\partial X^i(z)\,.\tag{11.70}$$

This is in fact the correct identification; $\partial X^i(z)$ are conformal fields of dimension one and are the $[U(1)]^n$ Cartan-subalgebra currents. They satisfy the operator algebra

$$\partial X^i(z)\partial X^j(w) = -\frac{\delta^{ij}}{(z-w)^2} + \dots\tag{11.71}$$

which immediately leads to the correct commutator eq.(11.61) between H^i_m and H^j_n. We can also check that, using

$$H^i(z)\tilde{E}^{\alpha}(w) = \frac{\alpha^i}{(z-w)}\tilde{E}^{\alpha}(w) + \dots,\tag{11.72}$$

the correct algebra between the H_m^i and $\tilde{E}_n^\alpha$ is obtained. Finally, we have to compensate the factor $(-1)^{\alpha \cdot \beta}$ in eq.(11.69). This is done by introducing so-called cocycle factors (also called Klein factors). We define $E^\alpha(z)$ as:

$$E^\alpha(z) = c_\alpha \tilde{E}^\alpha(z). \tag{11.73}$$

The cocycles c_α have to satisfy:

$$\begin{aligned} c_\alpha \cdot c_\beta &= (-1)^{\alpha \cdot \beta} c_\beta \cdot c_\alpha \\ c_\alpha \cdot c_\beta &= \epsilon(\alpha, \beta) c_{\alpha + \beta}. \end{aligned} \tag{11.74}$$

They can be explicitly constructed from the boson zero modes. We omit to present this construction.

In conclusion, the level one Kac-Moody algebra of a simply laced group G with rank n possesses an explicit operator construction from n free bosonic fields $X^i(z)$:

$$\begin{aligned} H_m^i &= \oint_{C_0} \frac{dz}{2\pi i} z^m H^i(z), & H^i(z) &= i\partial X^i(z), \\ E_m^\alpha &= \oint_{C_0} \frac{dz}{2\pi i} z^m E^\alpha(z), & E^\alpha(z) &= c_\alpha :e^{i\alpha \cdot X(z)}: \, . \end{aligned} \tag{11.75}$$

The generators H_0^i, E_0^α obviously form a finite dimensional subalgebra isomorphic to the Lie algebra g of eq.(11.37). In this case the Cartan subalgebra generators H_0^i are simply given by the momentum operators p^i in eq.(11.63). The non-commuting E_0^α's are characterized by the momentum eigenvalues which are the roots α of the Lie algebra. It means that the allowed momentum eigenvalues of the bosons $X^i(z)$ are quantized, or, equivalently, the bosons $X^i(z)$ live on a (right-moving) n-dimensional torus.

Now we can also give an explicit construction of the vertex operators of the gauge boson states which results from the compactification of the bosonic coordinates. The vertex operators of the Cartan subalgebra $[U(1)]^n$ gauge bosons are just the currents $\partial X^i(z)$. The corresponding asymptotic states are given by

$$|i\rangle = \oint_{C_0} \frac{dz}{2\pi i} \frac{i}{z} \partial X^i(z)|0\rangle = \alpha^i_{-1}|0\rangle \qquad (11.76)$$

which is nothing but an internal oscillator excitation in agreement with our considerations in Chapter 10.

The vertex operators of the non-Abelian gauge bosons are again given by the corresponding currents $E^\alpha(z) = c_\alpha : e^{i\alpha \cdot X(z)} :$. Since $\alpha^2 = 2$ this vertex operator has conformal dimension $h = \frac{\alpha^2}{2} = 1$ as required for a physical state. The gauge boson state corresponding to the soliton vacuum has the form

$$|\alpha\rangle = \oint_{C_0} \frac{dz}{2\pi i} \frac{1}{z} c_\alpha : e^{i\alpha \cdot X(z)} : |0\rangle \qquad (11.77)$$

This state carries quantized internal momentum corresponding to the roots of G. It is a winding state on the maximal torus of G defined by $\mathbf{R}/\Lambda_R$. The allowed quantized momenta of the $X^i(z)$ are however not restricted to lie in the root lattice of G but can a priori be any weight vector. So we consider vertex operators

$$V_\lambda(z) = c_\lambda : e^{i\lambda \cdot X(z)} : \qquad (11.78)$$

where the λ's are weight vectors of some (irreducible) representation $\underline{R}$ of G and c_λ is again a cocycle factor. Then this operator creates states which transform under the representation $\underline{R}$. The operator product algebra is

$$V_\lambda(z)V_{\lambda'}(w) = (z-w)^{\lambda \cdot \lambda'} V_{\lambda+\lambda'}(w) + \ldots \qquad (11.79)$$

The full operator algebra is completely determined by the addition rules of the different conjugacy classes. The requirement for a closed operator algebra implies that the quantized momenta λ are vectors of a Lie algebra lattice of G. Additional requirements from string theory, like locality and modular invariance, imply that the lattice is integral and self-dual. This will be discussed further in these lectures.

11.4 Fermionic construction of the current algebra – Bosonization

In the previous section we presented the bosonic construction of the level one Kac-Moody algebra of a simply laced group G of rank n. It was motivated by the fact that both n free bosons and also the level one Kac-Moody algebra provide a Virasoro algebra with central charge $c = n$. On the other hand, a two-dimensional "real" chiral (Majorana-Weyl) fermion corresponds to a conformal field theory with $c = \frac{1}{2}$. Thus we expect that one can realize the level one Kac-Moody algebra by $2n$ fermions. Furthermore, we want to establish that the conformal field theory of $2n$ fermions with specific boundary conditions is equivalent to a conformal field theory of n bosons compactified on a torus. The quantum equivalence between fermions and bosons in two dimensions was discovered by Coleman [10] and Mandelstam [11].

Consider a system of $2n$ two-dimensional real fermions $\psi^i(z)$ ($i = 1, \ldots, 2n$) all having either periodic or antiperiodic boundary conditions and transforming as a vector of $SO(2n)$. The operator products are

$$\psi^i(z)\psi^i(w) = -\frac{\delta^{ij}}{z - w} + \ldots \tag{11.80}$$

The fields $\psi^i(z)$ transform as a vector of $SO(2n)$. We can then build the fermion bilinears

$$J^a(z) = \frac{1}{2} : \psi^i(z)\, T^a_{ij}\, \psi^j(z) :, \qquad a = 1, \ldots, 2n^2 - n \tag{11.81}$$

where the antisymmetric $2n \times 2n$ matrices T^a_{ij} are the generators of $SO(2n)$ in the fundamental representation. An explicit representation of the $SO(2n)$ generators in the vector representation is $(T^{kl})_{ij} = \delta^k_i \delta^l_j - \delta^k_j \delta^l_i$ and the currents become $J^{kl}(z) =: \psi^k \psi^l(z) :$. Using eq.(11.80) it is not difficult to show that the fermionic currents $J^a(z)$ in eq.(11.81) obey the level one Kac-Moody algebra of $D_n \sim SO(2n)$ as given in eq.(11.19).

This suggests that the bilinears of $2n$ free fermions are identical to the currents eq.(11.75) constructed from n free bosons with momentum vectors being root vectors of D_n. This is basically the statement of bosonization. It means that the conformal field theories of n free bosons and $2n$ free fermions are equivalent in the sense that the conformal properties of their operators and all correlation functions are identical. However, this is only true if the bosons are compactified on a special torus. E.g. a single boson is only equivalent to two free fermions if the radius of the circle the boson is compactified on takes special values.

Let us work out the bosonization prescription in more detail. For this purpose it is useful to convert the real basis for the fermions ψ^i to a complex basis. Define

$$\Psi^{\pm i}(z) = \frac{1}{\sqrt{2}}(\psi^{2i-1} \pm i\psi^{2i})(z), \qquad i = 1, \ldots, n \tag{11.82}$$

The Cartan subalgebra currents of D_n are then given by

$$J^{i,-i}(z) =\, :\Psi^i(z)\Psi^{-i}(z): \tag{11.83}$$

and the non-commuting D_n currents are

$$J^{\pm i, \pm j}(z) =\, :\Psi^{\pm i}(z)\Psi^{\pm j}(z): \qquad (i < j). \tag{11.84}$$

Bosonization then consists of identifying $J^{i,-i}(z)$ with the derivative of the bosons $X^i(z)$

$$:\Psi^i(z)\Psi^{-i}(z): = i\partial X_i(z) \tag{11.85}$$

and the non-commuting currents with the operators $E^\alpha(z)$

$$:\Psi^{\pm i}(z)\Psi^{\pm j}(z): = c_{\pm i, \pm j} :e^{i(\pm X_i \pm X_j)(z)}: \qquad (i < j). \tag{11.86}$$

We recognize that the corresponding root vector of D_n is given by $\alpha = \pm e_i \pm e_j$.

One can also give a bosonic representation of the fermion $\Psi^i(z)$ itself. Since the fermions transform as a vector of D_n they are bosonized according to

$$\Psi^{\pm i}(z) = c_{\pm i} \; : e^{\pm i X_i(z)} : \tag{11.87}$$

where the quantized bosonic momentum is now a vector weight of D_n : $\lambda = \pm e_i$. This gives the correct conformal dimension $h = \frac{\lambda^2}{2} = \frac{1}{2}$ for the fermions $\Psi^{\pm i}(z)$.

As already discussed in Chapter 8, the Hilbert space of the fermionic theory splits into two sectors, namely the NS and R sector. In the NS sector, the fermions have antiperiodic boundary conditions on the cylinder which means that they are periodic on the complex plane. For R states the situation is reversed. This change of boundary conditions is due to the Jacobian factor in eq.(4.10) for $h = \frac{1}{2}$.

All states in the NS sector are (even or odd rank) tensors of $SO(2n)$. Therefore, in bosonic language, the corresponding vertex operators are of the form

$$V_\lambda(z) = c_\lambda \; : e^{i\lambda \cdot X(z)} : \tag{11.88}$$

where λ is either in the (0) or (V) conjugacy class of D_n. The currents eq.(11.86) or the fermions eq.(11.87) are of this type. Since the mutual scalar products between the (0) and (V) conjugacy classes are integer (see table 11.2), the operator algebra eq.(11.79) in the NS sector is local, i.e. contains no branch cuts.

On the other hand, states in the R sector are build from the vacua $|S^\alpha\rangle$ $(|S^{\dot\alpha}\rangle)$ which transform as spinors (anti-spinors) of D_n. Thus the vertex operators in the R sector are of the form eq.(11.88) but now with λ being a weight either in the (S) or (C) conjugacy class of D_n. We can also give an explicit construction for the spinorial vacuum $|S^\alpha\rangle$ $(|S^{\dot\alpha}\rangle)$. It is created by an operator eq.(11.88) with λ_α $(\lambda_{\dot\alpha})$ being an (anti) spinor weight of D_n as displayed in eqs.(11.54,55). This operator has conformal dimension $h = \frac{\lambda^2}{2} = \frac{n}{8}$.

Now, inspection of table 11.2 shows that the operator algebra eq.(11.79) in the R sector is non-local. Furthermore, states in the NS sector corresponding to the (V) conjugacy class of D_n are non-local with respect to the R sector. The corresponding operator algebra contains branch cuts.

In conclusion, it is evident that using the techniques of bosonization the vertex operators of the fermionic conformal field theory become extremely simple. This is of great help in string theory as we will discuss in more detail later.

11.5 Unitary representations and characters of Kac-Moody algebras

Let us now return to the discussion at the beginning of this chapter about the representations of the combined chiral algebra $\mathcal{A} = \hat{v} \oplus \hat{g}$. Following Chapter 4 we define primary or highest weight states of $\mathcal{A}$ by

$$|\phi_i\rangle = \phi_i(0)|0\rangle \tag{11.89}$$

where i is a representation index of G and $\phi_i(z)$ is primary according to eq.(11.21). The vacuum is $SL(2, \mathbf{C})$ and G invariant. The highest weight states satisfy

$$
\begin{aligned}
L_n|\phi_i\rangle &= 0, \quad n > 0, \\
J_n^a|\phi_i\rangle &= 0, \quad n > 0, \\
L_0|\phi_i\rangle &= h_i|\phi_i\rangle, \\
J_0^a|\phi_i\rangle &= (T^a)_i{}^j|\phi_j\rangle.
\end{aligned}
\tag{11.90}
$$

Alternatively, one could also write the highest weight condition in terms of the modes of $H^i(z)$ and $E^\alpha(z)$. To each primary field there exists an infinite number of descendant fields of the form

$$L_{-k_1} \cdots L_{-k_m} J^{a_1}_{-l_1} \cdots J^{a_n}_{-l_n} |\phi_i\rangle, \qquad k_i, l_i > 0. \tag{11.91}$$

The conformal dimensions of the descendant states are $h_i + \sum_{i=1}^{m} k_i + \sum_{i=1}^{n} l_i$. We denote the totality of fields in a given highest weight representation by $[\phi_i]_{\mathcal{A}}$. The $[\phi_i]_{\mathcal{A}}$ are also called current algebra families.

The conformal dimension of the primary field is given by

$$h = \frac{C_2^R}{C_2 + 2k} \tag{11.92}$$

where C_2^R is the second Casimir of the representation $\underline{R}$ under which $|\phi\rangle$ transforms:

$$(T^a)_i{}^j (T^a)_j{}^k = C_2^R \delta_i^k. \tag{11.93}$$

Eq.(11.92) is easily derived using the explicit expression for L_0 given in eq.(11.22b).

Let us now determine the restrictions on k following from unitarity. To do this consider the $SU(2)$ subalgebra of $\hat{g}$ generated by $E_1^{-\alpha}$, E_{-1}^{α} and $(k - \alpha \cdot \boldsymbol{H}_0)$ where α is any root. Again, if we identify these generators with J_+, J_- and $2J_3$ we have the angular momentum algebra and know that unitarity requires that the eigenvalues of $(k - \alpha \cdot \boldsymbol{H}_0)$ have to be integer. Acting on a primary state $|m\rangle$ with weight m we find $k - \alpha \cdot m \in \mathbf{Z}$ or, since $\alpha \cdot m \in \mathbf{Z}$

$$k \in \mathbf{Z}. \tag{11.94}$$

Since $|m\rangle$ is primary, $E_{+1}^{-\alpha} |m\rangle = 0$ and

$$\|E_{-1}^{\alpha} |m\rangle\|^2 = \langle m | E_1^{-\alpha} E_{-1}^{\alpha} |m\rangle = (k - \alpha \cdot m) \||m\rangle\|^2. \tag{11.95}$$

Positivity of the Hilbert space then gives $k \geq \alpha \cdot m$. The right hand side of this inequality is maximized if we chose α to be the highest root ψ and m to be the highest weight m_0 of the given representation. We then have

$$k \geq \psi \cdot m_0. \tag{11.96}$$

Clearly, for a given k only a finite number of highest weights satisfies this criterion.[5] For $k = 1$ they are those belonging to the lowest dimensional representation in each conjugacy class. So, in spite of the fact that we are considering conformal field theories with c larger than one, the combined Kac-Moody and Virasoro algebra has only a finite number of primary fields or current algebra families. Therefore, we call them also rational conformal field theories. Of course, the irreducible representations of the combined algebra are highly reducible under the Virasoro algebra. A current algebra family contains an infinite number of conformal fields which all transform under some representation of G. On the other hand, any current algebra family is generated from a specific lowest dimensional representation; states in all other representations are obtained by acting with the E^{α}_{-n}.

As usual, the operator product algebra between the primary fields is determined by the conformal dimensions and the fusion rules among the fields:

$$\phi_i \times \phi_j = N_{ij}{}^l \phi_l. \qquad (11.97)$$

For arbitrary level k the fusion coefficients $N_{ij}{}^l$ are quite difficult to determine. Of course, the fusion rules obey the decomposition rules of the tensor products between two irreducible representations. Therefore, the $N_{ij}{}^l$ are necesssarily zero if the corresponding Clebsch-Gordon coefficients vanish. However, the converse is not true which makes a systematic discussion unfeasable.

For this reason we will concentrate on the simplest case $k = 1$. We have already seen that one can explicitly construct the level one currents and also the states transforming under a specific representation by n free bosons compactified on the weight lattice of G. Now, eq.(11.96) which

---

[5] The corresponding highest weight representations are also called integrable representations.

provides the number of current algebra families is satisfied only for the lowest dimensional representation in each conjugacy class of G. Therefore, the number of primary fields is identical to the number of conjugacy classes. The highest weight state has the form:

$$|\phi_i\rangle = \oint_{C_0} \frac{dz}{2\pi i} \frac{1}{z} :e^{i\lambda_i \cdot X(z)}: |0\rangle . \tag{11.98}$$

λ_i is the highest weight vector of the lowest representation in the ith conjugacy class. The fusion rules eq.(11.97) are simply determined by the addition rules for the different conjugacy classes resp. highest weight vectors. All non-vanishing coefficients $N_{ij}{}^l$ are one.

Let us finally discuss the generalized characters of $\mathcal{A}$. Recall that the characters are defined by

$$Ch_i(\tau) = \text{Tr}_{\phi_i} q^{L_0 - c/24} \tag{11.99}$$

where the trace is over all states of the current algebra family ϕ_i. This definition holds for all k. Again, for $k = 1$ these characters have a rather simple form. Since the L_0 eigenvalue of the field $\phi(z) = e^{i\lambda \cdot X(z)}$ is given by $h = \frac{\lambda^2}{2}$, the level one characters are

$$Ch_i(\tau) = \frac{1}{[\eta(\tau)]^n} \sum_{\lambda \in (i)} e^{i\pi\tau\lambda^2} = \frac{1}{[\eta(\tau)]^n} P_i(\tau). \tag{11.100}$$

The sum $P_i(\tau)$ is over all lattice vectors within the conjugacy class (i). The factor $[\eta(\tau)]^{-n}$, where n is the rank of G, takes into account the contribution of the L_{-k}'s and $q^{-c/24}$ $(c = n)$.

For the D_n algebras at level one we can give a closed expression for the lattice sums. From the explicit expressions for the simple roots and the representative weights, one can construct all weights in each of the four conjugacy classes. One then finds that the lattice sums corresponding to the 0,V,S,C conjugacy classes of D_n are given by:

$$P_0(\tau) = \frac{1}{2}\Big\{[\theta_3(0|\tau)]^n + [\theta_4(0|\tau)]^n\Big\}$$

$$P_V(\tau) = \frac{1}{2}\Big\{[\theta_3(0|\tau)]^n - [\theta_4(0|\tau)]^n\Big\}$$

$$P_S(\tau) = \frac{1}{2}\Big\{[\theta_2(0|\tau)]^n + i^n[\theta_1(0|\tau)]^n\Big\}$$

$$P_C(\tau) = \frac{1}{2}\Big\{[\theta_2(0|\tau)]^n - i^n[\theta_1(0|\tau)]^n\Big\}$$

$$(11.101)$$

The θ-functions were defined in Chapter 9. It is illustrative to compare these expressions with the partition functions of the fermionic string theory (cf. Chapter 9). Again, the equivalence between n bosons compactified on a D_n lattice and $2n$ periodic resp. antiperiodic fermions becomes manifest. Explicitly we obtain the following identities between the bosonic and fermionic partition functions[6]

$$\chi_{PP}(\tau) = (i)^n(Ch_S(\tau) - Ch_C(\tau))$$

$$\chi_{PA}(\tau) = \quad Ch_S(\tau) + Ch_C(\tau)$$

$$\chi_{AA}(\tau) = \quad Ch_0(\tau) + Ch_V(\tau)$$

$$\chi_{AP}(\tau) = \quad Ch_0(\tau) - Ch_V(\tau)$$

$$(11.102)$$

With the help of eq.(11.101) the partition function for the $SO(32)$ string, given at the end of the previous chapter, follows immediately.

Similarly, one can also express the level one characters of E_6, E_7 and E_8 by theta functions; this requires again the explicit forms of the simple roots and the representative weights. For E_8 they have been given in this chapter. For E_6 and E_7 we refer to the literature. One then finds for the characters

---

[6]These identities remain also true for higher genus partitition functions

$$P^{c=6}_{(E_6)0} = \frac{1}{\eta^6(\tau)}\left\{\theta_3(0|3\tau)\big[\theta_3(0|\tau)\big]^5 + \theta_4(0|3\tau)\big[\theta_4(0|\tau)\big]^5 + \theta_2(0|3\tau)\big[\theta_2(0|\tau)\big]^5\right\}$$

$$P^{c=6}_{(E_6)1} = \frac{1}{\eta^6(\tau)}\left\{\theta[^{1/6}_{\ 0}](0|3\tau)\big[\theta_2(0|\tau)\big]^5 + \theta[^{2/3}_{\ 0}](0|3\tau)\big[\theta_3(0|\tau)\big]^5 \right.$$
$$\left. + e^{-2\pi i/3}\,\theta[^{2/3}_{1/2}](0|3\tau)\big[\theta_4(0|\tau)\big]^5\right\}$$

$$P^{c=6}_{(E_6)\bar{1}} = \frac{1}{\eta^6(\tau)}\left\{\theta[^{5/6}_{\ 0}](0|3\tau)\big[\theta_2(0|\tau)\big]^5 + \theta[^{1/3}_{\ 0}](0|3\tau)\big[\theta_3(0|\tau)\big]^5 \right.$$
$$\left. - e^{-i\pi/3}\,\theta[^{1/3}_{1/2}](0|3\tau)\big[\theta_4(0|\tau)\big]^5\right\}$$

$$P^{c=7}_{(E_7)0} = \frac{1}{\eta^7(\tau)}\left\{\theta_2(0|2\tau)\big[\theta_2(0|\tau)\big]^6 + \theta_3(0|2\tau)\left(\big[\theta_3(0|\tau)\big]^6 + \big[\theta_4(0|\tau)\big]^6\right)\right\}$$

$$P^{c=7}_{(E_7)1} = \frac{1}{\eta^7(\tau)}\left\{\theta_3(0|2\tau)\big[\theta_2(0|\tau)\big]^6 + \theta_2(0|2\tau)\left(\big[\theta_3(0|\tau)\big]^6 - \big[\theta_4(0|\tau)\big]^6\right)\right\}$$

$$P^{c=8}_{(E_8)0} = \frac{1}{\eta^8(\tau)}\left\{\big[\theta_2(0|\tau)\big]^8 + \big[\theta_3(0|\tau)\big]^8 + \big[\theta_4(0|\tau)\big]^8\right\}.$$

$$\tag{11.103}$$

Let us conclude this chapter by discussing some of the modular properties of the level one Kac-Moody characters. This is important because they appear in string theory as one-loop partition functions of n bosonic coordinates compactified on a specific Lie algebra lattice. The Hilbert space then contains, in general, several current algebra families (or conjugacy classes) including the identity family which of course corresponds to the (0) conjugacy class.

The level one characters of the Kac-Moody algebra $\hat{g}$ build a finite dimensional representation of the modular group, i.e. the characters transform under under the generators of the modular group T $(\tau \to \tau + 1)$ and S $(\tau \to -\frac{1}{\tau})$ like $(i = 1, \ldots, N_C = \text{number of conjugacy classes})$:

$$Ch_i(\tau + 1) = T_{ij}Ch_j(\tau),$$
$$Ch_i(-\frac{1}{\tau}) = S_{ij}Ch_j(\tau). \tag{11.104}$$

The matrix T is easy to determine. From eq.(11.99) we immediately derive

$$T_{ij} = e^{-\pi i(\frac{c}{12} - \lambda_i^2)} \delta_{ij} \tag{11.105}$$

where λ_i is some vector in the conjugacy class (i). The S transformation, on the other hand, is more involved. We obtain for the level one characters the following result:

$$S: \quad Ch_i(-\frac{1}{\tau}) = \frac{1}{\sqrt{N_c}} \sum_{j=1}^{N_c} e^{2\pi i \lambda_i \cdot \lambda_j} Ch_j(\tau). \tag{11.106}$$

This is actually not difficult to derive. The weights in the i-th conjugacy class can be written as

$$\{\lambda^{(i)}\} = \bar{\lambda}^{(i)} + \Lambda_R \tag{11.107}$$

where $\bar{\lambda}^{(i)}$ is some representative weight and Λ_R the root lattice. We then write

$$\begin{aligned}
Ch_i(-\frac{1}{\tau}) &= \frac{1}{\eta^n(-1/\tau)} \sum_{\lambda \in \Lambda_R} e^{-\frac{i\pi}{\tau}(\bar{\lambda}^{(i)} + \lambda)^2} \\
&= \frac{1}{\eta^n(\tau)} \frac{1}{\sqrt{N_C}} \sum_{\lambda \in \Lambda_w} e^{2\pi i \lambda \cdot \bar{\lambda}^{(i)}} e^{i\pi\tau\lambda^2}
\end{aligned} \tag{11.108}$$

where we have used eq.(10.53) and the fact that $\Lambda_w = (\Lambda_R)^*$. With $\lambda^{(j)} \cdot \bar{\lambda}^{(i)} = \lambda^{(j)} \cdot \lambda^{(i)} \bmod 1$ we get eq.(11.106). It shows that the S transformation acts as a Fourier transformation on the characters $Ch_i(\tau)$: For the D_n algebra we obtain, using table 11.2, the following matrix:

$$S_{ij} = \frac{1}{2} \begin{pmatrix} 1 & 1 & 1 & 1 \\ 1 & 1 & -1 & -1 \\ 1 & -1 & e^{i\pi n/2} & -e^{i\pi n/2} \\ 1 & -1 & -e^{i\pi n/2} & e^{i\pi n/2} \end{pmatrix} \tag{11.109}$$

($i = 1$ corresponds to (0), $i = 2$ to (V), $i = 3$ to (S) and $i = 4$ to (C)). One can verify that $S^2 = 1$ for n even. For n odd, acting with S^2 interchanges the S and C conjugacy classes. This is a general feature: if complex representations are present, $S^2 \neq 1$ but $S^4 = 1$. For the A_n series S_{kl} is given by ($k, l = 1, \ldots, n+1$)

$$S_{kl} = \frac{1}{\sqrt{n+1}} e^{-\frac{2\pi i}{n+1}(k-1)(l-1)}. \tag{11.110}$$

Finally the E_6 characters transform like A_2 (E_6 and A_2 have the same fusion rules), the E_7 characters like the A_1 characters and for E_8 S and T are one. The E_8 character is a modular invariant in agreement with our result of Chapter 10 that the partition function of an even self-dual lattice is modular invariant.

REFERENCES

[1] V.G. Kac, *"Simple graded algebras of finite growth"*,
   Funct. Anal. Appl. **1** (1967) 328.

[2] R.V. Moody, *"Lie algebras associated with generalized Cartan matrices"*, *Bull. Amer. Math. Soc.* **73** (1967) 217.

[3] P. Goddard and D. Olive, *"Kac-Moody and Virasoro algebras"*, in 'Advanced Series in Mathematical Physics', World Scientific (1988).

[4] H. Sugawara, *"A field theory of currents"*, *Phys. Rev. Lett.* **170** (1968) 1659.

[5] P. Goddard and D. Olive, *"Algebras, lattices and strings"*, in 'Vertex Operators in Mathematics and Physics', eds. J. Lepowsky, S. Mandelstam, I. Singer, Springer-Verlag (1983).

[6] W. Lerche, A.N. Schellekens and N. Warner,
   "Lattices and Strings", *Phys. Rep.* **177** (1989) 1.

[7] H. Niemeier, Journal of Number Theory **5** (1973) 142.

[8] J. Frenkel and V.G. Kac, *"Basic representations of affine Lie algebras and dual resonance models"*, *Inv. Math.* **62** (1980) 23.

[9] G. Segal, *"Unitary representations of some infinite dimensional groups"*, *Comm. Math. Phys.* **80** (1981) 301.

[10] S. Coleman, *"Quantum sine-Gordon equation as the massive Thirring model"*, *Phys. Rev.* **D11** (1975) 2088.

[11] S. Mandelstam, *"Soliton operators for the quantized sine-Gordon equation"*, *Phys. Rev.* **D11** (1975) 3026.

Chapter 12

Conformal Field Theory III: Superconformal Field Theory

In Chapter 4 we have demonstrated the usefulness of conformal field theory as a tool for the bosonic string. In the same way as conformal symmetry was a remnant of the reparametrization invariance of the bosonic string in conformal gauge, super-conformal invariance is a remnant of local supersymmetry of the fermionic string in super-conformal gauge. This leads us to consider super-conformal field theory. In many aspects our discussion of super-conformal field theory parallels that of conformal field theory and we will treat those rather briefly. There are, however, new features which we will present in more detail.

The $n = 1$ superconformal algebra is due to Ramond [1] and Neveu and Schwarz [2]. The models with extended $n = 2$, $n = 4$ superconformal symmetry were formulated by Ademollo et al. [3]. An introduction to the subject and also a more complete list of references can be found in the reviews listed in Chapter 4 and in refs. [4, 5].

The generators of super-conformal transformations are the conserved energy-momentum tensor $T(z)$ and the conserved world-sheet supercurrent $T_F(z)$[1].

---

[1] As in Chapter 4 we will only consider the holomorphic part of the theory. Note that whereas both sectors of the theory are conformally invariant, it is possible that only one of them, say the holomorphic one, exhibits superconformal invariance. This is for instance the case in the heterotic string theory. We should, however, mention that superconformal invariance can also appear in the internal sector of the bosonic string.

The basic objects of super-conformal field theory are conformal (or primary) superfields

$$\Phi(z, \theta) = \phi_0(z) + \theta\phi_1(z) \tag{12.1}$$

where θ is a constant anticommuting Grassmann variable and ϕ_0 and ϕ_1 are conformal fields of opposite statistics.[2] If Φ is of weight h, ϕ_0 and ϕ_1 have conformal weights h and $(h + \frac{1}{2})$ respectively; i.e. under infinitesimal conformal transformations we get

$$\begin{aligned}
\delta_\xi \phi_0(z) &= [h\partial\xi + \xi\partial]\phi_0(z)\,, \\
\delta_\xi \phi_1(z) &= [(h + \tfrac{1}{2})\partial\xi + \xi\partial]\phi_1(z)\,.
\end{aligned} \tag{12.2}$$

Under two-dimensional supersymmetry transformations the two components of a superfield are transformed into each other according to

$$\begin{aligned}
\delta_\epsilon \phi_0(z) &= \frac{1}{2}\epsilon(z)\phi_1(z)\,, \\
\delta_\epsilon \phi_1(z) &= \frac{1}{2}\epsilon(z)\partial\phi_0(z) + h\partial\epsilon(z)\phi_0(z)
\end{aligned} \tag{12.3}$$

where the anticommuting analytic function $\epsilon(z)$ parametrizes infinitesimal holomorphic supersymmetry transformations. We have used the notation

$$\delta_\epsilon \phi(z) = [T_{F_\epsilon}, \phi(z)] \tag{12.4}$$

where

$$T_{F_\epsilon} = \oint \frac{\mathrm{d}z}{2\pi i}\epsilon(z)T_F(z) \tag{12.5}$$

are an infinity of conserved charges. T_F is the anticommuting generator of the superconformal algebra. Note that the supersymmetry transformation eq.(12.3) is the "square root" of conformal transformations eq.(12.2) in the sense that

---

[2] One can formulate the following discussion in terms of superfields and super-space variables which makes two-dimensional supersymmetry manifest. We will however use the component formalism.

$$[\delta_{\epsilon_1}, \delta_{\epsilon_2}]\phi_0(z) = [h\partial\xi + \xi\partial]\phi_0\,,$$

$$[\delta_{\epsilon_1}, \delta_{\epsilon_2}]\phi_1(z) = \left[(h + \tfrac{1}{2})\partial\xi + \xi\partial\right]\phi_1 \tag{12.6}$$

where

$$\xi = \tfrac{1}{2}\epsilon_2\epsilon_1.$$

We can again translate the transformation rules for conformal superfields to operator product expansions. Using the techniques of Chapter 4 we easily find

$$T(z)\phi_0(w) \sim \frac{h\phi_0(w)}{(z-w)^2} + \frac{\partial\phi_1(w)}{z-w} + \cdots$$

$$T(z)\phi_1(w) \sim \frac{(h+\tfrac{1}{2})\phi_1(w)}{(z-w)^2} + \frac{\partial\phi_1(w)}{z-w} + \cdots$$

$$T_F(z)\phi_0(w) \sim \frac{\tfrac{1}{2}\phi_1(w)}{z-w} + \cdots \tag{12.7}$$

$$T_F(z)\phi_1(w) \sim \frac{h\phi_0(w)}{(z-w)^2} + \frac{\tfrac{1}{2}\partial\phi_0(w)}{z-w} + \cdots$$

The superconformal algebra is specified by the operator products of the generators of superconformal transformations:

$$T(z)T(w) \sim \frac{\tfrac{3}{4}\hat{c}}{(z-w)^4} + \frac{2T(w)}{(z-w)^2} + \frac{\partial T(w)}{z-w} + \cdots \tag{12.8a}$$

$$T(z)T_F(w) \sim \frac{\tfrac{3}{2}T_F(w)}{(z-w)^2} + \frac{\partial T_F(w)}{z-w} + \cdots \tag{12.8b}$$

$$T_F(z)T_F(w) \sim \frac{\tfrac{1}{4}\hat{c}}{(z-w)^3} + \frac{\tfrac{1}{2}T(w)}{z-w} + \cdots \tag{12.8c}$$

This is the $n = 1$ superconformal algebra. (It is $n = 1$ because we have only one supercurrent. We will encounter extended superconformal algebras below.) The operator products can again be verified by commuting various combinations of conformal and superconformal transformations. Eq.(12.8b) simply states that T_F is a primary field of the Virasoro algebra of weight $h = 3/2$. Eq.(12.8c) reflects eq.(12.6). Note that the central charges in eqs.(12.8a) and (12.8c) are related. The reason for this will be given below. A central extension of eq.(12.8b) is forbidden by scale invariance and the Grassmann properties of T (even) and T_F (odd). We have normalized the central charge such that a free superfield $\mathcal{X}(z,\theta) = X(z) + \theta\psi(z)$, where $X(z)$ and $\psi(z)$ are free world-sheet bosons and fermions respectively, has $\hat{c} = 1$. (The central charge c in eq.(4.26) and $\hat{c}$ are related by $c = \frac{3}{2}\hat{c}$.) We will explicitly verify the algebra for this case below. Comparing eqs.(12.7) and (12.8) we find that, apart from the central charge terms, T_F and T are the two components of a $h = 3/2$ superfield.

As in the bosonic case we now expand T and T_F in modes and derive their algebra from the operator products. This will of course give the super-Virasoro algebra. The modes of $T(z)$ are defined as in Chapter 4. We expand T_F as

$$T_F(z) = \frac{1}{2} \sum_{r \in \mathbf{Z}+a} z^{-3/2-r} G_r \tag{12.9}$$

or

$$G_r = 2 \oint \frac{\mathrm{d}z}{2\pi i} T_F(z) z^{r+1/2}. \tag{12.10}$$

We have introduced the parameter a to distinguish NS and R sectors. Integer modings ($a = 0$) correspond to the R sector and half-integer modings ($a = \frac{1}{2}$) to the NS sector. From the reality of T_F we get the hermiticity condition

$$G_r^\dagger = G_{-r}. \tag{12.11}$$

Notice that $T_F(z)$ is single-valued on the complex plane in the NS sector and double-valued in the R sector. In general, the fermionic (anticommuting) components of NS superfields are single-valued on the plane whereas the fermionic components of R superfields are double-valued. That is $\phi_f^{NS}(e^{2\pi i}z) = +\phi_f^{NS}(z)$ and $\phi_f^{R}(e^{2\pi i}z) = -\phi_f^{R}(z)$. This is the reversed situation we had on the cylinder. The reason is that when we map a field of dimension h from the cylinder to the complex plane we have the Jacobian factor $(1/z)^h$ which changes the analyticity properties of world-sheet fermions which have half-integer conformal weights. This discussion will especially apply to the world-sheet fermions $\psi(z)$. Using the contour deformation trick of Chapter 4 we easily show that the operator algebra eq.(12.8) is equivalent to

$$[L_m, L_n] = (m - n)L_{m+n} + \frac{\hat{c}}{8}(m^3 - m)\delta_{m+n}$$

$$[L_m, G_r] = (\tfrac{1}{2}m - r)G_{m+r} \qquad (12.12)$$

$$\{G_r, G_s\} = 2L_{r+s} + \frac{\hat{c}}{2}[r^2 - \tfrac{1}{4}]\delta_{r+s}.$$

Note that since T_F is an anti-commuting field the operator product $T_F(z)T_F(w)$ leads to an anti-commutator of the modes. Only for $a = \frac{1}{2}$ (i.e. in the NS sector) exists a finite dimensional subalgebra, generated by $L_0, L_{\pm 1}$ and $G_{\pm\frac{1}{2}}$. This is the super-algebra $\widehat{SL}(2) \simeq OSp(1|2)$. From the Jacobi identity $[\{G_r, G_s\}, L_n] + \{[L_n, G_r], G_s\} - \{[G_s, L_n], G_r\} = 0$ we get that the central charge parameters $\hat{c}$ in eqs.(12.8a) and (12.8c) have to be the same.

We can now also easily work out the commutation relations of the L_n and G_r with the modes of the primary fields. Define

$$\phi_0(z) = \sum z^{-n-h}\phi_{0,n}\,,$$

$$\phi_1(z) = \sum z^{-n-h-1/2}\phi_{1,n}\,. \qquad (12.13)$$

Fields with integer conformal weight always have integer mode numbers and fields with half-integer weights have integer mode numbers if they are in the R sector and half-integer mode numbers if they are in the NS sector. We then get

$$[L_m, \phi_{0,n}] = [m(h - 1) - n]\phi_{0,m+n}\,,$$

$$[L_m, \phi_{1,n}] = [m(h - \tfrac{1}{2}) - n]\phi_{1,m+n}\,,$$

$$[\epsilon G_r, \phi_{0,n}] = \epsilon\phi_{1,r+n}\,, \qquad (12.14)$$

$$[\epsilon G_r, \phi_{1,n}] = \epsilon[(2h - 1)r - n]\phi_{0,r+n}\,,$$

where we have introduced a constant anti-commuting parameter ϵ to make ϵG_m a commuting quantity.

Let us now turn to the Hilbert space of the superconformal field theory. From the commutation relations

$$
\begin{aligned}
[L_0, L_m] &= -m L_m\,, \\
[L_0, G_r] &= -r G_r
\end{aligned}
\tag{12.15}
$$

we conclude that $L_m, G_m, m > 0$ are lowering operators. Ground states are the highest weight states $|h\rangle$ of the superconformal algebra; they are annihilated by all the lowering operators and have conformal weight h:

$$
\begin{aligned}
L_0|h\rangle &= h|h\rangle \\
L_m|h\rangle &= 0 \qquad && \text{for } m > 0 \\
G_r|h\rangle &= 0 \qquad && \text{for } r > 0.
\end{aligned}
\tag{12.16}
$$

We will treat the action of G_0 in the R sector below.

We have already seen in Chapter 4 that unitarity requires $\hat{c} \geq 0$ and $h \geq 0$. This can be refined for the superconformal case by considering $(r > 0)$

$$
\begin{aligned}
\langle h|G_r G_{-r}|h\rangle &= \langle h|\{G_r, G_{-r}\}|h\rangle \\
&= 2\langle h|L_0|h\rangle + \frac{1}{2}\hat{c}(r^2 - \tfrac{1}{4})\langle h|h\rangle \\
&= \left\{2h + \frac{\hat{c}}{2}(r^2 - \tfrac{1}{4})\right\}\langle h|h\rangle \geq 0
\end{aligned}
\tag{12.17}
$$

from which we find that

$$
\begin{aligned}
h &\geq 0 \qquad && \text{(NS)}, & (12.18a) \\
h &\geq \frac{\hat{c}}{16} \qquad && \text{(R)}. & (12.18b)
\end{aligned}
$$

For $\hat{c} \geq 1$ these are indeed the only restrictions imposed by unitarity (cf. $c \geq 1$ in the conformal case). For $\hat{c} < 1$ one gets again only a discrete set of allowed $\hat{c}$ and h values, namely [6]

$$\hat{c} = 1 - \frac{8}{m(m+2)}$$

$$h_{p,q} = \frac{[(m+2)p - mq]^2 - 4}{8m(m+2)} + \frac{1}{32}(1 - (-1)^{p-q}) \qquad (12.19)$$

$$m = 2, 3, \ldots, \qquad 1 \le p < m, \ 1 \le q < m+2$$

where $p - q \in 2\mathbf{Z}$ in the NS sector and $p - q \in 2\mathbf{Z} + 1$ in the R sector.

Let us now look at the two sectors separately starting with the NS sector. In analogy with the conformal case we define the $\widehat{SL}(2, \mathbf{C})$ invariant in- and out-vacua $|0\rangle$ and $\langle 0|$ to be the states annihilated by the generators of $\widehat{SL}(2, \mathbf{C})$. Clearly this vacuum is in the NS sector as it satisfies $L_0|0\rangle = 0$. Regularity of T and T_F at the origin and infinity requires

$$
\begin{array}{llll}
L_n|0\rangle = 0 & n \ge -1, & \langle 0|L_n = 0 & n \le 1 \\[2mm]
G_r|0\rangle = 0 & r \ge -\dfrac{1}{2}, & \langle 0|G_r = 0 & r \le \dfrac{1}{2}.
\end{array}
\qquad (12.20)
$$

The correspondence between highest weight states and conformal superfields $\Phi(z, \theta)$ of conformal weight h is made as follows:

$$
\begin{aligned}
\phi_0(0)|0\rangle &= |h\rangle \\
\phi_1(0)|0\rangle &= \left(\hat{G}_{-1/2}\phi_0\right)(0)|0\rangle = G_{-1/2}|h\rangle
\end{aligned}
\qquad (12.21)
$$

where the second equation follows from $[\epsilon G_{-1/2}, \phi_0(z)] = \epsilon \phi_1(z)$ and we have defined (cf. Chapter 4)

$$\hat{G}_r \phi_0(z) = 2 \oint \frac{dw}{2\pi i} T_F(w)\phi_0(z)(w - z)^{r+1/2}. \qquad (12.22)$$

It is straightforward to show that $|h\rangle$ satisfies the highest weight conditions eq.(12.16) and for $\phi_1(0)|0\rangle$ we easily find

$$
\begin{array}{ll}
L_0\left(G_{-1/2}|h\rangle\right) = (h + \dfrac{1}{2})\left(G_{-1/2}|h\rangle\right), & L_m\left(G_{-1/2}|h\rangle\right) = 0 \qquad m > 0, \\[3mm]
G_{1/2}\left(G_{-1/2}|h\rangle\right) = 2h|h\rangle, & G_m\left(G_{-1/2}|h\rangle\right) = 0 \qquad m > \dfrac{1}{2}.
\end{array}
$$

$$(12.23)$$

Note that the relation

$$G^2_{-\frac{1}{2}} = L_{-1} \qquad (12.24)$$

is the global supersymmetry algebra on the complex plane. It is the global algebra since $G_{-\frac{1}{2}} = 2 \oint \frac{dz}{2\pi i} T_F(z)$ and $L_{-1} = \oint \frac{dz}{2\pi i} T(z)$ and it is the supersymmetry algebra since L_{-1} is the translation operator on the plane (cf. Chapter 4). Since the vacuum satisfies $G_{-\frac{1}{2}}|0\rangle = 0$, NS supersymmetry is unbroken.

In the R sector we can define a global supersymmetry charge on the cylinder; it is simply G_0 as can be most easily seen by referring to Chapter 7. Then the global supersymmetry algebra on the cylinder is

$$G^2_0 = L_0 - \frac{\hat{c}}{16}. \qquad (12.25)$$

Using the transformation law of the energy-momentum tensor under finite transformations, we find, for the map from the cylinder to the plane (cf. eq.(4.41))

$$(L_0)_{\text{cyl.}} = (L_0)_{\text{plane}} - \frac{\hat{c}}{16} \qquad (12.26)$$

and $(L_0)_{\text{cyl.}}$ is the translation operator on the cylinder. Note that this shift in L_0 in the R sector is the one described in Chapter 8 which was necessary to bring the NS and R super-Virasoro algebras into identical form. This is automatic here as the operator products in eq.(12.8) are the same for the two sectors.

Clearly, R supersymmetry can only be unbroken if there exists a ground state which satisfies $G_0|h\rangle = 0$, implying that $L_0|h\rangle = \frac{\hat{c}}{16}|h\rangle$.[3] This is in agreement with eq.(12.18b) if we take into account eq.(12.26). Since $[G_0, L_0] = 0$, highest weight states come in orthogonal pairs

$$|h^+\rangle \qquad \text{and} \qquad |h^-\rangle \equiv G_0|h^+\rangle. \qquad (12.27)$$

(It is easy to see that if $|h^+\rangle$ is a highest weight state then so is $|h^-\rangle$.) If $|h^+\rangle$ is the R ground state with $h = \frac{\hat{c}}{16}$ then $|h^-\rangle$ is a null state:

---

[3] Note that in the $n = 1$ discrete series R supersymmetry is unbroken for even m only.

$\langle h^- | h^- \rangle = \langle h^+ | G_0^2 | h^+ \rangle = 0$. It decouples since it is perpendicular to all descendants of $|h^+\rangle$. So we can formally set $|h^-\rangle = 0$ in the case of unbroken supersymmetry.

Since the Virasoro algebra is a subalgebra of the R-algebra, above highest weight states must be created from the vacuum by ordinary conformal fields, the so-called spin fields $S^\pm(z)$; i.e.

$$|h^+\rangle = S^+(0)|0\rangle \qquad \text{and} \qquad |h^-\rangle = S^-(0)|0\rangle. \qquad (12.28)$$

This, together with eq.(12.27) implies that $\hat{G}_0 S^+(z) = S^-(z)$ from which we derive

$$T_F(w)S^+(z) \sim \frac{1}{2} \frac{1}{(w-z)^{3/2}} S^-(z) + \text{ less singular.} \qquad (12.29)$$

Likewise $\hat{G}_0 S^-(z) = (h - \frac{\hat{c}}{16})S^+(z)$ leads to

$$T_F(w)S^-(z) \sim \frac{1}{2}(h - \frac{\hat{c}}{16})\frac{1}{(w-z)^{3/2}} S^+(z) + \text{ less singular.} \qquad (12.30)$$

We see that the spin fields introduce branch cuts into the operator algebra. Also, since they transform the NS ground state into the R ground state they interpolate between the two sectors. This means that if we take a NS fermion and carry it around the spin field it feels the branch cut and changes sign; i.e.

$$\phi_f^{\text{NS}}(ze^{2\pi i})S^\pm(0) = -\phi_f^{\text{NS}}(z)S^\pm(0) \qquad (12.31)$$

which means that $\phi_f^{\text{NS}}(z)S^\pm(0)$ has an expansion in half-integer powers of z and $z^{1/2}\phi_f^{\text{NS}}(z)S^\pm(0)$ is single valued on the plane. We define the operators

$$\phi_{f,n}^R S^\pm(z) = \oint \frac{dw}{2\pi i} \phi_f^{\text{NS}}(w)(w-z)^{n+h-1}S^\pm(z), \qquad n + h \in \mathbf{Z} + \frac{1}{2} \qquad (12.32)$$

where h is the conformal weight of $\phi_f^{\text{NS}}(w)$. The $\phi_{f,n}^R$ are the modes of the fermionic components of a R superfield. Thus one should not think of the NS and R superfields as two separate sets of superfields but rather as one superfield whose fermionic component gets modified in the presence of a spin field. The superfields themselves act diagonally on states in the two sectors of the Hilbert space, i.e.

$$\begin{pmatrix} |\text{NS}\rangle' \\ |\text{R}\rangle' \end{pmatrix} = \begin{pmatrix} \phi & 0 \\ 0 & \phi \end{pmatrix} \begin{pmatrix} |\text{NS}\rangle \\ |\text{R}\rangle \end{pmatrix} \tag{12.33}$$

whereas spin fields act off-diagonally

$$\begin{pmatrix} |\text{NS}\rangle' \\ |\text{R}\rangle' \end{pmatrix} = \begin{pmatrix} 0 & S \\ S & 0 \end{pmatrix} \begin{pmatrix} |\text{NS}\rangle \\ |\text{R}\rangle \end{pmatrix}. \tag{12.34}$$

In both sectors the descendants of a given ground state are obtained by acting with the lowering operators of the superconformal algebra. The n-th level of an irreducible representation is spanned by the following vectors with L_0 eigenvalue $h+n$

$$G_{-r_1} G_{-r_2} \ldots L_{-m_1} L_{-m_2} \ldots |h\rangle \tag{12.35}$$

where

$$0 < r_1 < r_2 \ldots \quad \text{and} \quad 0 < m_1 \le m_2 \ldots$$

$$n = \sum r_i + \sum m_i , \qquad r_i \in \mathbf{Z} + \frac{1}{2} \quad (\text{NS}), \qquad r_i \in \mathbf{Z} \quad (\text{R}) .$$

The condition on the fermionic oscillators takes into account that $G_m^2 = L_{2m}$.

Let us now return to string theory. The world-sheet bosons and fermions form two-dimensional superfields of dimension 1/2:

$$\mathcal{D}\mathcal{X}^\mu(z, \theta) = \psi^\mu(z) + \theta \partial X^\mu(z). \tag{12.36}$$

with action[4]

$$S = \frac{1}{4\pi} \int d^2z \left(\partial X^\mu \bar{\partial} X^\mu - \psi^\mu \bar{\partial} \psi^\mu - \bar{\psi}^\mu \partial \bar{\psi}^\mu \right) \tag{12.37}$$

In the NS sector ψ^μ has the mode expansion

---

[4]This is the same action as given in eq.(7.32) after going to Euclidean coordinates. ∂X and ψ are now anti-hermitian.

$$i\psi^\mu(z) = \sum_{n\in\mathbf{Z}+\frac{1}{2}} b^\mu_n z^{-n-\frac{1}{2}} \tag{12.38}$$

with $(b^\mu_n)^\dagger = b^\mu_{-n}$. Their basic operator product is

$$\psi^\mu(z)\psi^\nu(w) \sim -\frac{g^{\mu\nu}}{z-w} + \text{finite} \tag{12.39}$$

which, by now familiar arguments, is equivalent to

$$\{b^\mu_m, b^\nu_n\} = \delta_{m+n}g^{\mu\nu}. \tag{12.40}$$

We can also calculate the propagator. Using $b^\mu_n|0\rangle = 0$ for $n \geq 1/2$, we get

$$\begin{aligned}
\langle\psi^\mu(z)\psi^\nu(w)\rangle_{\text{NS}} &= -\sum_{m,n\in\mathbf{Z}} z^{-n}w^{-m}\langle 0|b^\mu_{n+\frac{1}{2}}b^\nu_{m-\frac{1}{2}}|0\rangle \\
&= -\sum_{n\geq 0}\frac{1}{z}\left(\frac{w}{z}\right)^n g^{\mu\nu} \\
&= -\frac{g^{\mu\nu}}{z-w} \qquad \text{for } |z| > |w|.
\end{aligned} \tag{12.41}$$

The propagator is also easily derived from the free field action eq.(12.37) using standard field theory arguments. We should note that this is the propagator on the Riemann sphere. On higher genus Riemann surfaces the propagator is more complicated due to the global structure of the surfaces. The short distance limit will however be the same as in eq.(12.41). Using the relation

$$\frac{1}{2\pi}\partial_{\bar{z}}\frac{1}{z-w} = \delta^2(z-w) \tag{12.42}$$

we find that the propagator satisfies

$$\partial_{\bar{z}}\langle\psi^\mu(z)\psi^\nu(w)\rangle = -2\pi\delta^2(z-w)g^{\mu\nu}. \tag{12.43}$$

In the R sector the fields $\psi^\mu(z)$ have integer modings (cf. eq.(12.32)) and one can readily show that they also satisfy the anti-commutation relations eq.(12.40). This also follows from the fact that the short distance expansion for NS and R fermions should be the same since the branch cut in the R case cannot be felt locally. Of particular interest is the relation

$$\{b_0^\mu, b_0^\nu\} = g^{\mu\nu} \tag{12.44}$$

which is just the Dirac algebra and allows us to identify the ψ zero modes with Dirac matrices. Since $[L_0, b_0^\mu] = 0$, the b_0^μ act on the R ground state which has to be degenerate. The spin fields that create them are spinors of $SO(9,1)$ with conformal weight $h = \frac{\hat{c}}{16} = \frac{5}{8}$ ($\hat{c} = d = 10$, see below). The fact that spinors of $SO(9,1)$ have conformal weight $\frac{5}{8}$ can be easily seen by going to the Wick-rotated Lorentz group $SO(10)$ and bosonize them (cf. Chapter 11). The R ground states are then labeled by a spinor index of $SO(9,1)$:

$$|\alpha\rangle = S_\alpha(0)|0\rangle \tag{12.45}$$

and the b_0^μ act as

$$b_0^\mu|\alpha\rangle = \frac{1}{\sqrt{2}}(\Gamma^\mu)_\alpha{}^\beta|\beta\rangle \tag{12.46}$$

where Γ^μ are the $SO(9,1)$ Dirac matrices which satisfy $\{\Gamma^\mu, \Gamma^\nu\} = 2g^{\mu\nu}$.

We can now also calculate the propagator of ψ^μ in the R sector. Using $b_n^\mu|\alpha\rangle = 0$ for $n > 0$ and eq.(12.44) we get

$$
\begin{aligned}
\langle\psi^\mu(z)\psi^\nu(w)\rangle_{\mathrm{R}} &= -\sum_{m,n\in\mathbf{Z}} \langle b_n^\mu b_m^\nu\rangle_{\mathrm{R}} z^{-n-\frac{1}{2}} w^{-m-\frac{1}{2}} \\
&= -\frac{1}{\sqrt{zw}}\left\{\sum_{n>0} \langle b_n^\mu b_{-n}^\nu\rangle_{\mathrm{R}} z^{-n} w^n + \langle b_0^\mu b_0^\nu\rangle_{\mathrm{R}}\right\} \\
&= -\frac{1}{\sqrt{zw}}\left\{\sum_{n=1}^\infty \left(\frac{w}{z}\right)^n + \frac{1}{2}\right\} g^{\mu\nu} \\
&= -\frac{1}{2}\frac{1}{z-w}\left(\sqrt{\frac{z}{w}} + \sqrt{\frac{w}{z}}\right) g^{\mu\nu}, \qquad |z| > |w|.
\end{aligned}
\tag{12.47}
$$

Note that this is just the four-point function $\langle 0|S_\alpha(\infty)\psi^\mu(z)\psi^\nu(w)S_\alpha(0)|0\rangle$ (no sum over α). The propagators in the NS and the R sectors have the same short distance behaviour, as they should.

We still have to discuss the question of locality. As we have seen, the spin fields introduce branch cuts into the theory. A suitable string theory however has to have local correlation functions to give well defined S-matrix elements. To arrive at a local string theory one performs a generalized GSO projection. This is a consistent truncation of the spectrum such that the resulting superconformal field theory is local. In fact, not only is the truncation consistent but also required by modular invariance.

The energy-momentum tensor and supercurrent for the fermionic string are

$$T(z) = -\frac{1}{2}\partial X^\mu \partial X_\mu(z) - \frac{1}{2}\partial \psi^\mu \psi_\mu(z)$$
$$T_F(z) = -\frac{1}{2}\psi_\mu \partial X^\mu(z).$$
(12.48)

$T(z)$ follows from eq.(12.2) and (4.82). T_F is computed in a similar way. Under infinitesimal supersymmetry transformations eq.(12.3), the action changes as

$$\delta S = -\frac{1}{2\pi}\int \mathrm{d}^2z\, \bar\partial\epsilon\, T_F.$$
(12.49)

Since X and ψ are free fields, we can use Wick's theorem to evaluate operator products of composite operators such as T and T_F. (We have again dropped normal ordering symbols.) One easily verifies the algebra eq.(12.8) with $\hat c = \frac{2}{3}c = d$ where d is the range of the index μ.

The ghost system of the fermionic string is another example of a $n = 1$ superconformal field theory. It consists of the (anti-commuting) conformal ghosts b, c and the (commuting) superconformal ghosts β, γ. Together they form two conformal superfields $B = \beta + \theta\, b$ and $C = c + \theta\,\gamma$ with conformal weights $h = 3/2$ and $h = -1$ respectively. The action is (cf. Chapters 3 and 8)

$$S = \frac{1}{2\pi}\int \mathrm{d}^2z\,(b\bar\partial c + \beta\bar\partial\gamma).$$
(12.50)

We can generalize the ghost system by considering superfields B and C with conformal weights $\lambda - 1/2$ and $1 - \lambda$ ($\lambda \in \mathbf{Z}$) respectively. The action is still given by eq.(12.50). (We will consider these generalized b, c systems in great detail in Chapter 13.) The energy momentum tensor and supercurrent are calculated as before; one finds

$$T^{(\lambda)} = -\lambda b \partial c + (1 - \lambda)(\partial b)c - (\lambda - \tfrac{1}{2})\beta \partial \gamma + (\tfrac{3}{2} - \lambda)(\partial \beta)\gamma,$$

$$T_F^{(\lambda)} = \tfrac{1}{2}b\gamma + (1 - \lambda)(\partial \beta)c - (\lambda - \tfrac{1}{2})\beta \partial c. \tag{12.51}$$

We now use the operator products

$$c(z)b(w) \sim \gamma(z)\beta(w) \sim \frac{1}{z - w} \tag{12.52}$$

to show that above system generates a $n = 1$ superconformal algebra with central charge $c = -c^\lambda + c^{\lambda - \frac{1}{2}}$. Here $c^\lambda = 12\lambda^2 - 12\lambda + 2$. The first contribution, $-c^\lambda$, is that of the anti-commuting b, c system and the second, $c^{\lambda - \frac{1}{2}}$, that of the commuting β, γ system. For the fermionic string ($\lambda = 2$) we then find the ghost contribution to the anomaly $c^{\text{ghost}} = -26 + 11 = -15$. The matter fields come in chiral multiplets of the $n = 1$ superconformal symmetry. Each of them contributes $c = 3/2$ to the conformal anomaly. This means that we need ten of them for anomaly cancellation. Hence $d = 10$ as the critical dimension of the fermionic string. It is however by no means necessary to represent the $\hat{c} = 10$ algebra in terms of free superfields eq.(12.36). For a d-dimensional space-time interpretation it suffices to represent d of them as free superfields and have a $\hat{c} = 10 - d$ internal $n = 1$ superconformal theory (cf. Chapter 14).

One may now ask whether extended superconformal algebras play a role in string theory. The answer is that indeed they do. Let us begin with the case $n = 2$. The $n = 2$ gravity multiplet is $(g_{\alpha\beta}, \chi_\alpha, A_\alpha)$ where besides the graviton we have a gravitino which is, in contrast to the $n = 1$ case, a complex Dirac fermion, and a $U(1)$ gauge current. Since we have now gauge fields of spin $2, 3/2$ and 1, Faddeev-Popov quantization will give b, c systems with $\lambda = 2, 3/2$ and $\lambda = 1$ where the $\lambda = 3/2$ system is doubled, since a Dirac spinor is equivalent to two Majorana spinors. Their combined contribution to the conformal anomaly is $c^{\text{ghost}} = -26 + 2 \times 11 - 2 =$

-6. This has to be cancelled by the matter fields. The $n = 2$ matter multiplets are (X, ψ) where X is a complex scalar and ψ a spin 1/2 Dirac-Weyl fermion, which is equivalent to two real scalars and two Majorana-Weyl fermions, thus contributing $c = 3$ to the conformal anomaly. We then conclude that the critical dimension of the theory with local $n = 2$ superconformal invariance is $d = 2$. The situation becomes even worse if we consider the $n = 4$ case[5]. The $n = 4$ gravity multiplet is $(g_{\alpha\beta}, \chi^i_\alpha, A^i_{\alpha j})$. i is a $SU(2)$ index. The four gravitini form a complex doublet of $SU(2)$ and the $SU(2)$ currents transform in the adjoint. The ghost contribution to the anomaly is therefore $c^{\text{ghost}} = -26 + 4 \times 11 - 3 \times 2 = 12$. $n = 4$ matter multiplets $(4X, \psi)$ contain four real scalars and a complex doublet of spinors which gives a $c = 6$ contribution to the conformal anomaly. This leads to a critical dimension $d = -2$. This is clearly unacceptable for our purposes.

Nevertheless, extended superconformal algebras do play an important role in string theory, however only as global symmetries. The maximal gauged symmetries is $n = 1$. There does exist a deep connection between space-time symmetries and world-sheet symmetries. In particular, $N = 1$ space-time supersymmetry in four dimensions requires (global) $n = 2$ superconformal invariance on the world-sheet. We will show this in Chapter 14. Here we will only present a brief discussion of the extended $n = 2$ superconformal algebra.

The local symmetries for $n = 2$ are reparametrization invariance, supersymmetry and $U(1)$ invariance; they are generated by the energy-momentum tensor $T(z)$, the supercharge $T_F(z)$ and a $U(1)$ current $J(z)$. (Again, we only discuss the holomorphic sector of the theory.) The supercharge splits into two parts with $U(1)$ charges ± 1 respectively: $T_F =$

---

[5]The case $n = 3$ does not seem to be of interest for string theory.

$T_F^+ + T_F^-$. These generators satisfy the following operator algebra:

$$T(z)T(w) \sim \frac{\frac{1}{2}c}{(z-w)^4} + \frac{2T(w)}{(z-w)^2} + \frac{\partial T(w)}{z-w} + \dots$$

$$T(z)T_F^{\pm}(w) \sim \frac{\frac{3}{2}T_F^{\pm}(w)}{(z-w)^2} + \frac{\partial T_F^{\pm}}{z-w} + \dots$$

$$T(z)J(w) \sim \frac{J(w)}{(z-w)^2} + \frac{\partial J(w)}{z-w} + \dots$$

$$J(z)J(w) \sim \frac{\frac{1}{3}c}{(z-w)^2} + \dots \tag{12.53}$$

$$J(z)T_F^{\pm}(w) \sim \pm\frac{T_F^{\pm}(w)}{z-w} + \dots$$

$$T_F^+(z)T_F^-(w) \sim \frac{\frac{1}{12}c}{(z-w)^3} + \frac{\frac{1}{4}J(w)}{(z-w)^2} + \frac{\frac{1}{4}T(w) + \frac{1}{8}\partial J(w)}{z-w} + \dots$$

$$T_F^{\pm}(z)T_F^{\pm}(w) \sim \text{finite}.$$

It is easy to see that T and $T_F = T_F^+ + T_F^-$ form a $n=1$ superconformal algebra. Above operator algebra can be converted to (anti)commutators of the modes of the generators defined by

$$T(z) = \sum_{n\in\mathbf{Z}} z^{-n-2}L_n \qquad\qquad L_n = \oint \frac{dz}{2\pi i} z^{n+1} T(z)$$

$$T_F^{\pm}(z) = \sum_{n\in\mathbf{Z}} z^{-n-3/2\mp a}G_{n\pm a}^{\pm} \quad\Longleftrightarrow\quad G_{n\pm a}^{\pm} = \oint \frac{dz}{2\pi i} z^{n+\frac{1}{2}\pm a} T_F^{\pm}(z)$$

$$J(z) = \sum_{n\in\mathbf{Z}} z^{-n-1}J_n \qquad\qquad J_n = \oint \frac{dz}{2\pi i} z^n J(z)$$

$$\tag{12.54}$$

with hermiticity conditions

$$L_n^{\dagger} = L_{-n} \quad, \qquad G_{n+a}^{+\dagger} = G_{-n-a}^{-} \quad, \qquad J_n^{\dagger} = J_{-n}. \tag{12.55}$$

The operator algebra is then equivalent to

$$[L_n, L_m] = (n - m)L_{n+m} + \frac{c}{12}n(n^2 - 1)\delta_{n+m}\,,$$

$$[L_n, G^{\pm}_{m+a}] = (\frac{n}{2} - m \mp a)G^{\pm}_{n+m\pm a}\,,$$

$$[L_n, J_m] = -mJ_{n+m}\,,$$

$$[J_n, J_m] = \frac{c}{3}n\delta_{m+n}\,,$$

$$[J_n, G^{\pm}_{m\pm a}] = \pm G^{\pm}_{n+m\pm a}\,, \tag{12.56}$$

$$\{G^{+}_{n+a}, G^{-}_{m-a}\} = \frac{1}{4}L_{m+n} + \frac{1}{8}(n - m + 2a)J_{m+n} + \frac{c}{24}[(n + a)^2 - \tfrac{1}{4}]\delta_{n+m}\,,$$

$$\{G^{+}_{n+a}, G^{+}_{m+a}\} = \{G^{-}_{n-a}, G^{-}_{m-a}\} = 0.$$

The relation between the central charges in the $T(z)T(w)$, $T^{+}_F(z)T^{-}_F(w)$ and $J(z)J(w)$ operator products are again fixed by Jacobi identities.

Note that we have introduced the real parameter a. From the mode expansion of $T^{\pm}_F$ we find

$$T^{\pm}_F(e^{2\pi i}z) = -e^{\mp 2\pi i a}T^{\pm}_F(z)\,, \tag{12.57}$$

i.e. a labels the boundary conditions of the fermionic operators $T^{\pm}_F$. For $a \in \mathbf{Z}$ we are in the R sector whereas $a \in \mathbf{Z} + \frac{1}{2}$ corresponds to the NS sector. The algebras for a and $a + 1$ are clearly isomorphic and we can restrict a to $a \in [0, 1)$. Note that only for $a = 0$ and $a = \frac{1}{2}$ can we form a real $T_F = T^{+}_F + T^{-}_F$ with definite boundary conditions. In the $n = 1$ case we had only two sectors since there the supercurrent was real which allows only for periodic or anti-periodic boundary conditions. In the $n = 2$ case we can interpolate between the two sectors by varying a from 0 to $\frac{1}{2}$.

For $a = \frac{1}{2}$, i.e. in the NS sector, there exists again a finite dimensional subalgebra generated by $L_{0,\pm 1}$, $G^{\pm}_{+1/2}$, $G^{\pm}_{-1/2}$ and J_0. This is the algebra $OSp(2, 2)$.

For different values of a the corresponding representation spaces are obviously different. From an algebraic point of view the algebras are however equivalent for all values of a. To see this let us write the $\{G^+, G^-\}$ anti-commutator in the following form

$$\{G^-_{n+a+\eta}, G^-_{m-a-\eta}\} = \frac{1}{4}\left[L_{n+m} + \eta J_{n+m} + \frac{c}{6}\eta^2 \delta_{n+m}\right]$$

$$+ \frac{1}{8}(n - m + 2a)\left[J_{n+m} + \frac{1}{3}\eta\delta_{n+m}\right] + \frac{c}{24}\left[(n+a)^2 - \frac{1}{4}\right]\delta_{n+m}.$$
$$(12.58)$$

One now verifies that the generators

$$L^\eta_n = L_n + \eta J_n + \frac{c}{6}\eta^2\delta_n \,,$$

$$G^{\eta\pm}_{n\pm a} = G^\pm_{n\pm(a+\eta)} \,,$$
$$(12.59)$$

$$J^\eta_n = J_n + \frac{1}{3}\eta\delta_n$$

satisfy the $n = 2$ superconformal algebra. Therefore the algebras are equivalent for all a. Under the change of a the conformal weights h and $U(1)$ charges q of any state change by

$$h \to h^\eta = h + \eta q + \frac{c}{6}\eta^2 \,,$$

$$q \to q^\eta = q + \frac{c}{3}\eta \,.$$
$$(12.60)$$

This is called the spectral flow [7]. For instance, the $OSp(2,2)$ invariant vacuum $|0\rangle$ with $(h, q) = (0, 0)$ is always in the NS sector. Now the spectral flow takes it to a state $(\eta = 1/2)$ with $(h, q) = (\frac{c}{24}, \frac{c}{6})$.

As usual, we now define highest weight states by the conditions

$$G^\pm_r|\phi\rangle = 0 \qquad\qquad r > 0 \,,$$

$$L_n|\phi\rangle = J_n|\phi\rangle = 0 \qquad\qquad n > 0 \,, \qquad (12.61)$$

$$L_0|\phi\rangle = h|\phi\rangle \,, \qquad J_0|\phi\rangle = q|\phi\rangle \,.$$

Here q is the $U(1)$ charge of the state. Actually, the conditions for L_n and J_n $(n > 0)$ already follow from $G^\pm_r|\phi\rangle = 0$ and the use of the algebra. Let us consider the constraints from unitarity. In the NS sector we have

$$
\begin{aligned}
0 \;\le\; &\langle\phi|G^{\pm}_{1/2}G^{\mp}_{-1/2}|\phi\rangle \\[4pt]
=\; &\langle\phi|\{G^{\pm}_{+1/2},G^{\mp}_{-1/2}\}|\phi\rangle \\[4pt]
=\; &\frac{1}{4}\langle\phi|L_0\pm\tfrac{1}{2}J_0|\phi\rangle \\[4pt]
=\; &\frac{1}{4}(h\pm\tfrac{1}{2}q)\langle\phi|\phi\rangle ,
\end{aligned}
\tag{12.62}
$$

i.e. $h\ge\frac{1}{2}|q|$ in the NS sector. When the bound is satisfied, there are two states which satisfy

$$
G^{\pm}_{-1/2}|h\mp\tfrac{1}{2}q\rangle = 0.
\tag{12.63}
$$

To each highest weight state there corresponds a primary field $\phi^{(h,q)}$ of the $n=2$ algebra. For the fields with $h=\frac{1}{2}|q|$, eq.(12.63) is equivalent to

$$
T^{\pm}_F(z)\phi^{(\mp\frac{1}{2}q,q)}(w)\sim\text{finite}.
\tag{12.64}
$$

These fields are called chiral (anti-chiral) primary fields. For a (anti-)chiral primary field the action of only one of the two supercharges results in a new field. It is not hard to show that any state with $h=\frac{1}{2}|q|$ is (anti-) chiral primary and satisfies $h\le\frac{c}{6}$.

In the R sector we find, as in the $n=1$ case, that $h\ge\frac{c}{24}$ and states with $h=\frac{c}{24}$ satisfy $G^{\pm}_0|\phi\rangle=0$. Also, states that follow from chiral primary fields through the spectral flow, satisfy $G^{\pm}_{-1}|\phi\rangle=0$.

Since the spectral flow connects the NS with the R sector, and states in these sectors of the fermionic string are space-time bosons and fermions respectively, we have here an indication of the connection between $n=2$ world-sheet and $N=1$ space-time supersymmetry. We will elaborate on this in more detail in Chapter 14.

To close this discussion of the $n=2$ algebra, let us give two simple realizations. The first one consists of two free real $n=1$ superfields $\mathcal{X}^{1,2}(z,\theta)=X^{1,2}(z)+\theta\psi^{1,2}(z)$. Let us define the complex fields

$$
X^{\pm}=\frac{1}{\sqrt{2}}(X^1\pm iX^2)\quad\text{and}\quad \psi^{\pm}=\frac{1}{\sqrt{2}}(\psi^1\pm i\psi^2).
\tag{12.65}
$$

The energy momentum tensor and supercurrent are then

$$T(z) = -\partial X^+ \partial X^-(z) - \frac{1}{2}\left(\partial \psi^+ \psi^-(z) + \partial \psi^- \psi^+(z)\right),$$
$$T_F(z) = T_F^+(z) + T_F^-(z)$$

(12.66)

where

$$T_F^+ = -\frac{1}{2}\psi^+ \partial X^-(z) \qquad \text{and} \qquad T_F^- = -\frac{1}{2}\psi^- \partial X^+(z).$$

(12.67)

We can now define a $U(1)$ current

$$J(z) = \psi^- \psi^+(z).$$

(12.68)

It is then straightforward to verify that T, $T_F^{\pm}$ and J generate a $c = 3$, $n = 2$ superconformal algebra. This example is easily generalized to the case of $2k$ free superfields from which we can build k complex ones. This leads to a $c = 3k$, $n = 2$ algebra. In the string context this example arises in $2k$-dimensional torus compactifications. The second example consists of only one real free boson:

$$T(z) = -\frac{1}{2}\partial\phi\partial\phi(z),$$

$$T_F^{\pm}(z) = \frac{1}{2\sqrt{3}}e^{\pm i\sqrt{3}\phi(z)},$$

(12.69)

$$J(z) = \frac{i}{\sqrt{3}}\partial\phi(z),$$

which generate a $c = 1$, $n = 2$ superconformal algebra.

In a similar way one can discuss the $n = 4$ algebra which is relevant for $N = 2$ space-time supersymmetric string theories. We will however not do it here and refer instead to the literature.

REFERENCES

[1] P. Ramond, *"Dual theory for free fermions"*, Phys. Rev. **D3** (1971) 2415.

[2] A. Neveu and J.H. Schwarz, *"Factorizable dual model of pions"*, Nucl. Phys. **B31** (1971) 86.

[3] M. Ademollo et al, *"Dual string with $U(1)$ colour symmetry"*, Nucl. Phys. **B111** (1976) 77; *"Dual string models with non-Abelian colour and flavour symmetries"*, Nucl. Phys. **B114** (1976) 297; *"Supersymmetric strings and color confinement"*, Phys. Lett. **62B** (1976) 105.

[4] S. Shenker, *"Introduction to two dimensional conformal and superconformal field theory"*, in proceedings of Workshop on 'Unified String Theories', Santa Barbara 1985, eds. M. Green and D. Gross, World Scientific.

[5] D. Friedan, *"Notes on string theory and two dimensional conformal field theory"*, in proceedings of Workshop on 'Unified String Theories', Santa Barbara 1985, eds. M. Green and D. Gross, World Scientific.

[6] D. Friedan, Z. Qiu and S. Shenker, *" Superconformal invariance in two dimensions and the tricritical Ising model"*, Phys. Lett. **151B** (1985) 37.

[7] A. Schwimmer and N. Seiberg, *"Comment on the $N = 2$, $N = 3$, $N = 4$ superconformal algebras in two dimensions"*, Phys. Lett. **184B** (1987) 191.

Chapter 13

Bosonization of the Fermionic String – Covariant Lattices

We will now reexamine the 10-dimensional fermionic (spinning) string using the bosonic language. The aim of this bosonic formulation is the construction of the covariant fermion emission vertex operators, as they were discovered by Friedan, Martinec and Shenker [1] and by Knizhnik [2]. This will in turn lead to the introduction of the so-called covariant lattices [3–6].

Recall the action of the fermionic string in superconformal gauge:

$$S = \frac{1}{4\pi} \int d^2z \left(\partial X^\mu \bar\partial X_\mu - \psi^\mu \bar\partial \psi_\mu - \bar\psi^\mu \partial \bar\psi_\mu \right) \tag{13.1}$$

where we have only written the matter (X^μ, ψ^μ) part. We will turn to the ghost part below. These fields generate a superconformal field theory with $\hat c = 10$ ($c = 15$) where from now on we discuss only the right-moving, holomorphic part of the theory. The two-dimensional supercharge, also called supercurrent, is given by:

$$T_F(z) = -\frac{1}{2} \partial X_\mu(z) \psi^\mu(z). \tag{13.2}$$

Now, applying the techniques of bosonization as introduced in the previous chapter, we replace the ten real fermions $\psi^\mu(z)$ ($\mu = 1, \ldots, 10$) by five chiral bosons $\phi^i(z)$ ($i = 1, \ldots, 5$) with momentum eigenvalues being lattice vectors of the D_5 weight lattice. The bosonization is performed by converting the ten real fermions $\psi^\mu(z)$ to the complex Cartan-Weyl basis:

$$\Psi^{\pm i}(z) = \frac{1}{\sqrt{2}} (\psi^{2i-1} \pm i\psi^{2i})(z) \qquad i = 1, \ldots, 5. \tag{13.3}$$

The action for the complex fermions is

$$S = -\frac{1}{4\pi} \int \mathrm{d}^2 z (\Psi^{+i} \bar{\partial} \Psi^{-i} + \Psi^{-i} \bar{\partial} \Psi^{+i}). \tag{13.4}$$

The part of the generators of the Wick rotated Lorentz group $SO(10)$ which are built from the world-sheet fermions are bosonized according to

$$
\begin{aligned}
J^{i,-i}(z) &=: \Psi^i \Psi^{-i}(z) := i\partial\phi^i(z) \\
J^{\pm i, \pm j}(z) &=: \Psi^{\pm i} \Psi^{\pm j} :=: e^{\pm i\phi^i \pm i\phi^j}(z) : \quad (i < j)
\end{aligned}
\tag{13.5}
$$

where the complex fermions themselves are expressed as

$$\Psi^{\pm i}(z) =: e^{\pm i\phi^i(z)} : . \tag{13.6}$$

The states of the spinning string theory are created by vertex operators which contain the five bosons $\phi^i(z)$. Let us concentrate on expressions built by exponentials of these bosons; possible derivative terms $\partial\phi^i(z)$ play only a trivial role in the following. Also, we will not discuss the X^μ-dependent part of the vertex operators.

In the NS sector the states are space-time bosons. The ground state is the NS vacuum $|0\rangle$ which is a tachyon as discussed in Chapter 8. The first excited state is the massless ten-dimensional vector $|\psi^\mu\rangle = b^\mu_{-1/2}|0\rangle$ with the corresponding vertex operator $\psi^\mu(z)$. Thus, the vector vertex operator in the bosonized version of the theory is simply given by eq.(13.6). In general, states in the NS sector are described by vertex operators[1]

$$V_\lambda(z) =: e^{i\lambda\cdot\phi(z)} : \tag{13.7}$$

where λ are D_5 lattice vectors in the 0 or V conjugacy class. E.g. for the tachyon $\lambda = 0$ and for the $SO(10)$ vector $\lambda = (0,\ldots,\pm 1,\ldots,0)$.

On the other hand, states in the R sector of the theory have fermionic statistics. They are created again by vertex operators of the form eq.(13.7) but now with λ being a lattice vector of the S or C conjugacy classes

---

[1] Here and in the following we will drop cocycle factors. They are however necessary to produce manifestly covariant results, as e.g. in eq.(13.8) below.

of D_5. As ground states there are the two massless spinors of opposite chirality denoted by $|S^\alpha\rangle$ and $|S^{\dot\alpha}\rangle$ with corresponding D_5 weights $\lambda = (\pm\frac{1}{2}, \pm\frac{1}{2}, \pm\frac{1}{2}, \pm\frac{1}{2}, \pm\frac{1}{2})$ with an even and odd number of minus signs respectively.

The vertex operator of the massless vector has conformal dimension $h = \frac{\lambda^2}{2} = \frac{1}{2}$. Remember on the other hand that vertex operators of physical massless states must have $h = 1$. Similarly, the massless spinors belong to vertex operators with conformal dimension $\lambda^2 = \frac{5}{8}$ in contradiction to the physical state condition. (Note that for massless states $e^{ik\cdot X}$ has vanishing conformal weight.) This discrepancy clearly indicates that the vertex operators of eq.(13.7) with $\lambda \in 0, V, S, C$ of D_5 are not the full vertex operators of the fermionic string theory and have to be complemented by an additional piece.

The incorrect conformal dimension is however not the only serious drawback. Consider the operator algebra between the vector ψ^μ and the spinor fields S^α, $S^{\dot\alpha}$ in a $SO(10)$ covariant basis:

$$\psi^\mu(z)S^\alpha(w) = \frac{1}{\sqrt{2}} \frac{(\Gamma^\mu)^\alpha{}_{\dot\beta}}{(z-w)^{1/2}} S^{\dot\beta}(w) + \dots$$

$$S^\alpha(z)S^\beta(w) = \frac{1}{\sqrt{2}} \frac{(\Gamma_\mu)^{\alpha\beta}}{(z-w)^{3/4}} \psi^\mu(w) + \dots \tag{13.8}$$

$$S^\alpha(z)S^{\dot\beta}(w) = \frac{C^{\alpha\dot\beta}}{(z-w)^{5/4}} + \frac{1}{2} \frac{(\Gamma_\mu\Gamma_\nu)^{\alpha\dot\beta}}{(z-w)^{1/4}} \psi^\mu \psi^\nu(w) + \dots.$$

Here $C^{\alpha\dot\beta}$ is the $SO(10)$ charge conjugation matrix and $(\Gamma_\mu)^\alpha{}_{\dot\beta}$ are ten-dimensional Dirac matrices. There are several ways to derive these operator products. One uses bosonization and an explicit form for the cocycle factors. Another is to use $SO(10)$ invariance.

These equations show that $S^\alpha(z)$ $(S^{\dot\alpha}(z))$ creates a branch cut which renders the theory non-local. (Locality of the operator product algebra is however necessary e.g. to get well-defined scattering amplitudes.) This is due to the D_5 inner-product rule $(V) \cdot (S) = \frac{1}{2} + n$, $n \in \mathbf{Z}$ (cf. table 11.2). Furthermore, S^α does not anticommute with itself since $(S) \cdot (S) = 5/4 + n$. Also note the branch cut in the operator product between S^α and $S^{\dot\beta}$ due to $(S) \cdot (C) = 3/4 + n$.

In summary, without further modifications the fermionic string theory is non-local and therefore ill-defined. This is simply a reflection of the properties of the D_5 weight lattice. Therefore one can expect that the complete vertex operators of the fermionic string requires a modification of the D_5 lattice. The missing piece will be provided by the superconformal ghost system. Before demonstrating this, let us first present a general discussion of first order systems such as the b, c and β, γ ghost systems. Their bosonization will be of particular importance [1, 7].

13.1 First order systems

We introduce a common notation and consider b and c being conjugate fields with first order action

$$S = \frac{1}{2\pi} \int \mathrm{d}^2 z \, b\bar\partial c. \tag{13.9}$$

The field b has conformal weight λ and c has weight $1 - \lambda$. The action is then conformally invariant. The statistics of b and c is parametrized by ϵ: $\epsilon = 1$ for Fermi statistics and $\epsilon = -1$ for Bose statistics. The equations of motion and their solutions are

$$\begin{aligned}
\bar\partial b &= 0, &&\text{i.e. } b = b(z), \\
\bar\partial c &= 0, &&\text{i.e. } c = c(z).
\end{aligned} \tag{13.10}$$

The propagator is

$$\langle c(z)b(w)\rangle = \frac{1}{z-w} \tag{13.11}$$

and the basic operator products are

$$\begin{aligned}
c(z)b(w) &= \frac{1}{z-w} + \dots \\
b(z)c(w) &= \frac{\epsilon}{z-w} + \dots
\end{aligned} \tag{13.12}$$

The bb and cc products are non-singular. We have the mode expansions

$$\begin{aligned}
b(z) &= \sum_{n\in\delta-\lambda+\mathbf{Z}} z^{-n-\lambda}b_n \\
c(z) &= \sum_{n\in\delta+\lambda+\mathbf{Z}} z^{-n-(1-\lambda)}c_n
\end{aligned} \tag{13.13}$$

with the following hermiticity conditions

$$b_n^\dagger = \epsilon b_{-n} \ , \quad c_n^\dagger = c_{-n} \tag{13.14}$$

and (anti) commutator

$$[c_m, b_n]_\epsilon = \delta_{m+n}. \tag{13.15}$$

For the case of half-integer λ, there are two sectors, the R sector specified by $\delta = \frac{1}{2}$ and the NS sector with $\delta = 0$.[2] The action of the modes on the SL_2 invariant vacuum is

$$\begin{aligned}
b_n|0\rangle &= 0 \qquad \text{for } n \geq 1 - \lambda \\
c_n|0\rangle &= 0 \qquad \text{for } n \geq \lambda
\end{aligned} \tag{13.16}$$

The energy-momentum tensor is

$$\begin{aligned}
T &= -\lambda b\partial c + (1-\lambda)\partial bc = \\
&= \frac{1}{2}(\partial bc + b\partial c) + \frac{1}{2}\epsilon Q\partial(bc)
\end{aligned} \tag{13.17}$$

where we have introduced

$$Q = \epsilon(1 - 2\lambda). \tag{13.18}$$

---

[2] We can also have $\delta = \frac{1}{2}$ for integer λ corresponding to a twisted sector.

The significance of Q as a background charge will become clear below. The Virasoro operators are

$$L_n = \sum_m \left(m - (1-\lambda)n\right) b_{n-m} c_m.$$ (13.19)

The central charge is easily found by computing the operator product between $T(z)$ and $T(w)$:

$$c = -\epsilon(12\lambda^2 - 12\lambda + 2) = \epsilon(1 - 3Q^2)$$ (13.20)

and it is straightforward to verify that

$$T(z)b(w) = \frac{\lambda b(w)}{(z-w)^2} + \frac{\partial b(w)}{z-w} + \text{finite},$$ (13.21)

and

$$T(z)c(w) = \frac{(1-\lambda)c(w)}{(z-w)^2} + \frac{\partial c(w)}{z-w} + \text{finite}.$$ (13.22)

as expected. In table 13.1 we collect the values of the various parameters for the conformal and super-conformal ghost systems and complex NSR fermions.

Table 13.1 Familiar first order systems

	ϵ	λ	Q	c
b, c	1	2	-3	-26
β, γ	-1	$\frac{3}{2}$	2	11
$\psi^{\pm i}$	1	$\frac{1}{2}$	0	1

The action eq.(13.9) is invariant under a chiral $U(1)$ with current

$$j(z) = -b(z)c(z) = \sum_n z^{-n-1} j_n$$ (13.23)

where

$$j_n = \sum_m \epsilon c_{n-m} b_m.$$ (13.24)

The operator products of b and c with j reflect the fact that they have $U(1)$ charges -1 and $+1$ respectively:

$$j(z)b(w) = \frac{-1}{z-w}\, b(w) + \ldots$$
$$j(z)c(w) = \frac{1}{z-w}\, c(w) + \ldots \qquad (13.25)$$

The operator product of the chiral current and the energy-momentum tensor is anomalous:

$$T(z)j(w) = \frac{Q}{(z-w)^3} + \frac{j(w)}{(z-w)^2} + \frac{\partial j(w)}{z-w} + \ldots \qquad (13.26)$$

and only for $Q = 0$, $(\lambda = \frac{1}{2})$ is j a true conformal field with $h = 1$. This implies that the $U(1)$ current itself is not conserved. One can show that the anomalous conservation law is $\bar{\partial}j(z) = -\frac{1}{2}Q\bar{\partial}\partial\sigma$ where σ is the Weyl degree of freedom of the metric $\mathrm{d}s^2 = 2e^\sigma \mathrm{d}z\mathrm{d}\bar{z}$. With $R = -2e^{-\sigma}\bar{\partial}\partial\sigma$ we find

$$\bar{\partial}j(z) = \frac{1}{4}Q\sqrt{h}R, \qquad (13.27)$$

where h is the determinant of the two-dimensional metric and R is the corresponding curvature scalar. The anomaly in the $U(1)$ current is related to the existence of b, c zero modes, their number being determined by the Riemann-Roch theorem:

$$N_c - N_b = \epsilon Q(g-1) = (1 - 2\lambda)(g-1) \qquad (13.28)$$

where g is the genus of the Riemann surface. The right hand side is the amount by which the ghost charge is not conserved. This can be seen by integrating eq.(13.27). Note that the situation is very similar to the $U(1)$ anomaly in gauge theory. There the anomalous divergence of the chiral $U(1)$ current is also given by a topological quantity which measures the difference between the number of massless left- and right-handed fermions, i.e. the zero modes of the chiral Dirac operator.

The operator product eq.(13.26) is equivalent to the anomalous commutators

$$[L_n, j(z)] = \frac{1}{2}Qn(n+1)z^{n-1} + (n+1)z^n j(z) + z^{n+1}\partial j(z) \quad (13.29)$$

and

$$[L_n, j_m] = \frac{1}{2}Qn(n+1)\delta_{n+m} - m j_{m+n}. \quad (13.30)$$

From eq.(13.29) it is appareant that $j(z)$ transforms covariantly under translations (L_{-1}) and dilatations (L_0) but not under the transformations generated by L_{+1}; i.e. $j(z)$ is not quasi-primary. It is easy to show using the hermiticity conditions eq.(13.14) that $j_n^\dagger = -j_{-n}$, $\forall n \neq 0$. The case $n = 0$ is delicate because of normal ordering ambiguities. We can use eq.(13.30) to find $j_0^\dagger$:

$$j_0^\dagger = -[L_{-1}, j_1]^\dagger = -[L_1, j_{-1}] = -j_0 - Q. \quad (13.31)$$

Then, if O_p is an operator with $U(1)$ charge p, i.e. $[j_0, O_p] = pO_p$ and $|q\rangle$ a state with $U(1)$ charge q, we find $p\langle q'|O_p|q\rangle = \langle q'|[j_0, O_p]|q\rangle = -(q' + q + Q)\langle q'|O_p|q\rangle$; i.e. we have to insert an operator with $U(1)$ charge $p = -(q + q' + Q)$ in order to get a non-vanishing result. We then normalize the states such that

$$\langle -q - Q|q\rangle = 1 \quad (13.32)$$

for the non-vanishing inner products.

The b, c system is bosonized by defining a chiral scalar field $\phi(z) = \epsilon \int^z j(z')\mathrm{d}z'$, or

$$j(z) = \epsilon\partial\phi(z) \quad (13.33)$$

with

$$\phi(z)\phi(w) \sim \epsilon \ln(z - w). \quad (13.34)$$

In terms of ϕ the action is

$$S = -\frac{1}{4\pi}\int \mathrm{d}^2 z\left(\epsilon\partial\phi\bar{\partial}\phi + \frac{Q}{2}\sqrt{h}R\phi\right) \quad (13.35)$$

from which the anomalous current conservation law eq.(13.27) follows as the equation of motion for ϕ. The energy momentum tensor derived from above action is

$$T^{(j)} = \epsilon\Big(\frac{1}{2}j^2 - \frac{1}{2}Q\partial j\Big). \tag{13.36}$$

As a check we can reproduce eq.(13.26).

Conformal fields $V(z)$ are given by exponentials of $\phi(z)$:

$$V(z) =: e^{q\phi(z)}: \tag{13.37}$$

where q is (half–) integer for the NS (R) sector. This will become clear below. (We again suppress possible derivative terms and will drop normal ordering symbols from now on.) Therefore, the allowed bosonic momenta are points in a D_1 lattice where integer lattice point belong to the 0 or V conjugacy classes and half-integer elements to the S and C conjugacy classes of D_1. The $U(1)$ charge of $e^{q\phi(z)}$ is easily determined

$$j(z)e^{q\phi(w)} = \frac{q}{z-w}e^{q\phi(w)} + \dots , \tag{13.38}$$

and its conformal dimension follows directly from

$$T(z)e^{q\phi(w)} = \Big[\frac{\frac{1}{2}\epsilon q(q+Q)}{(z-w)^2} + \frac{\partial}{z-w}\Big]e^{q\phi(w)} + \dots \tag{13.39}$$

The contribution to the conformal weight which is linear in q has its origin in the ghost number anomaly. The operator $e^{q\phi(z)}$ shifts the ghost charge of the vacuum by q units. It is the vertex operator for a state $|q\rangle$:

$$|q\rangle = \oint_{C_0} \frac{dz}{2\pi i} \frac{1}{z} e^{q\phi(z)}|0\rangle = e^{q\phi(0)}|0\rangle \tag{13.40}$$

which satisfies $j_0|q\rangle = q|q\rangle$. E.g according to eq.(13.32) one obtains

$$\langle 0|e^{-Q\phi(z)}|0\rangle = 1. \tag{13.41}$$

So far we have bosonized only the $U(1)$ current $j(z)$ which is a bilinear in the b,c fields. In the case of Fermi statistics ($\epsilon = +1$), the fields b,c themselves can be bosonized in a straightforward way – they are given by the exponentials of $\phi(z)$ (compare also with eq.(13.6)):

$$\epsilon = +1 : \qquad \begin{aligned} b(z) &= e^{-\phi(z)} \quad , \quad c(z) = e^{\phi(z)} \\[2mm] &\phi(z)\phi(w) \sim \ln(z - w). \end{aligned} \tag{13.42}$$

On the other hand, in case of Bose statistics the "bosonization" of $b(z)$, $c(z)$ is more complicated. In fact, the energy-momentum tensor $T^{(j)}$ is not complete for $\epsilon = -1$. If we calculate the anomaly in the $T^{(j)}T^{(j)}$ operator product we find

$$c^{(j)} = (1 - 3\epsilon Q^2) = \begin{cases} c & \epsilon = +1 \\ c + 2 & \epsilon = -1 \end{cases} \tag{13.43}$$

where c refers to the value of the anomaly given in eq.(13.20). This means that in the bosonic case the field ϕ does not give a complete description of the system. This can already be deduced from the fact the solitons $e^{\pm\phi}$ are always fermions and cannot accommodate Bose statistics. One requires extra fields for the "bosonization" of b, c. What is needed is a fermionic system with $c = -2$. Let us therefore define two conjugate free fermions $\eta(z)$, $\xi(z)$ of conformal weight 1 and 0 respectively which constitute a first order system with $\epsilon = 1$, $\lambda^{\eta\xi} = 1$, $Q^{\eta\xi} = -1$, $c^{\eta\xi} = -2$. Their operator product is

$$\eta(z)\xi(w) = \xi(z)\eta(w) = \frac{1}{z - w} + \ldots \tag{13.44}$$

Then the Bose fields b, c can be "bosonized" as

$$\epsilon = -1 : \qquad \begin{aligned} b(z) &= e^{-\phi(z)}\partial\xi(z) \quad , \quad c(z) = e^{\phi(z)}\eta(z) \\[2mm] &\phi(z)\phi(w) \sim -\ln(z - w). \end{aligned} \tag{13.45}$$

The correct operator products eq.(13.12) between b and c are easily verified using eq.(13.45).

The η, ξ system contains its own chiral $U(1)$ current which provides a second scalar field $\chi(z)$:

$$\eta(z)\xi(z) = \partial\chi(z) \quad , \quad \chi(z)\chi(w) = \ln(z - w)$$
$$\eta(z) = e^{-\chi(z)} \quad , \quad \xi(z) = e^{\chi(z)}. \tag{13.46}$$

Thus in terms of ϕ and χ, b and c are expressed as:

$$b(z) = e^{-\phi(z)+\chi(z)}\partial\chi(z) \quad , \quad c(z) = e^{\phi(z)-\chi(z)}. \tag{13.47}$$

It is important to note that the irreducible representations of the b, c algebra are built only from ϕ, η and $\partial\xi$; the zero mode field ξ_0 never appears in the b, c algebra. As long as we do not include ξ_0 we do not have to neutralize the background charge of the η, ξ system.

Let us now address the question of vacuum states. A hint that the SL_2 invariant vacuum $|0\rangle$ might not be the only possible one came already from our discussion of the conformal ghost system. We know that locality requires that $c_n|0\rangle = 0$ for $n \geq \lambda$. From $[L_0, c_n] = -nc_n$ we find that $L_0 c_n|0\rangle = -nc_n|0\rangle \neq 0$ for $n < \lambda$. In particular, for $0 < n < \lambda$, c_n lowers the energy of the vacuum $|0\rangle$. If c_n is a bosonic operator, we can apply it to $|0\rangle$ an arbitrary number of times thus lowering the vacuum energy by an arbitrary amount. We find that in the bosonic case the spectrum is unbounded from below. It is clearly also unbounded from above. Consequently, the question of the vacuum is ambiguous. In the Fermi case we can also build states with negative energy, but not without lower bound. Here the situation is familiar from Dirac theory; we can define different vacua depending on to what level the states are filled, and these vacua are stable due to the exclusion principle. In the Bose case the situation is unfamiliar and would be a disaster were it not for the fact that we are dealing with a free i.e. non-interacting theory which does not allow transitions from one vacuum to another.

We can now define an infinite number of vacua $|q\rangle$, which can be viewed as Bose/Fermi seas, by requiring

$$b_n|q\rangle = 0, \quad n > \epsilon q - \lambda$$
$$c_n|q\rangle = 0, \quad n \geq -\epsilon q + \lambda \tag{13.48}$$

where $q \in \mathbf{Z}$ for the NS sector and $q \in \mathbf{Z} + \frac{1}{2}$ for the R sector. Since in the fermi case the different vacua are distinguished by the occupation of a finite number of states, we can go from one vacuum to another by application of a finite number of creation or annihilation operators. This is not so in the bose case. Here a finite number of operators will never bring us from one vacuum to another. The q-vacua are in fact identical to the states $|q\rangle = e^{q\phi(0)}|0\rangle$ we have encountered before. It is straightforward to show that they satisfy eq.(13.48). For instance, for bose statistics ($\epsilon = -1$) we have $b_n|q\rangle = \oint \frac{dz}{2\pi i} z^{n+\lambda-1} e^{-\phi(z)} \partial\xi(z) e^{q\phi(0)}|0\rangle = \oint \frac{dz}{2\pi i} z^{n+\lambda+q-1} :e^{-\phi(z)} \partial\xi(z) e^{q\phi(0)}:|0\rangle$. Eq.(13.48) then follows from regularity of the normal ordered product at $z = 0$. We also see that when acting with an operator in the NS sector ($n + \lambda=$integer) on a state with half-integer q we get a branch cut. Hence these states belong to the R sector. From the expressions of j_n and L_n in terms of modes of b and c it follows that

$$j_n|q\rangle = L_n|q\rangle = 0 \quad , \quad n > 0. \tag{13.49}$$

The propagator receives a finite correction from the vacuum charge:

$$\langle -q - Q|c(z)b(w)|q\rangle \equiv \langle c(z)b(w)\rangle_q$$
$$= \sum_{n \leq \epsilon q - \lambda} \langle -q - Q|[c_{-n}, b_n]_\epsilon|q\rangle z^{-n-(1-\lambda)} w^{-n-\lambda}$$
$$= \left(\frac{z}{w}\right)^{\epsilon q} \frac{1}{z - w} \ . \tag{13.50}$$

The conformal properties of the current $j(z)$ and the energy momentum tensor are also modified. Using above propagator we find

$$\langle j(z)j(w)\rangle_q = \frac{\epsilon}{(z-w)^2},$$

$$\langle T(z)j(w)\rangle_q = \frac{Q}{(z-w)^3} + \frac{q}{z(z-w)^2}, \tag{13.51}$$

$$\langle T(z)T(w)\rangle_q = \frac{c^{b,c}}{2(z-w)^4} + \frac{\epsilon q(Q+q)}{zw(z-w)^2}.$$

Comparing this with eq.(13.26) and the TT operator product gives

$$\langle j(z)\rangle_q = \frac{1}{z}\langle j_0\rangle_q = \frac{q}{z},$$

$$\langle T(z)\rangle_q = \frac{1}{z^2}\langle L_0\rangle_q = \frac{1}{2}\epsilon q(q+Q)\frac{1}{z^2}, \tag{13.52}$$

i.e.

$$j_0|q\rangle = q|q\rangle,$$

$$L_0|q\rangle = \frac{1}{2}\epsilon q(q+Q)|q\rangle. \tag{13.53}$$

Eq.(13.53) also follows directly from eqs.(13.39) and (13.40). We see that the L_0 eigenvalue is bounded from below for fermions ($\epsilon+1$) and unbounded for bosons ($\epsilon = -1$), in agreement with our discussion above.

The SL_2 invariant vacuum has $h = 0$. There are two states which satisfy this condition, namely $|0\rangle$ and $|-Q\rangle$. However it is easy to show that $L_{-1}|q\rangle \neq 0$ for $q \neq 0$; i.e. $|0\rangle$ is the unique SL_2 invariant state.

13.2 Covariant vertex operators, BRST and picture changing

Let us now apply the bosonization of the superconformal ghosts β, γ to the construction of the vertex operators of the fermionic string theory. As already mentioned, the conformal fields of the NSR (ψ^μ) part of the fermionic string have to be completed by conformal fields of the β, γ system. This is in analogy to the bosonic string where physical states contain also ghost excitations. The fundamental reason for including the superconformal ghosts is again the requirement of BRST invariance of the physical states. We will see this in the following.

Let us consider states of the form $|\lambda\rangle_\psi \otimes |q\rangle_{\beta,\gamma}$ with corresponding vertex operators in bosonized form:

$$V_{\lambda,q}(z) = e^{i\lambda\cdot\phi(z)} e^{q\phi(z)} \tag{13.54}$$

where λ is a weight vector of D_5 and the ghost charge q is a D_1 "lattice vector". The conformal dimension of (13.54) is given by

$$h = \frac{1}{2}\lambda^2 - \frac{1}{2}q^2 - q + N. \tag{13.55}$$

(N counts possible oscillator excitations which we neglected in eq.(13.54).) The superghosts β, γ satisfy the same periodicity conditions as the world-sheet gravitino and therefore, due to the coupling of the gravitino to the supercurrent, the same boundary conditions as the NSR fermions ψ^μ. This implies that we must couple the R (NS) sector of the ψ^μ system to the R (NS) sector of the superconformal ghost system. This is to say that for $\lambda \in S, C$ of D_5, q must be half-integer, and for $\lambda \in 0, V$ of D_5, q is integer.

Let us first look at the NS sector of the theory. Here, massless vectors are characterized by D_5 lattice vectors $\lambda = (0,\ldots,\pm1,0,\ldots,0)$ and the corresponding vertex operators have conformal dimension $\frac{1}{2}$. Using formula eq.(13.55) we see that the full vertex operator eq.(13.54) describes a massless vector with $h = 1$ if the ghost charge of this state is chosen to be

$$q = -1. \tag{13.56}$$

We will call this the canonical choice for the ghost number and NS vertex operators with $q = -1$ are said to be in the canonical ghost picture. Thus, a general NS state in the canonical ghost picture is created by a vertex operator

$$V_{\lambda,-1}(z) = e^{i\lambda\cdot\phi(z)} e^{-\phi(z)} \quad \lambda \in 0, V \text{ of } D_5 \tag{13.57}$$

and has mass

$$m^2 = \frac{1}{2}\lambda^2 - \frac{1}{2}q^2 - q - 1 = \frac{1}{2}\lambda^2 - \frac{1}{2}. \tag{13.58}$$

Here we have again neglected oscillator contributions. The -1 contribution to the zero point energy is due to the requirement that physical vertex operators have conformal dimension $h = +1$ or, from the point of view of Chapter 5, due to the reparametrization ghosts. Together with the $\frac{1}{2}$ mass unit from the superconformal ghosts β, γ one obtains the correct normal ordering constant, the tachyon mass in the NS sector. The ground state in the NS sector is thus $e^{-\phi} c(0)|0\rangle$.

In the R sector massless spinors correspond to the weight vectors $\lambda = (\pm\frac{1}{2}, \pm\frac{1}{2}, \pm\frac{1}{2}, \pm\frac{1}{2}, \pm\frac{1}{2})$ with an even and odd number of minus signs for S_α and $S_{\dot\alpha}$ respectively. We now derive that the ghost charge q must be

$$q = -\frac{1}{2} \tag{13.59}$$

for the vertex operator eq.(13.54) of the massless spinors to have conformal dimension $h = 1$. Again, we call this the canonical ghost charge in the R sector. All vertex operators in this ghost picture are of the form

$$V_{\lambda, -\frac{1}{2}}(z) = e^{i\lambda \cdot \phi(z)} e^{-\frac{1}{2}\phi(z)} \quad \lambda \in \text{S}, \text{C of } D_5 \tag{13.60}$$

with mass

$$m^2 = \frac{1}{2}\lambda^2 - \frac{1}{2}q^2 - q - 1 = \frac{1}{2}\lambda^2 - \frac{5}{8}. \tag{13.61}$$

Now consider operator products between two different vertex operators as given in eq.(13.54):

$$V_{\lambda, q}(z) V_{\lambda', q'}(w) = (z - w)^{\lambda \cdot \lambda' - qq'} V_{\lambda + \lambda', q + q'}(w) + \ldots \tag{13.62}$$

Since the ghost charges add, it is appeareant that vertex operators in non-canonical ghost pictures appear in the operator product expansion. So let us discuss the meaning of states with arbitrary, non-canonical ghost charge q which are needed for closure of the operator algebra eq.(13.62). We will investigate this question using the BRST formalism. Thus, we have first to construct the BRST charge Q of the fermionic string theory. It gets contribution from the fields X, b, c as well as from their superpartners ψ, β, γ. Generalizing eq.(5.30) we get

$$Q = \oint \frac{dz}{2\pi i}\left\{c(T^{X,\psi} + \tfrac{1}{2}T^{b,c,\beta,\gamma}) - \gamma(T_F^{X,\psi} + \tfrac{1}{2}T_F^{b,c,\beta,\gamma})\right\} \tag{13.63}$$
$$= Q_0 + Q_1 + Q_2$$

where

$$
\begin{aligned}
Q_0 &= \oint \frac{dz}{2\pi i}(cT^{X,\psi,\beta,\gamma} + bc\partial c)\,, \\
Q_1 &= -\oint \frac{dz}{2\pi i}\gamma T_F^{X,\psi} &= \frac{1}{2}\oint \frac{dz}{2\pi i}e^{\phi-\chi}\psi_\mu\partial X^\mu\,, \\
Q_2 &= -\oint \frac{dz}{2\pi i}\frac{1}{4}b\gamma^2 &= -\oint \frac{dz}{2\pi i}\frac{1}{4}be^{2\phi-2\chi}\,.
\end{aligned}
\tag{13.64}
$$

The subscript on Q denotes the superconformal ghost charge. Q_0 is the bosonic BRST operator if we treat β,γ as extra matter fields. Q_1 generates world-sheet supersymmetry transformations with parameter given by the supersymmetry ghost γ. Finally, Q_2 is needed for the closure of the BRST algebra. One easily works out the BRST transformations of the various fields. We find

$$
\begin{aligned}
[Q, X^\mu(z)] &= c\partial X^\mu(z) - \tfrac{1}{2}\gamma\psi^\mu(z)\,, \\
\{Q, \psi^\mu(z)\} &= \left(\tfrac{1}{2}\partial c\psi^\mu(z) + c\partial\psi^\mu(z)\right) - \tfrac{1}{2}\gamma\partial X^\mu(z)\,, \\
\{Q, c(z)\} &= c\partial c(z) - \tfrac{1}{4}\gamma^2(z)\,, \\
[Q, \gamma(z)] &= -\tfrac{1}{2}\partial c\gamma(z) + c\partial\gamma(z)\,, \\
\{Q, b(z)\} &= T^{\text{tot}}(z)\,, \\
[Q, \beta(z)] &= -T_F^{\text{tot}}(z)\,, \\
[Q, T^{\text{tot}}(z)] &= \tfrac{1}{8}(d-10)\partial^3 c(z)\,, \\
\{Q, T_F^{\text{tot}}(z)\} &= \tfrac{1}{8}(d-10)\partial^2\gamma\,.
\end{aligned}
\tag{13.65}
$$

Here T^{tot} and T_F^{tot} are the energy-momentum tensor and the supercurrent for all the fields involved. On the matter fields Q acts as a combined conformal transformation with parameter c and a superconformal transformation with parameter $-\gamma$. One can verify that $Q^2 = 0$ for $d = 10$.

Now consider the vertex operator for a massless spinor in the canonical $q = -\frac{1}{2}$ ghost picture:

$$V_{-\frac{1}{2}}(z) = u_\alpha S^\alpha(z) e^{-\frac{1}{2}\phi(z)} e^{ik_\mu X^\mu(z)}. \tag{13.66}$$

u_α is the spinor wave function. Let us first show that $V_{-\frac{1}{2}}$ is BRST invariant i.e. $[Q, V_{-\frac{1}{2}}]$ vanishes up to a total derivative which is irrelevant upon integration over z. First we have (cf. eq.(5.37))

$$[Q_0, V_{-\frac{1}{2}}] = \partial(c V_{-\frac{1}{2}}). \tag{13.67}$$

Then, using eqs.(13.8) and (13.34) with $\epsilon = -1$, we obtain

$$e^{\phi - \chi} T_F(z) V_{-\frac{1}{2}}(w) \sim \frac{i}{2\sqrt{2}} (z - w)^{-1} (\not{k} u)^\alpha{}_{\dot\alpha} e^{\frac{1}{2}\phi - \chi} S^{\dot\alpha}(w) e^{ik_\mu X^\mu(w)}. \tag{13.68}$$

Thus $[Q_1, V_{-\frac{1}{2}}] = 0$ if we demand that u_α satisfies the on-shell condition $\not{k} u = 0$. Finally

$$e^{2\phi - 2\chi} b(z) V_{-\frac{1}{2}}(w) \sim (z - w) \, e^{\frac{3}{2}\phi - 2\chi} \, b \, u_\alpha S^\alpha e^{ik_\mu X^\mu}(w) \tag{13.69}$$

shows that $[Q_2, V_{-\frac{1}{2}}] = 0$. Therefore $V_{-\frac{1}{2}}$ is a BRST invariant vertex operator. In general, BRST invariance of physical state vertex operators requires that they satisfy on-shell conditions.

However we can create a second version of the fermion vertex operator with different ghost number which is also BRST invariant. This operator is defined as

$$V_{\frac{1}{2}} = 2[Q, \xi V_{-\frac{1}{2}}]. \tag{13.70}$$

The subscript $\frac{1}{2}$ denotes that this vertex operator now has ghost charge $\frac{1}{2}$. $V_{\frac{1}{2}}$ is obviously BRST invariant. One might think that since $V_{-\frac{1}{2}}$ is

BRST invariant and ξ has zero conformal dimension that $\xi V_{-\frac{1}{2}}$ is itself BRST invariant. This is however not the case since the β, γ algebra and consequently also Q only contain $\partial \xi$, but not the zero mode of ξ. We can then compute[3]

$$V_{\frac{1}{2}} = [Q_0 + Q_1 + Q_2, 2\xi V_{-\frac{1}{2}}] \tag{13.71}$$

$$= 2\partial(c\xi V_{-\frac{1}{2}}) + \frac{1}{\sqrt{2}} u_\alpha \left\{ e^{\frac{1}{2}\phi}(\partial X^\mu + \frac{i}{4}(k\cdot\psi)\psi^\mu)(\Gamma_\mu)^\alpha{}_{\dot\beta} S^{\dot\beta} \right.$$
$$\left. + \frac{1}{2} e^{\frac{3}{2}\phi - \chi} \eta \, b \, S^\alpha \right\} e^{ik\cdot X}.$$

The derivative term is the contribution from Q_0 and the term with ghost charge $\frac{3}{2}$ is generated by Q_2. Both of these terms will never contribute to correlation functions (at least at tree level); the former vanishes upon integration over z and the latter contains one b field which will not be absorbed by a c field. Also note that the zero mode of ξ does not contribute to $V_{+\frac{1}{2}}$, so its ghost part is also in the β, γ algebra. Therefore we define the physical fermion vertex operator in the $+\frac{1}{2}$ ghost picture as

$$V_{\frac{1}{2}}(z) = \quad [Q_1, 2\xi V_{-\frac{1}{2}}(z)]$$

$$= -2 \oint \frac{dw}{2\pi i} e^{\phi(w) - \chi(w)} T_F(w) \, e^{\chi(z)} V_{-\frac{1}{2}}(z)$$

$$= -2 \lim_{w \to z} e^{\phi(w)} T_F(w) V_{-\frac{1}{2}}(z) \tag{13.72}$$

$$= \frac{1}{\sqrt{2}} u_\alpha \left\{ e^{\frac{1}{2}\phi}(\partial X^\mu + \frac{i}{4}(k\cdot\psi)\psi^\mu)(\Gamma_\mu)^\alpha{}_{\dot\beta} S^{\dot\beta} \right\}.$$

The operator

$$P_{+1}(z) = -2e^{\phi(z)} T_F(z) = e^\phi \psi^\mu \partial X_\mu(z) \tag{13.73}$$

---

[3]In the derivation of eq.(13.71) we need the subleading term of the first of the operator products in eq.(13.8). One can show that in d dimensions

$$k_\mu u_\alpha \psi^\mu(z) S^\alpha(w) \sim \frac{1}{\sqrt{2}}(z - w)^{1/2} \frac{1}{1 - \frac{d}{2}} u_\alpha k_\mu \psi^\mu \psi^\nu (\Gamma_\nu)^\alpha{}_{\dot\beta} S^{\dot\beta}$$

for on-shell u_α.

is called picture changing operator. It carries one unit of ghost charge. Apart from the ghost part it acts like a two-dimensional supersymmetry transformation.

In summary, the picture changing operation provides a second version of the fermion vertex operator, now in the ghost picture $q = +\frac{1}{2}$. Moreover, one can use P_{+1} to generate further, equivalent, and also BRST invariant vertex operators like

$$V_{\frac{3}{2}}(z) = \lim_{w \to z} P_{+1}(w) V_{\frac{1}{2}}(z) \tag{13.74}$$

or in general

$$V_{q+1}(z) = \lim_{w \to z} P_{+1}(w) V_q(z). \tag{13.75}$$

This equation states that vertex operators with ghost charges differing by integral units represent the same physical state. The different copies are said to be in different ghost pictures. The limit in eq.(13.75) always exists for BRST invariant vertex operators. There is a copy of any canonical vertex operator in every ghost sector. As a second example, picture changing of the vertex operator of the massless vector in the -1 picture,

$$V_{-1}(z) = \zeta_\mu \psi^\mu(z) e^{-\phi(z)} e^{ik_\rho X^\rho(z)} \tag{13.76}$$

leads to its copy in the 0 picture:

$$V_0(z) = -\zeta_\mu (\partial X^\mu + i\psi^\mu (k \cdot \psi)) e^{ik_\rho X^\rho(z)}. \tag{13.77}$$

For the tachyon we derive

$$V_0(z) = -ik \cdot \psi e^{ik_\mu X^\mu(z)}. \tag{13.78}$$

One can immediately check that the mass formula eq.(13.55) gives the same answer in any ghost picture. P_{+1}, which has conformal weight 0, always acts in a way that the effect of changing the ghost charge q is compensated e.g. by oscillators ∂X_μ or extra ψ^μ factors. Thus the mass spectrum of the fermionic sting is in fact bounded from below unlike the spectrum of the β, γ theory. The reason is that only BRST invariant combinations are accepted as physical states. Also, the physical vertex operators, no matter in what ghost picture, never contain the zero mode of ξ; i.e. they are always in the β, γ algebra.

Note that it follows from eq.(12.7) that a NS field with zero superconformal ghost charge is BRST invariant if it is the upper component of a $h = \frac{1}{2}$ superfield. The lower component is provided by the vertex operator in the canonical ghost picture (apart from the factor $e^{-\phi}$).

The picture changing operation is important for the evaluation of correlation functions. We have seen that correlation functions in the SL_2 invariant vacuum will only be non-vanishing if we insert an operator with ghost charge $q = -Q = -2$. This means that to get non-vanishing scattering amplitudes we have to choose the ghost pictures for the vertex operators such that the total superconformal ghost charge adds up to -2. This is analogous to the situation we encountered when discussing the conformal ghosts in Chapter 6. There the vacuum carried three units of ghost charge and they had to be absorbed by ghost zero mode insertions. If we denote by σ the boson that arises from bosonizing the conformal ghosts b, c and by ϕ the boson from the β, γ system, we have[4]

$$\langle -Q^{b,c} - Q^{\beta,\gamma}|0\rangle = \langle 0|e^{3\sigma(0)-2\phi(0)}|0\rangle = 1. \qquad (13.79)$$

BRST invariance of the states $|0\rangle$ and $e^{3\sigma(0)-2\phi(0)}|0\rangle$ is easy to verify. For instance, since in the critical dimension Q commutes with the L_n's (cf. eq.(13.65)), we find that $L_{0,\pm 1}Q|0\rangle = 0$. Since the SL_2 invariant vacuum state is unique, we conclude $Q|0\rangle = 0$. Invariance of $e^{3\sigma(0)-2\phi(0)}|0\rangle$ is shown by computing $[Q, e^{3\sigma(0)-2\phi(0)}]$. Hence the vacuum expectation values of BRST invariant operators will be BRST invariant.

---

[4]If we include the zero mode of the η, ξ system, we also have to neutralize its background charge and get $\langle 0|e^{3\sigma-2\phi+\chi}|0\rangle = 1$. $e^{3\sigma-2\phi+\chi}|0\rangle$ is however not BRST invariant.

It is now important that correlation functions of physical vertex operators are independent of how we distribute the ghost charges among them as long as $\sum_i q_i = -2$ (at tree level, say). The reason is the following. Since none of the vertex operators depends on the zero mode ξ_0, we can insert it in the functional integral and integrate over it; i.e. we insert $1 = \int d\xi_0 \xi_0$. Since ξ_0 is a Grassmann variable, we can replace it by $\xi(z)$ for arbitrary z. (This follows from $\int \mathcal{D}\xi' d\xi_0 \xi_0 F(\xi') = \int \mathcal{D}\xi' d\xi_0\, \xi(z) F(\xi')$ where ξ' denotes the non-zero mode part of ξ.) So we can attach $\xi(z)$ to any of the vertex operators in the correlation function, say to $V_{q_1}(z_1)$. Now rewrite any of the other vertex operators, say $V_{q_2}(z_2)$, as $V_{q_2}(z_2) = 2 \oint \frac{dw}{2\pi i} \xi(w) j_{\mathrm{BRST}}(w) V_{q_2-1}(z_2)$. We deform the integration contour by pulling it off the back of the sphere. Due to BRST invariance it passes through all vertex operators except for $\xi(z_1) V_{q_1}(z_1)$ which becomes $V_{q_1+1}(z_1)$. Then $\int d\xi_0 \xi(w) = 1$. In summary, we have traded one unit of ghost charge between two vertex operators within a BRST invariant correlation function without changing its value.

13.3 The covariant lattice

Let us now return to the operator product expansion eq.(13.62). It strongly suggests to combine the D_5 weight vectors λ with the D_1 weights q to a six-dimensional vector $w = (\lambda, q)$. We can now write the operator algebra as

$$V_{w_1}(z) V_{w_2}(w) = (z - w)^{w_1 \cdot w_2} V_{w_1 + w_2}(w) + \dots \tag{13.80}$$

Closure of the algebra implies that w_1, w_2 are vectors of a six-dimensional Lorentzian lattice $D_{5,1}$ with metric signature $(+ + + + +, -)$. The minus sign is due to the ghost sector. This enlarged Lorentzian lattice is usually called covariant lattice since it describes the covariant vertex operators of the fermionic string. Since the only allowed vectors of $D_{5,1}$ decompose under $D_5 \otimes D_1$ as $w = (\lambda; q)$, where λ and q both belong to either the

NS sector or both to the R sector, the lattice $D_{5,1}$ contains four conjugacy classes: 0, V, S and C.

Now consider the weights of $D_{5,1}$ which correspond to the states of lowest mass in the canonical ghost picture:

$$w_{\text{tachyon}} = (0,0,0,0,0;-1) \qquad\qquad \in V \text{ of } D_{5,1}$$

$$w_{\text{vector}} = (0,\ldots,\pm 1,0,\ldots,0;-1) \qquad \in 0 \text{ of } D_{5,1}$$

$$w_{\text{spinor}} = (\pm\tfrac{1}{2},\pm\tfrac{1}{2},\pm\tfrac{1}{2},\pm\tfrac{1}{2},\pm\tfrac{1}{2},-\tfrac{1}{2}) \in C \text{ of } D_{5,1} \qquad \text{odd number of minus signs}$$

$$w_{\text{antispinor}} = (\pm\tfrac{1}{2},\pm\tfrac{1}{2},\pm\tfrac{1}{2},\pm\tfrac{1}{2},\pm\tfrac{1}{2};-\tfrac{1}{2}) \in S \text{ of } D_{5,1} \qquad \text{even number of minus signs}$$

$$(13.81)$$

We recognize that for states in the canonical ghost picture, all vectors in the V (S) conjugacy class of D_5 belong to the 0 (C) conjugacy class of $D_{5,1}$ and vice versa. The decomposition of the conjugacy classes of $D_{5,1}$ into those of $D_5 \otimes D_1$ is given in table 13.2.

Table 13.2 Decomposition of $D_{5,1}$ conjugacy classes

$D_{5,1}$	$D_5 \otimes D_1$
0	$(V,V) \oplus (0,0)$
V	$(V,0) \oplus (0,V)$
S	$(S,S) \oplus (C,C)$
C	$(C,S) \oplus (S,C)$

The mass of a state $|w\rangle$ can be written in terms of lattice vectors of $D_{5,1}$ as (remember the negative metric of this lattice)

$$m^2 = \frac{1}{2}w^2 + w \cdot e_6 - 1 \qquad (13.82)$$

where e_6 is the basis vector $(0,0,0,0,0;1)$. From the discussion above we understand the meaning of lattice vectors with different ghost charge q. They all correspond to copies of the same physical state but in different ghost pictures. However there is no one-to-one relation between $D_{5,1}$ lattice vectors and physical states. Only states in the canonical ghost picture are directly related to lattice vectors of $D_{5,1}$. In other ghost pictures there is no clear relation between lattice vectors and physical vertex operators. These are in general given by linear combinations of vertex operators of the form eq.(13.54). E.g. the relevant $D_{5,1}$ lattice vectors for the massless vector in the 0 picture are the null vector and the vectors $(\alpha; 0)$ where α is a root of D_5. The picture changing operation does not change the $D_{5,1}$ conjugacy class of a state. This is so because the picture changing operator P_{+1} corresponds to the $D_{5,1}$ lattice vector

$$w_{\mathrm{PC}} = (0, \ldots, \pm 1, \ldots, 0; +1) \tag{13.83}$$

which is a root of $D_{5,1}$. Picture changing acts on the lattice $D_{5,1}$ by simply adding w_{PC}.

Let us return to the question of locality, i.e. absence of branch cuts in the operator product algebra of the fermionic string. The exponents of $(z - w)$ are determined by the inner product rules of conjugacy classes of the Lorentzian lattice $D_{5,1}$ which are summarized in table 13.3.

Table 13.3 Mutual scalar products of $D_{5,1}$ conjugacy classes

	0	V	S	C
0	Z	Z	Z	Z
V		Z	$Z+\frac{1}{2}$	$Z+\frac{1}{2}$
S			Z	$Z+\frac{1}{2}$
C				Z

We recognize that we have almost reached our aim to obtain a local theory by extending D_5 to $D_{5,1}$. The massless spinor ($\in C$ of $D_{5,1}$) is now local with respect to the massless vector ($\in 0$ of $D_{5,1}$). The branch cut in the operator product $\psi^\mu(z) S^\alpha(w)$ is cancelled by the branch cut in $e^{-\phi(z)} e^{-\frac{1}{2}\phi(w)}$. However there are still some sources of non-locality. The tachyon ($\in V$ of

280

$D_{5,1}$) is non-local with respect to the spinor (and also to the antispinor) and also the spinor is non-local with respect to the antispinor ($\in$ S of $D_{5,1}$). Thus a projection is needed which eliminates half of the conjugacy classes of $D_{5,1}$ and makes the theory local. We see that the NS sector with 0 and V of $D_{5,1}$ leads to a local, closed operator algebra as well as the projection onto the 0 and C (or equivalently S) conjugacy classes of $D_{5,1}$. The latter projection is identical to the GSO projection introduced within the fermionic formulation of the spinning string. It leads to a space-time supersymmetric spectrum; the 0 and C conjugacy classes contain each others supersymmetric partners.

We now show that this projection on the lattice $D_{5,1}$ with 0 and C (or S) conjugacy classes is enforced by modular invariance of the one loop partition function of the fermionic string. This is in complete analogy to the fact that in the fermionic language the GSO projection was necessary when summing over a modular invariant combination of different spin structures.

The holomorphic part of the one loop partition function of the fermionic string in Hamiltonian description has the form

$$\chi(\tau) \sim \mathrm{Tr} e^{2\pi i \tau (L_0 - 1)} \Phi. \tag{13.84}$$

Φ is a phase factor which takes into account the correct space-time statistics, i.e. Φ is 1 for space-time bosons and -1 for space-time fermions. Let us first discuss the non-trivial part of eq.(13.84), namely the contribution of the zero modes of the bosons ϕ^i ($i = 1,\ldots,5$) and ϕ. They just give the sum over the lattice vectors of the Lorentzian lattice $D_{5,1}$:

$$\tilde{\chi}(\tau) = \sum_{w=(\lambda,q)\in D_{5,1}} e^{2\pi i \tau (\frac{1}{2}\lambda^2 - \frac{1}{2}q^2 - q)} e^{-2\pi i q}. \tag{13.85}$$

The factor $e^{-2\pi i q}$ ensures just the correct space-time statistics using the fact that q is (half) integer for space-time bosons (fermions).

However this expression for $\tilde{\chi}(\tau)$ cannot represent the physical partition function since the sum extends over arbitrarily high ghost numbers q.

In other words, eq.(13.85) sums over all possible equivalent ghost pictures. In addition we know also that the physical (light-cone) partition function should only count the transverse degrees of freedom. In the covariant lattices language these physical light-cone states in the canonical ghost picture can be characterized by decomposing $D_{5,1}$ to a part which describes the transverse Lorentz group $SO(8)$ and a two-dimensional part which describes the longitudinal, timelike and ghost degrees of freedom of any state:

$$D_{5,1} = D_4 \otimes D_{1,1} \quad , \qquad w = (u, x) \, ,$$
$$w \in D_{5,1}, \quad u \in D_4, \quad x \in D_{1,1} \, . \tag{13.86}$$

Then the physical state condition is to consider lattice vectors $w = (u, x)$ with fixed vector x_0 in the following way:

$$x_0 = (0, -1) \ \text{NS sector} \, ,$$
$$x_0 = (\frac{1}{2}, -\frac{1}{2}) \ \text{R sector} \, . \tag{13.87}$$

The trace in the partition function should then go only over those states satisfying eq.(13.87). To realize this constraint let us write a general vector $w \in D_{5,1}$ in a convenient form

$$w = (u, x_0) + \Delta = w_0 + \Delta \tag{13.88}$$

where Δ is the sum of two light-like (picture changing) vectors:

$$\Delta = m\Delta_1 + n\Delta_2 \, ,$$
$$\Delta_1 = (0, 0, 0, 0, 1, 1) \, , \tag{13.89}$$
$$\Delta_2 = (0, 0, 0, 1, 0, 1) \, .$$

Substituting eq.(13.88) into eq.(13.85) we are left with the following expression:

$$\tilde{\chi}(\tau) = \sum_{w_0} e^{2\pi i \tau (\frac{1}{2} w_0^2 + w_0 \cdot e_6)} e^{2\pi i w_0 \cdot e_6} \sum_{\Delta} e^{\pi i \tau \Delta^2} \tag{13.90}$$

where we have shifted Δ without affecting the infinite sum. The first part is simply the trace over all physical states. Therefore, to obtain the physical light-cone partition function, one has to divide $\tilde{\chi}(\tau)$ by

$$\Theta_{1,1}(\tau) = \sum_{\Delta} e^{\pi i \tau \Delta^2}. \tag{13.91}$$

$\Theta_{1,1}$ is the partition function of the two-dimensional Lorentzian even self-dual lattice $D_{1,1}$. Although the two functions $\Theta_{1,1}(\tau)$ and $\tilde{\chi}(\tau)$ are separately ill defined because of the Lorentzian metric, their ratio is nevertheless well defined and describes the physical partition function. Finally we also have to take into account the contribution of the bosonic X and ϕ oscillators. Then the complete (purely holomorphic) result is:

$$\chi(\tau) \sim \frac{1}{\eta(\tau)^{12}\Theta_{1,1}(\tau)} \sum_{w \in D_{5,1}} e^{2\pi i \tau \frac{1}{2}(w+e_6)^2} e^{2\pi i w \cdot e_6}. \tag{13.92}$$

Let us now check the modular invariance of $\chi(\tau)$. First consider the transformation $\tau \to \tau + 1$. Since $\eta(\tau)^{12}$ changes sign under this transformation while $\Theta_{1,1}(\tau)$ remains invariant we require that the lattice sum also changes sign. This implies that states with (half) integer q corresponding to (R) NS states have to be associated with (odd) even points on the lattice:

$$\frac{1}{2}w^2 - q \in \mathbf{Z}. \tag{13.93}$$

This is also what the spin statistics theorem demands. The operator product between vertex operators in the NS (R) sector ($\Phi = (\phi; -i\phi)$ is

$$e^{iw \cdot \Phi(z)} e^{-iw \cdot \Phi(w)} \sim (z - w)^{-w^2} + \dots \tag{13.94}$$

NS (R) states have to be (anti) commuting, i.e. $w^2 = 2n$ ($w^2 = 2n + 1$), $n \in \mathbf{Z}$. It is easily checked that all vectors in the S, C and V conjugacy classes of $D_{5,1}$ have odd (length)^2 whereas the vectors of the 0 conjugacy class have even (length)^2. Thus, requiring invariance under $\tau \to \tau + 1$ discards the V conjugacy class.

For the second transformation, $\tau \to -\frac{1}{\tau}$ we use that

$$\eta^{-1}(\tau) = (-i\tau)^{1/2}\eta^{-1}(-\frac{1}{\tau}),$$

$$\Theta_{1,1}^{-1}(\tau) = -i\tau\,\Theta_{1,1}^{-1}(-\frac{1}{\tau}). \tag{13.95}$$

The second relation follows since $\Theta_{1,1}$ is the lattice sum of the even self-dual lattice $D_{1,1}$. Using now the Poisson resummation formula of Chapter 10, we find

$$\chi(\tau) = \frac{\tau^4}{\text{vol}(D_{5,1})} \, \frac{1}{\eta^{12}(-\frac{1}{\tau})\Theta_{1,1}(-\frac{1}{\tau})} \sum_{w \in D_{5,1}^*} e^{-\frac{2\pi i}{\tau}\frac{1}{2}(w+e_6)^2} e^{-2\pi i w \cdot e_6}.$$

(13.96)

Thus, apart from a factor $e^{4\pi i q}$, which is irrelevant for q only integer or half-integer, $\chi(\tau)$ is invariant under $\tau \to -\frac{1}{\tau}$ if the covariant lattice $D_{5,1}$ is self-dual. (The factor τ^4 will be compensated by the transformation of $(\text{Im}\,\tau)^{-4}$ (cf. eq.(10.44).)

In summary, modular invariance implies that $D_{5,1}$ must be an odd self-dual Lorentzian lattice which contains only the 0 and S (or 0 and C) conjugacy classes. Then the spinning string is automatically local as discussed above. These two conjugacy classes contain as lowest states a massless vector and massless spinor. They build the on-shell degrees of freedom of a ten-dimensional supermultiplet and one can show that also the massive states can be arranged into supermultiplets. Thus the requirement of self-duality of the covariant lattice $D_{5,1}$ is equivalent to the GSO projection.

The covariant lattice description also allows for a straightforward derivation of the light-cone partition function of the fermionic string in terms of Jacobi theta functions. Since the lattice $D_{5,1}$ contains the 0 and S conjugacy classes, the light-cone partition function is given by the difference between the lattice sums of the V and C conjugacy classes of the D_4 weight lattice. The $V \in D_4$ conjugacy class is obtained from the $0 \in D_{5,1}$ by truncation to light-cone states according to the physical state selection rule eq.(13.87). Likewise this gives $C \in D_4$ from $S \in D_{5,1}$. The relative minus sign takes into account spin-statistics. Thus, using the expressions eq.(11.101) for the sums over the lattice vectors of the D_n-weight lattice, the contribution of the world-sheet fermions to the light-cone partition function becomes:

$$\chi(\tau) \sim \frac{1}{2}\frac{1}{\eta(\tau)^4}[\theta_3^4(0|\tau) - \theta_4^4(0|\tau) - \theta_2^4(0|\tau)]. \qquad (13.97)$$

This expression is identical to the one obtained from the sum over all spin structures of the world-sheet fermions in the fermionic description (see Chapter 9) and vanishes due to the triality relation among the V, C and S conjugacy classes of D_4 encoded in the first of the identities in eq.(9.11). This reflects the underlying space-time supersymmetry: the contribution of space-time bosons and space-time fermions cancels.

The covariant Lorentzian lattice $D_{5,1}$ with 0 and S conjugacy classes is very similar to the root lattice of E_8 which can also be thought of being the weight lattice of D_8, again with 0 and S conjugacy classes. In both cases the restriction to these two conjugacy classes implies self-duality. Therefore, due to this analogy, we may call the self-dual lattice $D_{5,1}$ also $E_{5,1}$. Both, E_8 and $E_{5,1}$, contain spinorial generators which, for E_8, correspond to length$^2 = 2$ vectors and commute; for $E_{5,1}$ however they correspond to length$^2 = 1$ vectors and therefore anticommute. They generate the space-time supersymmetry algebra (cf. the discussion in the next chapter). In fact, it turns out to be useful to consider instead of the Lorentzian lattice $E_{5,1}$ the Euclidean covariant lattice E_8. This is equivalent to replacing the superconformal ghost lattice D_1 by a D_3 lattice while simultaneously changing the signature of the metric of the lattice.[5] In other words, we replace the Lorentzian covariant lattice $D_{5,1}$ by the Euclidean covariant lattice D_8. The requirement of modular invariance implies in both cases that only the 0 and S conjugacy classes must be present such that we are dealing with the lattices $E_{5,1}$ resp. E_8. To describe the states in the canonical ghost picture we are forced to decompose E_8 (D_8) to $D_5^{\text{Lorentz}} \otimes D_3^{\text{ghost}}$ and consider only vectors of D_3^{ghost} with fixed entries. Conventionally we choose them as

---

[5] These lattice maps will be discussed in more detail in the next chapter.

$$w = (u, v) \in D_8 , \qquad u \in D_5^{\text{Lorentz}} , \qquad v \in D_3^{\text{ghost}} ,$$

$$v_0 = (\tfrac{1}{2}, \tfrac{1}{2}, -\tfrac{1}{2}) , \qquad\qquad \text{R sector} , \tag{13.98}$$

$$v_0 = (0, 0, -1) , \qquad\qquad \text{NS sector} .$$

Note that $\frac{1}{2} v_0^2$ is exactly the superconformal ghost contribution to the conformal weight in the canonical ghost picture. Furthermore, if we are only interested in the physical light-cone states we have to decompose D_8 to $D_4 \otimes D_4$ and consider states which have fixed entries in second D_4:

$$x_0 = (\tfrac{1}{2}, \tfrac{1}{2}, \tfrac{1}{2}, -\tfrac{1}{2}) \; \text{R sector} ,$$
$$x_0 = (0, 0, 0, -1) \; \text{NS sector} . \tag{13.99}$$

This means that we have replaced the Lorentzian lattice $D_{1,1}$, which describes the longitudinal, timelike and superconformal ghost degrees of freedom, by the Euclidean lattice D_4. The physical light-cone partition function can now be written as a lattice sum over the D_8 (E_8) lattice counting only those vectors satisfying eq.(13.99) and taking into account the correct spin statistics assignment. Since under $D_8 \rightarrow D_4$ the conjugacy classes are interchanged according to V $\leftrightarrow$ 0 and S $\leftrightarrow$ C, the truncation to physical light cone states acts on the theta functions as $\theta_3^8 \rightarrow \theta_3^4$, $\theta_4^8 \rightarrow -\theta_4^4$ and $\theta_2^8 \rightarrow -\theta_2^4$. Given the known expression for the E_8 partition function, $\chi_{E_8}(\tau) \sim \frac{1}{2} \frac{1}{\eta(\tau)^8} [\theta_2^8(0|\tau) + \theta_3^8(0|\tau) + \theta_4^8(0|\tau)]$ one immediately derives eq.(13.97) as the physical light-cone partition function. (Note that one also has to drop the contribution of four oscillators.) At this point it is important to realize that the replacement of the Lorentzian lattice $D_{5,1}$ by the Euclidian lattice D_8 does not mean that the non-unitary ghost Hilbert space is contained in the positive definite Hilbert space of the D_8 Kac-Moody algebra. So far this procedure is just a convenient technical trick since the Euclidean lattices are much nicer to handle. Both descriptions lead, with the conditions eq.(13.87) and eq.(13.99) respectively, to the correct light-cone degrees of freedom.

REFERENCES

[1] D. Friedan, E. Martinec and S. Shenker, *"Covariant quantization of superstrings"*, Phys. Lett. **160B** (1985) 55; *"Conformal invariance, supersymmetry and string theory"*, Nucl. Phys. **B271** (1986) 93.

[2] V.G. Knizhnik, *"Covariant fermionic vertex in superstrings"*, Phys. Lett. **160B** (1985) 403.

[3] J. Cohn, D. Friedan, Z. Qiu and S. Shenker, *"Covariant quantization of superstring theories: the spinor field of the Ramond-Neveu-Schwarz model"*, Nucl. Phys. **B278** (1986) 577.

[4] W. Lerche and D. Lüst, *"Covariant heterotic strings and odd self-dual lattices"*, Phys. Lett. **187B** (1987) 45.

[5] W. Lerche, D. Lüst and A.N. Schellekens, *"Ten-dimensional heterotic strings from Niemeier lattices"*, Phys. Lett. **B181** (1986) 71; Erratum Phys. Lett. **B184** (1987) 419.

[6] W. Lerche, D. Lüst and A.N. Schellekens, *"Chiral four-dimensional heterotic strings from self-dual lattices"*, Nucl. Phys. **B287** (1987) 477.

[7] D. Friedan, *"Notes on string theory and two dimensional conformal field theory"*, in proceedings of Workshop on 'Unified String Theories', Santa Barbara 1985, eds. M. Green and D. Gross, World Scientific.

Chapter 14

Heterotic Strings in Ten and Four Dimensions

In this chapter we present constructions which describe string theories in four space-time dimensions. It should be clear from the beginning that these are the only string theories consistent with our empirical observation of living in four (almost) flat space-time dimensions and that only they have a chance to make contact with low-energy phenomenology.

14.1 Ten-dimensional heterotic strings

As a warm-up exercise let us first study ten-dimensional heterotic strings in the covariant lattice description. In Chapter 10 we have discussed the original version of the heterotic string which has a space-time supersymmetric spectrum and gauge groups $E_8 \times E_8$ or $SO(32)$. Subsequently, additional heterotic string theories were discovered in [1, 2] and reformulated in the covariant lattice approach in [3, 4].

The holomorphic (right-moving) fermionic string in its bosonized version is characterized by the lattice $(D_{5,1})_R$ corresponding to the world-sheet fermions $\psi^\mu(z)$ and superconformal ghosts β, γ. The lattice vectors $w_R = (\lambda_R, q) \in D_{5,1}$ describe the Lorentz transformation properties and superconformal ghost charge of the right-moving part of any string state. To obtain the heterotic string theory we have to combine the right-moving fermionic string with the left-moving bosonic string. As discussed in Chapter 10, it consists of ten bosonic space-time coordinates $X^\mu(\bar{z})$ and, in addition, of 16 "compactified" bosonic variables $X^I(\bar{z})$ ($I = 1, \ldots, 16$). The corresponding quantized momenta build a 16-dimensional Euclidean lattice $(\Gamma_{16})_L$ whose lattice vectors we denote by w_L. Thus the (soliton) vertex operators of the heterotic string theory have the general form (neglecting contributions from bosonic oscillators):

$$V_{w_L;\lambda_R,q}(\bar{z},z) = e^{iw_L\cdot X(\bar{z})} e^{i\lambda_R\cdot\phi(z)} e^{q\phi(z)}. \tag{14.1}$$

Here and in the following we have dropped normal ordering symbols and cocycle factors. The operator product expansion of two such vertex operators

$$V_{w_{L_1};\lambda_{R_1},q_1}(\bar{z},z) V_{w_{L_2};\lambda_{R_2},q_2}(\bar{w},w) \tag{14.2}$$

$$= (\bar{z}-\bar{w})^{w_{L_1}\cdot w_{L_2}} (z-w)^{\lambda_{R_1}\lambda_{R_2}-q_1 q_2} V_{w_{L_1}+w_{L_2};\lambda_{R_1}+\lambda_{R_2},q_1+q_2}(\bar{w},w) + \dots$$

shows that the condition for locality, i.e. the absence of branch cuts, reads

$$-w_{L_1}\cdot w_{L_2} + \lambda_{R_1}\cdot\lambda_{R_2} - q_1 q_2 \in \mathbf{Z}. \tag{14.3}$$

This suggests to combine the 16-dimensional left-moving lattice Γ_{16} and the six-dimensional right-moving lattice $D_{5,1}$ to a Lorentzian lattice

$$\Gamma_{16;5,1} = (\Gamma_{16})_L \otimes (D_{5,1})_R \tag{14.4}$$

(the semicolon separates left- from right-movers) with lattice vectors $w = (w_L;\lambda_R,q)$ where the inner product $w_1\cdot w_2$ is defined with the metric $\mathrm{diag}[(-1)^{16},(+1)^5,(-1)]$. Locality demands that this lattice be integral with respect to this Lorentzian metric.

Combining the left- and right-moving sectors, the partition function of ten-dimensional heterotic string theories is essentially given by the sum over all lattice vectors of $\Gamma_{16;5,1}$:

$$\chi(\bar{\tau},\tau) = \mathrm{Tr}\, e^{-i\pi\bar{\tau}(\bar{L}_0-1)} e^{i\pi\tau(L_0-1)} \Phi \tag{14.5}$$

$$= \frac{(\mathrm{Im}\tau)^{-4}}{\eta^{24}(\bar{\tau})\eta^{12}(\tau)\Theta_{1,1}(\tau)} \sum_{w\in\Gamma_{16;5,1}} e^{-2i\pi\bar{\tau}\frac{1}{2}w_L^2} e^{2i\pi\tau(\frac{1}{2}\lambda_R^2-\frac{1}{2}q^2-q-\frac{1}{2})} e^{-2\pi i q}.$$

Again, as in the fermionic string theory, modular invariance forces $\Gamma_{16;5,1}$ to be an odd self-dual Lorentzian lattice. This can be proven by Poisson resummation.

The requirement of self-duality can be trivially satisfied if both $(\Gamma_{16})_L$ and $(D_{5,1})_R$ are self-dual separately, i.e. $\Gamma_{16;5,1}$ is a direct product of two self-dual lattices of which $(D_{5,1})_R$ must be odd. Then $(\Gamma_{16})_L$ must be either the root lattice of $E_8 \times E_8$ or the weight lattice of $\mathrm{Spin}(32)/\mathbf{Z}_2$ implying $E_8 \times E_8$ or $SO(32)$ as the two possible gauge groups. On the other hand, self-duality of $(D_{5,1})_R$ implies that the spectrum is space-time supersymmetric in ten dimensions.

However this is by far not the most general case – it is possible to obtain a self-dual lattice $\Gamma_{16;5,1}$ without selfdual sublattices $(\Gamma_{16})_L$ and $(D_{5,1})_R$. Then eq.(14.4) does not represent a direct product decomposition. Instead, $\Gamma_{16;5,1}$ is specified by non-trivial correlations between the various conjugacy classes of $(\Gamma_{16})_L$ and $(D_{5,1})_R$ given by the glue vectors as explained in Chapter 11. Thus, in this case there is a non-trivial interplay between the left- and right-moving degrees of freedom. This implies that in the fermionic description of the right-moving spinning string the GSO projections have to be modified. This destroys space-time supersymmetry. Analogously, the left-moving gauge group will be different from $E_8 \times E_8$ or $SO(32)$.

Let us classify all possible odd self-dual lattices $\Gamma_{16;5,1}$ which lead to a sensible heterotic string theory in ten dimensions. In general, the classification of Lorentzian self-dual lattices $\Gamma_{p,q}$ of a given dimension and metric $\mathrm{diag}[(+1)^p, (-1)^q]$ is meaningless since they are unique up to Lorentz rotations in $\mathbf{R}^{p,q}$. However, if we add the requirement of having a sensible space-time interpretation, a classification becomes possible. Because of Lorentz invariance all states are classified (off-shell) according to $SO(10)$ representations and we have to demand that $(D_{5,1})_R$ builds the right-moving part of $\Gamma_{16;5,1}$. The non-trivial question is now how the four conjugacy classes 0, V, S and C $\in (D_{5,1})_R$ are coupled to the conjugacy classes of Γ_{16}.

We analyze this problem by converting the Lorentzian lattices into Euclidean ones. Any even self-dual lattice consisting of one or several D_n factors can be replaced by another even self-dual lattice by changing the dimension of any D_n factor by multiples of eight and keeping all conjugacy classes the same, i.e. $D_n \to D_{n+p8}$ ($p \in \mathbf{Z}$). Such a transformation changes the (length)^2 of all vectors only modulo 2 and all mutual scalar products modulo 1, so that it does not affect self-duality. For example, the self-dual lattice D_8 with 0 and S conjugacy classes (i.e. the E_8 root lattice) can be mapped to the D_{16} weight lattice with the same conjugacy classes which is, as we know from Chapter 11, the self-dual weight lattice of Spin(32)/$\mathbf{Z}_2$. On the other hand, changing the dimension by multiples of four, maps an even self-dual lattice to an odd self-dual lattice and vice versa. Finally, one may even subtract multiples of eight or four to make the dimension of a D_n factor negative. This can be interpreted as change of signature, i.e. as a map of an Euclidean self-dual lattice to a Lorentzian self-dual lattice or vice versa. This is what happened in the previous chapter; when we replaced $D_{5,1}$ ($E_{5,1}$) by D_8 (E_8) – we changed the dimension by minus four units, thus converting an odd self-dual Lorentzian lattice to an even self-dual Euclidean lattice. The reason for doing so is the possibility of classifying Euclidean lattices as discussed in Chapter 11.

We are now ready to apply these techniques to the lattice $\Gamma_{16;5,1}$ which describes the heterotic string theories in ten dimensions. First we map $(D_{5,1})_R$ to $(D_8)_R$ obtaining the Lorentzian even self-dual lattice $\Gamma_{16;8} = (\Gamma_{16})_L \otimes (D_8)_R$. This lattice can in turn be mapped to those 24-dimensional Euclidean even self-dual lattices Γ_{24} which can be decomposed as

$$\Gamma_{24} = \Gamma_{16} \otimes D_8 \,. \tag{14.6}$$

Thus, our aim is to find all even self-dual Euclidean 24-dimensional lattices which allow for this decomposition. Each solution will be completely specified by the Lie algebra lattice Γ_{16} together with the glue vectors to D_8.

In table 11.3 we have listed all possible even self-dual lattices of dimension 24, the so-called Niemeier lattices. Seven of them contain D_8 as regular subalgebra and therefore lead to a heterotic string theory in ten dimensions. These are displayed in the first column of table 14.1. They lead to eight different heterotic string theories since D_8 can be embedded in two different ways in $E_8 \otimes D_{16}$. In all the other cases there is only one possible regular embedding of D_8. The algebras which commute maximally with D_8 build the left lattice Γ_{16} and are displayed in the third column of the table. The appearing conjugacy classes of Γ_{16} as well as their coupling to the D_8 conjugacy classes are shown in the last four columns.

The root vectors of Γ_{16} give rise to massless gauge bosons of the heterotic string theory and the corresponding gauge group, which is of course always of rank 16, can be read off from the third column of the table. We recognize that the first two models are the two original supersymmetric heterotic string theories. Here Γ_{16} is self-dual and so is D_8 which is thus the root lattice of E_8. All other theories are not supersymmetric. Only one other model, the last one in the table, is tachyon free and has gauge group $SO(16) \times SO(16)$. In all models the tachyon, if present, comes from the V conjugacy class of $(D_8)_R$. For the $SO(16) \times SO(16)$ model, however, the V conjugacy class is coupled to the (V,S) and (S,V) conjugacy classes of $(D_8 \otimes D_8)_L$. The lowest states within these two conjugacy classes have $m_L^2 = \frac{1}{2}$ such that the right-moving tachyonic state with $m_R^2 = -\frac{1}{2}$ does not satisfy the left-right level matching constraint. In the five remaining non-supersymmetric models it does.

In summary, via bosonization one obtains eight different modular invariant heterotic string theories in ten dimensions. There exists one additional tachyonic model with rank eight gauge group E_8 which cannot be described within the covariant lattice formalism. The reason is that this model in-

Table 14.1 Ten dimensional heterotic strings

Niemeier lattice		Heterotic string				
roots	weights	algebra	D_8-sector			
			(0)	(V)	(S)	(C)
E_8^3	$(0,0,0)$	$E_8 \times E_8$	$(0,0)$	$-$	$(0,0)$	$-$
$E_8 \times D_{16}$	$(0,S)$	D_{16}	$(0),(S)$	$-$	$(0),(S)$	$-$
$E_8 \times D_{16}$	$(0,S)$	$E_8 \times D_8$	$(0,0)$	$(0,V)$	$(0,S)$	$(0,C)$
D_{24}	(S)	D_{16}	(0)	(V)	(S)	(C)
D_{12}^2	$[V,S]$	$D_4 \times D_{12}$	$(0,0)$	$(V,0)$	(S,V)	(S,C)
	(C,C)		(V,S)	$(0,S)$	(C,C)	(C,V)
$D_{10} \times E_7^2$	$(S,1,0)$	$D_2 \times E_7^2$	$(0,0,0)$	$(V,0,0)$	$(S,1,0)$	$(C,1,0)$
	$(C,0,1)$		$(V,1,1)$	$(0,1,1)$	$(C,0,1)$	$(S,0,1)$
	$(V,1,1)$					
$D_9 \times A_{15}$	$(S,4k+2)$	$D_1 \times A_{15}$	$(0,4k)$	$(V,4k)$	$(S,4k+2)$	$(C,4k+2)$
	$(0,4k)$					
D_8^3	$[S,V,V]$	$D_8 \times D_8$	$(0,0)$	(S,V)	(V,V)	$(C,0)$
	$[C,0,C]$		(C,C)	(V,S)	(S,S)	$(0,C)$
	(S,S,S)					

volves real fermions with different spin structures which however cannot be bosonized with the methods described in Chapter 11 and therefore do not lead to a covariant lattice. For lattice models, the Kac-Moody algebra corresponding to the gauge group is always at level one. The theory with gauge group E_8 has a E_8 Kac-Moody algebra at level two and can thus not be represented by free bosons in the way described in Chapter 11.

14.2 Four-dimensional heterotic strings in the covariant lattice approach

It is now straightforward to generalize this method of using bosonic covariant lattices to construct chiral heterotic string theories also in four space-time dimensions. The covariant lattice construction of four-dimensional heterotic strings was first discussed in [5]. A comprehensive review of this construction scheme is given in reference [6]. We should emphasize, however, that this is not the only way to obtain four-dimensional heterotic string theories. The first "realistic" proposal in the context of four-dimensional string theories is the compactification of the ten-dimensional heterotic string on Calabi-Yau manifolds [7]. This corresponds in general to a highly nontrivial conformal field theory not immediately related to free fields. Soon after the so-called orbifold construction [8] was considered where one uses free bosons with twisted boundary conditions. Finally, in the fermionic construction [9, 10] one uses only free world-sheet fermions. All these construction schemes have in common that they can lead to chiral fermions unlike the simple toroidal compactification. An overview and further references about four-dimensional string theories can be found in [11]. In the next section we will discuss some general features of four-dimensional heterotic strings which are independent of the way they are constructed.

Going from ten to d dimensions only d bosonic fields $X^\mu(\bar{z}, z)$ ($\mu = 1, \ldots, d$) play the role of space-time coordinates. The d-dimensional center-of-mass momenta k^μ, which are canonically conjugate to the center-of-mass positions, have continuous eigenvalues. We are left with $26 - d$ left-moving bosonic fields $X^I(\bar{z})$ ($I = 1, \ldots 26 - d$) and $10 - d$ right-moving fields $X^J(z)$ ($J = 1, \ldots 10 - d$). For these variables we assume that the holomorphic and antiholomorphic fields move independently on a $(26 - d)$-dimensional left and a $(10 - d)$-dimensional right torus respectively. This implies that the corresponding momentum eigenvalues w_L, w_R are quantized and generate a $26 - d$ dimensional lattice $(\Gamma_{26-d})_L$ and a $10 - d$ dimensional lattice $(\Gamma_{10-d})_R$. (For general choice of background values of the underlying torus the left- and right-moving momenta do not necessarily build a lattice. We will however discard this case for simplicity.)

Following our strategy of "maximal bosonization" we replace the ten right-moving world-sheet fermions $\psi^I(z)$ $(I = 1, \ldots, 10)$ by five free bosonic fields. However only d (d even) of these fermions are the two-dimensional world-sheet superpartners of the bosonic coordinates $X^\mu(z)$. These d fermionic fields generate a level one $SO(d)$ Kac-Moody algebra which corresponds to the Wick rotated Lorentz group $SO(d-1,1)$. Via bosonization these d fermions $\psi^\mu(z)$ are replaced by $\frac{d}{2}$ bosons $\phi^i(z)$ $(i = 1, \ldots, \frac{d}{2})$. To obtain a sensible space-time interpretation we have to insist that the corresponding zero mode momentum vectors are elements of the weight lattice of the Lorentz group $SO(d)$: $\lambda_R \in D_{\frac{d}{2}}$. Then, just like in ten dimensions, states in the R sector are characterized by $\lambda_R \in$ S,C of $D_{\frac{d}{2}}$. These states are d-dimensional fermions. The NS sector with d-dimensional space-time bosons is obtained from $\lambda_R \in$ 0,V of $D_{\frac{d}{2}}$.

As before, we also bosonize the superconformal ghosts by introducing a free boson $\phi(z)$. The ghost charge is either half-integer or integer. The canonical choice for space-time fermions is $q = -\frac{1}{2}$ and for space-time bosons $q = -1$. Again, we combine the D_1 ghost lattice with the $SO(d)$ weight lattice $D_{\frac{d}{2}}$ to form the covariant lattice $D_{\frac{d}{2},1}$ with lattice vectors (λ_R, q) and signature $((+1)^{\frac{d}{2}}, -1)$.

Thus we are now left with $10 - d$ additional, internal world-sheet fermions $\tilde{\psi}^J(z)$ $(J = 1, \ldots, 10 - d)$ which we replace by $\frac{1}{2}(10 - d)$ bosons $\tilde{\phi}^i(z)$ $(i = 1, \ldots, \frac{1}{2}(10 - d))$. Now, for these bosons it is not necessary that the corresponding bosonic momenta $\tilde{\lambda}$ are vectors of a $D_{5-\frac{d}{2}}$ lattice since we have given up the notion of ten-dimensional space-time. We are treating these fields, just like the bosons $X^J(z)$, as internal degrees of freedom which are needed to ensure conformal invariance, i.e. the vanishing of the central charge of the Virasoro algebra. So let us denote the lattice with vectors $\tilde{\lambda}$ by $\Gamma_{5-\frac{d}{2}}$. In fact, there is no reason to treat $X^J(z)$ and $\tilde{\phi}^i(z)$ differently. The distinction between compactified bosons and bosonized internal fermions is meaningless. Let us therefore combine the internal right-moving bosonic degrees of freedom to $X_R(z) = (X^1(z), \ldots, X^{10-d}(z), \tilde{\phi}^1(z), \ldots, \tilde{\phi}^{5-\frac{d}{2}}(z))$ with corresponding momentum vectors $w_R = (w_R^1, \ldots, w_R^{10-d}, \tilde{\lambda}^1, \ldots, \tilde{\lambda}^{5-\frac{d}{2}})$; they build a $(15 - \frac{3}{2}d)$-dimensional lattice $(\Gamma_{15-\frac{3}{2}d})_R = \Gamma_{10-d} \otimes \Gamma_{5-\frac{d}{2}}$.

We would like to emphasize that the d-dimensional heterotic string constructed in this way cannot, in general, be regarded as a compactification of the 10-dimensional heterotic string in the sense that the string is moving on a $(10 - d)$-dimensional internal compact manifold, even when taking into account possible background fields. This is due to the asymmetric treatment of the left- and right-moving fields and also to the fact that we did not treat the internal bosonized world-sheet fermions and compactified bosons in any different way. On the contrary, these theories provide, in general, truly four-dimensional string theories where only d bosons and world-sheet fermions play the role of (flat) space-time coordinates and their two-dimensional superpartners. The other bosonic fields are only needed to cancel the conformal anomaly and could be replaced by a more general internal conformal field theory (see next section).

The covariant vertex operators of the d-dimensional string states can be written as (neglecting bosonic oscillators and space-time momentum dependence):

$$V_{w_L; w_R, \lambda_R, q}(\bar{z}, z) = e^{i w_L \cdot X_L(\bar{z})} e^{i w_R \cdot X_R(z)} e^{i \lambda_R \cdot \phi(z)} e^{q \phi(z)} \quad (14.7)$$

$$w_L \in (\Gamma_{26-d})_L, \quad w_R \in (\Gamma_{15-\frac{3}{2}d})_R, \quad (\lambda_R, q) \in D_{\frac{d}{2}, 1}.$$

The conformal dimension of this operator is given by

$$\bar{h} = h_L = \frac{1}{2} w_L^2,$$

$$h = h_R = \frac{1}{2} w_R^2 + \frac{1}{2} \lambda_R^2 - \frac{1}{2} q^2 - q. \quad (14.8)$$

It follows that the mass of a d-dimensional string state created by this vertex operator is given by the following expressions where N_L and N_R count the number of left-and right-moving space-time as well as internal oscillators:

$$m_L^2 = \frac{1}{2} w_L^2 + N_L - 1,$$

$$m_R^2 = \frac{1}{2} w_R^2 + \frac{1}{2} \lambda_R^2 - \frac{1}{2} q^2 - q + N_R - 1, \qquad (14.9)$$

$$m^2 = m_L^2 + m_R^2.$$

Physical states have to satisfy $m_L^2 = m_R^2$. The operator product expansion between two vertex operators eq.(14.7) reads:

$$V_{w_{L_1};w_{R_1},\lambda_{R_1},q_1}(\bar{z},z) V_{w_{L_2};w_{R_2},\lambda_{R_2},q_2}(\bar{w},w) \qquad (14.10)$$

$$= (\bar{z}-\bar{w})^{w_{L_1}\cdot w_{L_2}} (z-w)^{w_{R_1}\cdot w_{R_2}+\lambda_{R_1}\cdot\lambda_{R_2}-q_1 q_2}$$

$$\times V_{w_{L_1}+w_{L_2};w_{R_1}+w_{R_2},\lambda_{R_1}+\lambda_{R_2},q_1+q_2}(\bar{w},w) + \dots$$

It follows, in complete analogy with the heterotic string theories in ten dimensions, that it is convenient to combine $(w_L; w_R, \lambda_R, q)$ to form vectors of a $(42 - 2d)$-dimensional Lorentzian lattice $\Gamma_{26-d;15-d,1} = (\Gamma_{26-d})_L \otimes (\Gamma_{15-\frac{3}{2}d} \otimes D_{\frac{d}{2},1})_R$. Again, modular invariance of the partition function forces $\Gamma_{26-d;15-d,1}$ to be an odd self-dual Lorentzian lattice.

For an odd self-dual lattice $\Gamma_{26-d;15-d,1}$ to represent a physically sensible heterotic string theory additional constraints have to be imposed. One requirement, namely that all states are classified according to representations of the Lorentz group $SO(d)$, is already satisfied if we demand that $(D_{\frac{d}{2},1})_R$ is part of $\Gamma_{26-d;15-d,1}$. This requirement was essentially sufficient to classify all heterotic string theories in ten dimensions. However in lower dimensions this is not the end of the story. Remember that for decoupling of ghosts the right-moving fermionic string must possess a two-dimensional (local) world-sheet supersymmetry. However this is lost in lower dimensions when we treat the internal bosons and the internal fermions on the same footing. In other words, having bosonized all right-moving internal degrees of freedom, which leads to the lattice $(\Gamma_{15-\frac{3}{2}d})_R$, we have to find a way to realize the right-moving world-sheet supersymmetry entirely in terms of $15 - \frac{3}{2}$ internal bosonic fields. In contrast to the ten-dimensional string theory, the world-sheet supersymmetry will now be manifest only in a more complicated, in general non-linearly realized way. Since the realization of the two-dimensional supersymmetry is one of the key points in the construction of the lower dimensional string theories, let us be more precise.

The supercurrent of the fermionic string theory has to satisfy the following operator algebra:

$$T(z)T_F(w) = \frac{\frac{3}{2}T_F(w)}{(z-w)^2} + \frac{\partial T_F(w)}{z-w} + \dots$$
$$T_F(z)T_F(w) = \frac{\frac{5}{2}}{(z-w)^3} + \frac{\frac{1}{2}T(w)}{z-w} + \dots \tag{14.11}$$

Eq.(14.11) shows that $T_F(z)$ belongs to a superconformal field theory with central charge $c = 15$ which can be realized in the simplest case by 10 free bosons plus 10 free fermions. This kind of linear realization of world-sheet supersymmetry appears in the fermionic string theory in 10 dimensions. However in lower dimensional theories there are only d free bosons $X^\mu(z)$ together with their superpartners $\psi^\mu(z)$. These fields contribute $\frac{3}{2}d$ units to the central charge of the superconformal algebra. The corresponding space-time supercurrent takes it standard form:

$$T_F^{\text{space-time}}(z) = -\frac{1}{2}\psi^\mu(z)\partial X_\mu(z). \tag{14.12}$$

The missing $15 - \frac{3}{2}d$ units to the central charge are provided by the internal fields $X_R(z)$. The corresponding internal supercurrent $T_F^{\text{int}}(z)$ must be built entirely from the bosons $X_R(z)$ and it must satisfy

$$T_F^{\text{int}}(z)T_F^{\text{int}}(w) = \frac{\frac{1}{6}(15 - \frac{3}{2}d)}{(z-w)^3} + \frac{\frac{1}{2}T^{\text{int}}(w)}{z-w} + \dots \tag{14.13}$$

where $T^{\mathrm{int}}(w)$ is the internal energy momentum tensor. The most general ansatz for T_F^{int}, which is a conformal field of dimension $3/2$ and built entirely from the $15 - \frac{3}{2}d$ free bosons $X_R(z)$, is

$$T_F^{\mathrm{int}}(z) = \sum_t A(t) e^{it \cdot X_R(z)} + i \sum_l B(l) \cdot \partial X_R(z) e^{il \cdot X_R(z)} \qquad (14.14)$$

with

$$t^2 = 3 \quad , \qquad l^2 = 1 \quad , \qquad B \cdot l = 0.$$

The coefficients A and B have to be determined such that eq.(14.13) is satisfied. One can show that a necessary condition to arrive at theories with chiral fermions is $B = 0$. Since these are the most interesting theories from a phenomenological point of view, we will limit our discussion to this case.

The $(\mathrm{length})^2 = 3$ vectors t play a very important role in the construction of lower dimensional heterotic string theories. This becomes clear if we consider the picture changing operator $P_{+1}(z)$ which is itself a sum of a space-time and internal part:

$$P_{+1}(z) = P_{+1}^{\mathrm{s.t.}}(z) + P_{+1}^{\mathrm{int}}(z) = -2\big(e^{\phi} T_F^{\mathrm{s.t.}}(z) + e^{\phi} T_F^{\mathrm{int}}(z)\big). \qquad (14.15)$$

Because of BRST invariance, $P_+^{\mathrm{int}}(z)$ must act properly on all states, i.e. it maps a physical state to its picture changed image. Thus consider a state of the form eq.(14.7) characterized by a lattice vector $(w_L; w_R, \lambda_R, q) \in (\Gamma_{26-d})_L \otimes (\Gamma_{15-\frac{3}{2}d} \otimes D_{\frac{d}{2},1})_R$. The operator product with $P_{+1}^{\mathrm{int}}(z)$ then reads:

$$P_{+1}^{\mathrm{int}}(z) \, V_{w_L; w_R, \lambda_R, q}(\bar{w}, w)$$
$$= -2 \sum_t A(t)(z-w)^{t \cdot w_R - q} V_{w_L; w_R + t, \lambda_R, q+1}(\bar{w}, w) + \dots \qquad (14.16)$$

We find that the picture changed state is characterized by lattice vectors $(w_L; w_R + t, \lambda_R, q + 1)$ (if $w_R \cdot t = -1(-\frac{1}{2})$ in the NS (R) sector for states in the canonical ghost picture). Furthermore, because of the requirement of locality we have to demand that every lattice vector $w_R \in (\Gamma_{15-\frac{3}{2}d})_R$ satisfies $w_R \cdot t \in \mathbf{Z}$ (NS) and $w_R \cdot t \in \mathbf{Z} + \frac{1}{2}$ (R). Since we know on the other hand that $\Gamma_{26-d;15-d,1}$ is a self-dual lattice, it immediately follows that $(0; t, 0, 1) \in (\Gamma_{26-d})_L \otimes (\Gamma_{15-\frac{3}{2}d})_R \otimes (D_{\frac{d}{2},1})_R$.

The upshot of this discussion is that the vectors t must themselves be lattice vectors of the right-moving internal lattice $(\Gamma_{15-\frac{3}{2}d})_R$. They always appear in connection with the V conjugacy class of $(D_{\frac{d}{2},1})_R$, since $P_{+1}^{\mathrm{int}}(z)$ has ghost charge 1. The vectors t are called constraint vectors. Any sensible heterotic string theory, whose right-moving part possesses world-sheet supersymmetry, requires the existence of these constraint vectors.

In summary, the classification of d-dimensional heterotic strings within the covariant lattice construction amounts to determine all Lorentzian odd self-dual lattices $\Gamma_{22-d;15-d,1}$ which contain $(D_{\frac{d}{2},1})_R$ and allow for a proper realization of the internal supercurrent. For theories with chiral fermions the internal supercurrent must have the form of eq.(14.14) with constraint vectors t, $t^2 = 3$. Only a finite but huge number of Lorentzian lattices satisfies this constraint and these lattices can be shown to possess only a finite number of conjugacy classes, i.e. they correspond to rational conformal field theories.

Let us now discuss some general features of the spectrum of lower dimensional heterotic string theories. We restrict the discussion to the most interesting case, namely the four-dimensional theories. Generalization to other even dimensions is straightforward. In four dimensions, the covariant lattice has the structure

$$\Gamma_{22;11,1} = (\Gamma_{22})_L \otimes (\Gamma_9 \otimes D_{2,1})_R \tag{14.17}$$

In the following we are using the mass formula eq.(14.9). In the canonical ghost pictures $q = -1$ (NS) and $q = -\frac{1}{2}$ (R) massless states must satisfy

$\frac{1}{2}w_L^2 + N_L = 1$ and $\lambda_R^2 + w_R^2 = 1$, $N_R = 0$ (NS) or $\lambda_R^2 + w_R^2 = 5/4$, $N_R = 0$ (R). Let us first concentrate on the left sector. One possibility to obtain massless states is if $w_L^2 = 0$ and $N_L = 1$. There are two different kinds of oscillator contributions. First, a space-time oscillator $\partial X^\mu(\bar{z})$, from which the corresponding state will inherit a four-dimensional vector index μ. Second, we have 22 internal oscillators $\partial X^I(\bar{z})$ ($I = 1, \ldots, 22$) which correspond to the commuting Cartan subalgebra generators of $[U(1)]_L^{22}$. These two types of oscillator excitations are present in any covariant lattice model such that the rank of the gauge group is always 22. However the gauge symmetry can be extended to a non-Abelian group G via the Frenkel-Kac mechanism if the covariant lattice contains vectors $(w_L; 0_R)$, $w_L^2 = 2$. These vectors are called roots of the left lattice and the corresponding vertex operators generate, together with the internal oscillators, a level one Kac-Moody algebra $\hat{g}$. Thus G appears as non-Abelian gauge group in four dimensions.

Consider now the right-moving part of the theory. For space-time bosons (NS sector) the superconformal ghosts contribute already $\frac{1}{2}$ unit to m_R^2. It follows that λ_R can be either 0 or a vector weight of D_2 to obtain a massless state. In the former case the corresponding right-moving state is a Lorentz scalar and the internal lattice vector w_R must obey $w_R^2 = 1$ for masslessness. If both $\lambda_R = w_R = 0$ we obtain a right-moving tachyon of $m_R^2 = -\frac{1}{2}$. If λ_R is a vector weight of D_2, i.e. $\lambda_R = (\pm 1, 0), (0, \pm 1)$, and $w_R = 0$ we have a massless four-dimensional vector. For space-time fermions (R sector) the superconformal ghosts contribute $3/8$ units to the right mass. Now massless spinors of positive chirality are obtained if $\lambda_R = \pm(\frac{1}{2}, \frac{1}{2})$ and the internal lattice vector satisfies $w_R^2 = \frac{3}{4}$. Analogously, spinors of negative chirality have $\lambda_R = \pm(\frac{1}{2}, -\frac{1}{2})$ and again $w_R^2 = \frac{3}{4}$.

Now let us combine the left- and right-moving sectors to discuss the possible massless states in a four-dimensional heterotic string theory. It is important to note that the occurrence of most of these states is model dependent, i.e. depends on whether the various combinations of conjugacy classes, discussed in the following, are present in an explicit covariant lattice $\Gamma_{22;11,1}$. Also note that to every lattice vector its negative lattice vector is also present. This lattice automorphism corresponds in four dimensions to CPT conjugation. It changes the chirality of every spinor if we reexpress it in the canonical ghost picture. So any fermion in the representation $\underline{R}$ of the gauge group G is automatically accompanied by its CPT conjugate in the representation $\underline{\overline{R}}$. Together they count as one Weyl or Majorana spinor. The possible states are then:

a) Graviton, antisymmetric tensor, dilaton

This sector exits model independently in any four-dimensional string theory. There are no left-moving lattice excitations but one left-moving space-time oscillator. On the right-moving lattice we have $w_R = 0$, λ_R a vector weight of D_2 and $q = -1$. The corresponding vertex operator in the canonical ghost picture is

$$V(\bar{z}, z) = \epsilon_{\mu\nu}\, \partial X^\mu(\bar{z})\psi^\nu(z)e^{-\phi(z)}e^{ik_\rho X^\rho(\bar{z}, z)}. \qquad (14.18)$$

$\epsilon_{\mu\nu}$ denotes the polarization tensor. It is symmetric and traceless for the graviton $h_{\mu\nu}$, antisymmetric for $B_{\mu\nu}$ (which is a pseudo scalar in four dimensions after a duality transformation) and the trace part for the dilaton. BRST invariance requires $k^\mu \epsilon_{\mu\nu} = 0$.

b) Left gauge bosons

The right-moving sector is identical to the gravity sector. For the left-moving sector we have to consider two types of states. First, the 22 internal oscillators which lead to the gauge vector bosons of the Cartan subalgebra $[U(1)]_L^{22}$. Like the states in *a)* these states are present in any covariant lattice theory. Their vertex operator is almost identical to eq.(14.18). One only replaces the space-time oscillator $\partial X^\mu(\bar{z})$ by an internal oscillator $\partial X^I(\bar{z})$ ($I = 1, \ldots, 22$) and $\epsilon_{\mu\nu}$ by a polarization vector ϵ_ν. Second, $\partial X^I(\bar{z})$ can be replaced by root vectors of $(\Gamma_{22})_L$ with $w_L^2 = 2$ corresponding to non-Abelian gauge bosons of the gauge group G. Their vertex operators contain a factor $e^{iw_L \cdot X(\bar{z})}$ which replaces $\partial X^\mu(\bar{z})$ in eq.(14.18). The corresponding Kac-Moody algebra is always of level one.

c) Right gauge bosons

These types of states get their vector index from a left-moving space-time oscillator $\partial X^\mu(\bar{z})$. Thus, the right-moving part must contribute a Lorentz scalar with $w_R^2 = 1$, $\lambda_R = 0$, $q = -1$. Since there is no left-moving lattice excitation w_L connected with w_R, $w_R^2 = 1$, these states correspond to new "roots" $(w_R, 0, -1)$ of the lattice $(\Gamma_9 \otimes D_{2,1})_R$. This then implies that $D_{2,1}$ is part of a larger algebra $D_{n,1}$ $n > 2$. Then space-time spinors (see next paragraph) must come from the decomposition of $D_{n,1}$ spinors and appear automatically in non-chiral pairs. Thus chiral fermions arise only from covariant lattices which do not lead to right gauge bosons. In this sense the right gauge bosons are not interesting from a phenomenological point of view.

d) (Chiral) fermions

For these states λ_R is a spinor weight of D_2 and $w_R^2 = 3/4$ ($q = -1/2$). Since we are considering spin $\frac{1}{2}$ particles, the left-moving part must only contribute internal, gauge degrees of freedom. Therefore we are led to consider lattice vectors w_L with $w_L^2 = 2$. These vectors are in general not roots of $(\Gamma_{22})_L$, which will always be accompanied by internal oscillators and space-time oscillators, leading to gauginos and gravitinos, respectively (see *(f)* below). The w_L are then weight vectors of particular representations of the generally non-semi-simple gauge group G. In order to obtain chiral fermions these weights should correspond to complex representations of G (e.g. spinor representations of $SO(4n - 2)$). In addition one must ensure that the lattice $\Gamma_{22;11,1}$ does not contain conjugacy classes with the same vector $(w_L; w_R)$ coupled to the antispinor weight of D_2. These states would act as mirror particles and destroy the chiral structure.

e) Scalars

Massless scalars have $\lambda_R = 0$ and $w_R^2 = 1$ ($q = -1$). In general, they transform non-trivially under the left gauge group and are also characterized by weights $w_L^2 = 2$.

f) Gravitinos - space-time supersymmetry

To obtain gravitinos we need space-time spinors together with $w_R^2 = 3/4$ and $w_L = 0$, $N_L = 1$ where the left-moving oscillator carries a space-time vector index μ. The presence of the lattice vectors $(0; w_R, \lambda_R, -\frac{1}{2})$ with $w_R^2 = 3/4$ and λ_R a spinor of D_2 implies that $(D_{2,1})_R$ is part of a bigger algebra in analogy to the case of the right gauge bosons. The vectors $(w_R, \lambda, -\frac{1}{2})$ are just spinorial roots of this bigger algebra and the only regular embeddings which yield such roots are the embeddings of $D_{2,1}$ into the exceptional algebras $E_{3,1}$, $E_{4,1}$ and $E_{5,1}$ which are the Lorentzian analogs of the Euclidean algebras E_6, E_7 and E_8. If we map $D_{2,1}$ to D_5 the condition for the presence of gravitinos is that there are $(\text{length})^2 = 2$ vectors with spinor components in D_5. The only regular embeddings of D_5 with this property are those into E_6, E_7 and E_8 leading to one, two and four gravitinos respectively. The Euclidean exceptional algebras can be obtained from the Lorentzian ones by replacing D_1^{ghost} by D_3^{ghost} as explained before. We will come back to this in the next section where we will also discuss in detail the connection between exceptional groups and space-time supersymmetry which is a model-independent feature of all four-dimensional strings, not restricted to the covariant lattice construction. As a consequence of space-time supersymmetry, any physical fermion is accompanied by a physical boson. Adding or subtracting the gravitino lattice vector to a particular state provides its supersymmetric partner. Note that supersymmetric partners necessarily have different vectors w_R. However the total number of these vectors appearing together with a particular vector w_L does of course coincide for the fermions and bosons. We can now also replace the left-moving space-time oscillators of the gravitino vertex operator by an internal oscillator or a root of $(\Gamma_{22})_L$ and get the supersymmetric partners of the gauge bosons, the gauginos.

Let us now present an explicit example of a four-dimensional heterotic string theory which has $N = 1$ space-time supersymmetry, possesses chiral fermions and is tachyon free. We start with the Euclidean even self-dual 24-dimensional Niemeier lattice $D_7 \otimes A_{11} \otimes E_6$ with glue vector in the $(\underline{64}, \underline{12}, \underline{27})$ representation. The lattice can be decomposed to

$$
\begin{aligned}
\Gamma_{24} &= D_7 \otimes D_3^3 \otimes U(1)^2 \otimes E_6 \\
&= D_7 \otimes D_3^3 \otimes U(1)^3 \otimes D_5.
\end{aligned}
\tag{14.19}
$$

As explained before, we can map this lattice to a Lorentzian odd self-dual lattice $\Gamma_{22;11,1}$. To do this we make the replacements $(D_7)_R \rightarrow (D_3)_L \otimes (D_2)_R$, $(D_3)_R \rightarrow (D_7)_L \otimes (D_2)_R$, $(D_3)_R \rightarrow (D_7)_L \otimes (D_2)_R$, $(D_3)_R \rightarrow (D_5)_L$ and $(D_5)_R \rightarrow (D_{2,1})_R$. The last map corresponds to replacing D_3 by the ghost lattice D_1 while simultaneously reversing the signature (cf. also Chapter 13). It follows that the $U(1)$ factor from the decomposition of E_6 in eq.(14.19) combines with $D_{2,1}$ to $E_{3,1}$ as described above. We then get

$$
\begin{aligned}
\Gamma_{22;11,1} &= [D_3 \otimes D_7^2 \otimes D_5]_L \otimes [D_2^3 \otimes U(1)^2 \otimes E_{3,1}]_R \\
&= [D_3 \otimes D_7^2 \otimes D_5]_L \otimes [D_2^3 \otimes U(1)^3 \otimes D_{2,1}]_R
\end{aligned}
\tag{14.20}
$$

from which we immediately read off the rank 22 gauge group

$$
G = SO(6) \times [SO(14)]^2 \times SO(10).
\tag{14.21}
$$

We can now also decompose the glue vector according to eq.(14.19) and then generate all conjugacy classes of the lattice Γ_{24} or $\Gamma_{22;11,1}$. This is most economically done by computer. The following conjugacy classes contain massless spin 1/2 fermions (the semi-colon separates the left from the right lattice):

D_3	D_7	$D_{\bar 7}$	D_5	D_2	D_2	D_2	$U(1)$	$U(1)$	$U(1)$	$D_{2,1}$
(0,	V,	V,	0;	0,	0,	0,	$\frac{\sqrt{3}}{6}+\frac{1}{2}$,	$\frac{\sqrt{3}}{6}-\frac{1}{2}$,	$\frac{\sqrt{3}}{6}$,	S)
(0,	V,	0,	V;	0,	0,	0,	$\frac{\sqrt{3}}{6}-\frac{1}{2}$,	$\frac{\sqrt{3}}{6}+\frac{1}{2}$,	$\frac{\sqrt{3}}{6}$,	S)
(0,	0,	V,	V;	0,	0,	0,	$-\frac{\sqrt{3}}{3}$,	$-\frac{\sqrt{3}}{3}$,	$\frac{\sqrt{3}}{6}$,	S)
(V,	V,	0,	0;	0,	0,	0,	$-\frac{\sqrt{3}}{3}$,	$-\frac{\sqrt{3}}{3}$,	$\frac{\sqrt{3}}{6}$,	S)
(V,	0,	V,	0;	0,	0,	0,	$\frac{\sqrt{3}}{6}-\frac{1}{2}$,	$\frac{\sqrt{3}}{6}+\frac{1}{2}$,	$\frac{\sqrt{3}}{6}$,	S)
(V,	0,	0,	V;	0,	0,	0,	$\frac{\sqrt{3}}{6}+\frac{1}{2}$,	$\frac{\sqrt{3}}{6}-\frac{1}{2}$,	$\frac{\sqrt{3}}{6}$,	S)
(S,	0,	0,	C;	C,	0,	0,	$-\frac{1}{4}-\frac{\sqrt{3}}{12}$,	$\frac{1}{4}-\frac{\sqrt{3}}{12}$,	$\frac{\sqrt{3}}{6}$,	S)
(C,	0,	0,	C;	S,	0,	0,	$-\frac{1}{4}-\frac{\sqrt{3}}{12}$,	$\frac{1}{4}-\frac{\sqrt{3}}{12}$,	$\frac{\sqrt{3}}{6}$,	S)

Their CPT conjugate states are in the conjugacy classes with the negative lattice vectors. For the $U(1)$ factors we have given the charges. Note that this model possesses chiral fermion. The last entry $S \in D_{2,1}$ denotes just one light-cone degree of freedom. If we now look for conjugacy classes with massless scalars, we find the same list as above with only the last two columns replaced by $(\frac{\sqrt{3}}{6}, S) \to (-\frac{\sqrt{3}}{3}, V)$, each of them also representing one physical degree of freedom (apart from the multiplicity given by the $(D_2^3)_R$ part which provides a kind of family replication of the massless fields). This is just a reflection of space-time supersymmetry. We will discuss this situation in detail in the next section. For the discussion there it will be useful to keep this particular example in mind. For non-supersymmetric theories we will in general get different numbers of fermions and bosons characterized by different lattice vectors. We will refer to the scalars and fermions above as matter fields as they have spin 0 and 1/2 respectively.

There are three more conjugacy classes which lead to massless physical fields, namely

$$
\begin{array}{ccccccccccc}
D_3 & D_7 & D_7 & D_5 & D_2 & D_2 & D_2 & U(1) & U(1) & U(1) & D_{2,1} \\
(0, & 0, & 0, & 0; & 0, & 0, & 0, & 0, & 0, & 0, & 0) \\
(0, & 0, & 0, & 0; & 0, & 0, & 0, & 0, & 0, & \tfrac{\sqrt{3}}{2}, & \mathrm{C}) \\
(0, & 0, & 0, & 0; & 0, & 0, & 0, & 0, & 0, & -\tfrac{\sqrt{3}}{2}, & \mathrm{S})
\end{array}
$$

The trivial conjugacy class contains the graviton, antisymmetric tensor and dilaton as well as the gauge bosons. The remaining two conjugacy classes contain both the gauginos and the gravitino. It is now important to observe that if we take any of the massless scalars and add the gravitino lattice vector, we get a fermion, the supersymmetric partner of the boson. This is also trivially satisfied for the gravity sector. We learn that the gravitino acts as the supersymmetry charge. It is important to realize that supersymmetric partners differ only in their space-time and superconformal ghost quantum numbers and the $U(1)$ charge coming from the decomposition of the E_6 factor of the original lattice. It is the appearance of $E_6 \supset U(1) \otimes D_5$ or $E_{3,1} \supset U(1) \otimes D_{2,1}$ in the right lattice which is responsible for the presence of space-time supersymmetry. We can also add to the scalars twice the gravitino lattice vector and end up in conjugacy classes with $(\tfrac{2}{3}\sqrt{3}, 0) \in (U(1) \times D_{2,1})$. This leads to massless states with only exist in the $q = 0$ ghost picture and which are not BRST invariant. These are the auxiliary fields of the chiral $N = 1$ multiplets.

Given the conjugacy classes it is now trivial to write down the representations in which the matter fields transform under the gauge group G:

$$G: \quad SO(6) \otimes SO(14) \otimes SO(14) \otimes SO(10)$$

$$
\begin{array}{cccc}
\underline{1} & \underline{14} & \underline{14} & \underline{1} \\
\underline{1} & \underline{14} & \underline{1} & \underline{10} \\
\underline{1} & \underline{1} & \underline{14} & \underline{10} \\
\underline{6} & \underline{14} & \underline{1} & \underline{1} \\
\underline{6} & \underline{1} & \underline{14} & \underline{1} \\
\underline{6} & \underline{1} & \underline{1} & \underline{10} \\
\underline{4} & \underline{1} & \underline{1} & \underline{\overline{16}} \\
\underline{\overline{4}} & \underline{1} & \underline{1} & \underline{\overline{16}}
\end{array}
$$

The 24 constraint vectors t which build the two-dimensional supercurrent in this particular model are

$$
\begin{array}{cccccc}
U(1) & U(1) & U(1) & D_2 & D_2 & D_2
\end{array}
$$

$$
\begin{array}{l}
(\pm(\tfrac{\sqrt{3}}{3}, \quad \tfrac{\sqrt{3}}{3}, \quad \tfrac{\sqrt{3}}{3}),\ \pm 1,0,\ \pm 1,0,\ 0,0) \\[4pt]
(\pm(\tfrac{1}{2} - \tfrac{\sqrt{3}}{6}, \quad -\tfrac{1}{2} - \tfrac{\sqrt{3}}{6}, \quad \tfrac{\sqrt{3}}{3}),\ 0,\pm 1,\ 0,0,\ \pm 1,0) \\[4pt]
(\pm(-\tfrac{1}{2} - \tfrac{\sqrt{3}}{6}, \quad \tfrac{1}{2} - \tfrac{\sqrt{3}}{6}, \quad \tfrac{\sqrt{3}}{3}),\ 0,0,\ 0,\pm 1,\ 0,\pm 1)
\end{array}
$$

In order to satisfy the internal superconformal algebra eq.(14.13), the coefficients $A(t)$ must satisfy $A(t)^2 = \tfrac{1}{16}$.

14.3 General aspects of four-dimensional heterotic string theories

In the previous section we have presented one particular way to construct four-dimensional heterotic string theories, the covariant lattice construction, which uses only free (chiral) world-sheet bosons. This construction already gives an enormous number of consistent four-dimensional heterotic string theories. This shows that whereas in ten dimensions there existed only a few theories, their number proliferates rapidly as we go to lower dimensions. Moreover, there are different ways to construct heterotic string theories in lower dimensions and there is an overlap between them. However, all these constructions are examples of conformal field theories which describe the internal string degrees of freedom. A specific conformal field theory has to pass certain consistency requirements such as the correct value of the central charge, modular invariance, etc. So let us discuss properties of four-dimensional heterotic string theories which follow from the general structure of the internal conformal field theory. In addition, one can use "phenomenological" input like space-time supersymmetry or a chiral fermion spectrum to constrain the internal conformal field theory further. This will be discussed at the end of this section.

Any four-dimensional heterotic string theory possesses four space-time string coordinates $X^\mu(\bar{z}, z)$. In addition there are also the right-moving world-sheet fermions $\psi^\mu(z)$. Finally, there are the conformal ghosts $\bar{b}(\bar{z})$, $b(z)$, $\bar{c}(\bar{z})$ and $c(z)$ as well as the superconformal ghosts $\beta(z)$ and $\gamma(z)$. These fields build the external conformal field theory. The left-moving part is a conformal field theory with central charge $\bar{c}^{\text{ext}} = -22$ and the right-moving part a superconformal field theory with $c^{\text{ext}} = -9$. Conformal invariance requires that the internal world-sheet degrees of freedom cancel the conformal anomaly of the external fields, i.e. we need $\bar{c}^{\text{int}} = 22$, $c^{\text{int}} = 9$:

$$\bar{T}^{\text{int}}(\bar{z})\bar{T}^{\text{int}}(\bar{w}) = \frac{11}{(\bar{z}-\bar{w})^4} + \frac{2\bar{T}^{\text{int}}(\bar{w})}{(\bar{z}-\bar{w})^2} + \frac{\bar{\partial}\bar{T}^{\text{int}}(\bar{w})}{\bar{z}-\bar{w}} + \ldots$$

$$T^{\text{int}}(z)T^{\text{int}}(w) = \frac{\frac{9}{2}}{(z-w)^4} + \frac{2T^{\text{int}}(w)}{(z-w)^2} + \frac{\partial T^{\text{int}}(w)}{z-w} + \ldots \tag{14.22}$$

In addition, decoupling of the spurious states or, equivalently, Lorentz invariance in the light-cone gauge, requires the existence of a (local) $n = 1$ internal right-moving world-sheet supersymmetry generated by the internal supercurrent $T_F^{\text{int}}(z)$ which has to satisfy

$$T_F^{\text{int}}(z)T_F^{\text{int}}(w) = \frac{\frac{3}{2}}{(z-w)^3} + \frac{\frac{1}{2}T^{\text{int}}(w)}{z-w} + \ldots \tag{14.23}$$

It is crucial to realize that the left-moving and right-moving internal conformal field theories can be chosen independent of each other; only locality and modular invariance link the left- and right-moving sectors in a definite way. Moreover, it is not required that the internal conformal field theory admits an interpretation as a compactification on some manifold.

It is again convenient to replace the four external world-sheet fermions $\psi^\mu(z)$ by bosonic fields $\phi^i(z)$ $(i = 1, 2)$ and the superconformal ghosts by a scalar $\phi(z)$. Then all states are characterized by vectors (λ_R, q) of the covariant lattice $(D_{2,1})_R$ and the general form of vertex operators is

$$V(\bar{z}, z) = (\bar{\partial}^{n_L} X^\mu(\bar{z}))^{m_L} (\partial^{n_{1R}} X^\nu(z))^{m_{1R}} (\partial^{n_{2R}} \phi^i(z))^{m_{2R}}$$
$$\times\, e^{i\lambda_R \cdot \phi(z)} e^{q\phi(z)} V_{\text{int}}(\bar{z}, z) e^{ik_\mu X^\mu(\bar{z}, z)} . \tag{14.24}$$

$V_{\text{int}}(\bar{z}, z)$ are the conformal fields of the internal conformal field theory with conformal weight $(h_{\text{int}}, \bar{h}_{\text{int}})$. The mass of the corresponding state is given by

$$m_L^2 = N_L + \bar{h}_{\text{int}} - 1,$$

$$m_R^2 = \frac{1}{2}\lambda_R^2 - \frac{1}{2}q^2 - q + N_{1R} + N_{2R} + h_{\text{int}} - 1, \tag{14.25}$$

$$m^2 = m_L^2 + m_R^2.$$

$N = nm$ are oscillator excitation numbers. We require that the internal conformal field theory be unitary which implies that $\bar{h}_{\text{int}}, h_{\text{int}} \geq 0$. Furthermore, we have to insist that the theory is local. This requirement provides a link between the external and the internal part of the theory. Consider the operator product

$$V_1(\bar{z}, z) V_2(\bar{w}, w)$$

$$\sim (\bar{z} - \bar{w})^{-\bar{h}^1_{\text{int}} - \bar{h}^2_{\text{int}} + \bar{h}^3_{\text{int}}} (z - w)^{-h^1_{\text{int}} - h^2_{\text{int}} + h^3_{\text{int}} + \lambda_{1R} \cdot \lambda_{2R} - q_1 q_2} \tag{14.26}$$

$$\times\, e^{i(\lambda_{1R} + \lambda_{2R}) \cdot \phi(w)} e^{(q_1 + q_2)\phi(w)} V^3_{\text{int}}(\bar{w}, w) + \dots$$

where we have neglected the bosonic oscillators which always give rise to integer powers of $(z - w)$. Also, the $e^{ik \cdot X}$ factors give $|z - w|^{k_1 \cdot k_2}$ which does not have branch cuts. Locality now demands that

$$\bar{h}^1_{\text{int}} + \bar{h}^2_{\text{int}} - \bar{h}^3_{\text{int}} - h^1_{\text{int}} - h^2_{\text{int}} + h^3_{\text{int}} + \lambda_{1R} \cdot \lambda_{2R} - q_1 q_2 \in \mathbf{Z}. \qquad (14.27)$$

The complete Hilbert space, containing the left-moving and right-moving, the internal and external states, must obey above condition. Modular invariance of the one-loop partition function constrains the possible combinations of internal and external highest weight states further. However it is very difficult to analyze the constraints of modular invariance for a general model.

Let us now discuss the possible spectrum of the four-dimensional heterotic string theories.

Graviton sector

This sector is completely model independent and present in any four-dimensional string theory. This comes from the fact that any unitary internal conformal field theory contains the identity operator with $\bar{h}_{\text{int}} = h_{\text{int}} = 0$. Also, the 0 conjugacy class of $(D_{2,1})_R$ must always be present and it leads in the NS sector with canonical ghost charge $q = -1$ to a right-moving space-time vector with $\lambda_R = (\pm 1, 0), (0, \pm 1)$. Finally, $\partial X^\mu(\bar{z})$ is always possible. This combination of conformal fields yields the graviton $h_{\mu\nu}$, the anti-symmetric tensor $B_{\mu\nu}$ and the dilaton vertex operators eq.(14.18).

Gauge sector

Here we make the (reasonable) assumption that any gauge symmetry has its origin in an internal Kac-Moody algebra of level k as is the case for all known closed string theories. The affine Kac-Moody algebra eq.(11.16) can be realized either on the left-moving side by currents $J^a(\bar{z})$ $(a = 1, \ldots, \dim G)$ or on the right-moving side by currents $J^a(z)$. In the former case the vertex operators of the corresponding gauge bosons are given by (see also the discussion in the next chapter about the normalization of these vertex operators)

$$V_\mu^a(\bar{z}, z, k) = J^a(\bar{z})\psi_\mu(z)e^{-\phi(z)}e^{ik_\nu X^\nu(\bar{z},z)} . \tag{14.28}$$

In the latter case, super-BRST invariance demands that $J^a(z)$ is the upper component of a two-dimensional superfield, i.e. can be obtained via picture changing from a dimension $\frac{1}{2}$ field $\hat{J}^a(z)$. Then the right-moving gauge boson vertex operators in the canonical ghost picture is

$$V_\mu^a(\bar{z}, z, k) = \partial X_\mu(\bar{z})\hat{J}^a(z)e^{-\phi(z)}e^{ik_\nu X^\nu(\bar{z},z)} , \tag{14.29}$$

where in the 0-ghost picture we have

$$V_\mu^a(\bar{z}, z, k) = \partial X_\mu(\bar{z})J^a(z)e^{ik_\nu X^\nu(\bar{z},z)} . \tag{14.30}$$

Vertex operators of the form eq.(14.30) with $J^a(z)$ not being the upper component of a two-dimensional superfield correspond to "auxiliary" non-propagating gauge fields which are important for the description of space-time superfields in supersymmetric heterotic string theories. They are not BRST invariant.

Let us consider again the left gauge bosons of eq.(14.28). The internal Kac-Moody algebra $\hat{g}$ contributes to the central charge of the internal conformal field theory with

$$\bar{c} = \frac{2k \dim G}{C_2 + 2k}. \tag{14.31}$$

Clearly this value must not exceed the total internal central charge $\bar{c} = 22$. This gives a limit on the dimension of the gauge group dependent on the level k and the Casimir C_2. If k is one, the rank of the gauge group must be less or equal to 22 (Cf. eq.(11.25)). The existence of a level one Kac-Moody algebra also implies that the internal conformal field theory contains a certain number of free bosons or fermions which provide an explicit realization of the internal Kac-Moody currents. Thus, for these cases the corresponding part of the internal conformal field theory has a very simple structure.

Massless fermions

Now assume that the theory contains N_F massless fermions Ψ_i ($i = 1,\ldots,N_F$). Massless space-time fermions have to be right-moving spinors ($\lambda_R = (\pm\frac{1}{2},\pm\frac{1}{2})$) and $q = -\frac{1}{2}$. The associated internal conformal fields $G^{\Psi_i}(\bar{z},z)$ must be operators of dimension $(\bar{h},h)_{\text{int}} = (1,\frac{3}{8})$. Then the complete fermion vertex operator in the canonical ghost picture is

$$V^{\Psi_i}(\bar{z},z,k) = G^{\Psi_i}(\bar{z},z)S_\alpha e^{-\frac{\phi}{2}}(z)e^{ik_\mu X^\mu(\bar{z},z)}. \qquad (14.32)$$

Therefore, the number N_F of massless fermions is determined by the number of internal fields $G^{\Psi_i}(\bar{z},z)$ which are coupled to the spinors of $D_{2,1}$. For the fermions to be chiral the fields $G^{\Psi_i}(\bar{z},z)$ must transform under a complex representation of the gauge group G, and it must be ensured that the spinors of negative chirality are not coupled to the same fields $G^{\Psi_i}(\bar{z},z)$ as the spinors of positive chirality. BRST invariance of the fermion vertex operator requires that

$$T_F^{\text{int}}(z)G^{\Psi_i}(\bar{w},w) \sim \frac{\hat{G}^{\Psi_i}(\bar{w},w)}{(z-w)^{1/2}} + \text{finite} \qquad (14.33)$$

where $\hat{G}^{\Psi_i}$ are conformal fields of dimension $(\bar{h},h) = (1,\frac{11}{8})$. The presence of the branch cut tells us that G^{Ψ_i} is in the R sector of the internal superconformal field theory.

Constraints from superconformal field theory also exclude the presence of fermionic tachyons. The contribution to the conformal dimension of a R state in the canonical ghost picture is $\geq \frac{5}{8}$ and the internal conformal field theory contributes $h_{\text{int}} \geq \frac{c_{\text{int}}}{24} = \frac{3}{8}$ (cf. eq.(12.18)) so that $h \geq 1$. The contribution from $e^{ik\cdot X}$ must then be $\frac{1}{2}k^2 \leq 0$. This especially excludes the presence of tachyons in space-time supersymmetric theories.

Massless scalars

Massless space-time scalars Φ_i $(i = 1, \ldots, N_S)$ have $\lambda_R = 0$ and $q = -1$. Therefore their existence implies the presence of internal fields $G^{\Phi_i}(\bar{z}, z)$ with conformal dimension $(\bar{h}, h)_{\text{int}} = (1, \frac{1}{2})$. The corresponding vertex operators in the -1 ghost picture are

$$V^{\Phi_i}(\bar{z}, z, k) = G^{\Phi_i}(\bar{z}, z)e^{-\phi(z)}e^{ik_\mu X^\mu(\bar{z}, z)}. \qquad (14.34)$$

BRST invariance tells us that operator products between $T_F^{\text{int}}(z)$ and $G^{\Phi_i}(\bar{w}, w)$ have the structure

$$T_T^{\text{int}}(z)G^{\Phi_i}(\bar{w}, w) \sim \frac{\hat{G}^{\Phi_i}(\bar{w}, w)}{z - w} + \text{finite} \qquad (14.35)$$

where $\hat{G}^{\Phi_i}(\bar{w}, w)$ are internal conformal fields of dimension $(\bar{h}, h)_{\text{int}} = (1, 1)$. This means that $G^{\Phi_i}(\bar{z}, z)$ and $\hat{G}^{\Phi_i}(\bar{z}, z)$ are the two components of a two-dimensional superfields with respect to the right-moving world-sheet supersymmetry. Then the scalar vertex operator in the 0 ghost picture is given by

$$V(\bar{z}, z, k) = [-iG^{\Phi_i}(\bar{z}, z)k_\mu\psi^\mu(z) + \hat{G}^{\Phi_i}(\bar{z}, z)]e^{ik_\nu X^\nu(\bar{z}, z)} \qquad (14.36)$$

where the first term comes from picture changing with the space-time supercurrent $T_F^{\text{s.t}} = -\frac{1}{2}\psi_\mu\partial X^\mu(z)$.

Space-time Supersymmetry and Exceptional Groups

Without wanting to emphasize the "phenomenological" importance of space-time supersymmetry (vanishing of the cosmological constant or the possible solution of the hierarchy problem) let us assume that there are N space-time supersymmetries originating from the right-moving sector. The corresponding supercharges are denoted by Q_α^A $(A = 1, \ldots, N)$. They are the contour integrals (zero-mode part) of the holomorphic part of the gravitino vertex operators at zero space-time momentum:

$$Q_{\alpha(q)}^A = \oint \frac{dz}{2\pi i} V_{\alpha(q)}^A(z) \quad , \qquad \bar{Q}_{\dot{\alpha}A(q)} = \oint \frac{dz}{2\pi i} \bar{V}_{\dot{\alpha}A(q)}(z). \qquad (14.37)$$

314

In the canonical ghost picture we have

$$V^A_{\alpha(-\frac{1}{2})}(z) = S_\alpha e^{-\phi/2} \Sigma^A(z) \quad , \qquad \bar{V}_{\dot\alpha A(-\frac{1}{2})}(z) = S_{\dot\alpha} e^{-\phi/2} \Sigma_A(z). \quad (14.38)$$

$S_\alpha(z)$ and $S_{\dot\alpha}(z)$ are spin fields of the Lorentz group $SO(4)$ characterized by the spinor weights of D_2:

$$
\begin{aligned}
S_\alpha(z) &= e^{i\lambda_\alpha \cdot \phi(z)} \quad , & \lambda_\alpha &= \pm(\frac{1}{2}, \frac{1}{2}) , \\
S_{\dot\alpha}(z) &= e^{i\lambda_{\dot\alpha} \cdot \phi(z)} \quad , & \lambda_{\dot\alpha} &= \pm(\frac{1}{2}, -\frac{1}{2}) .
\end{aligned}
\qquad (14.39)
$$

The fields $\Sigma^A(z)$ are the (degenerate) Ramond ground states of dimension $h_{\text{int}} = \frac{3}{8}$ of the internal superconformal field theory with $c = 9$.

Now the space-time supersymmetry algebra ($Z^{AB} = -Z^{BA}$ are central charges, $C_{\alpha\beta}$ the charge conjugation matrix and $(\gamma^\mu)_{\alpha\dot\beta}$ the four-dimensional Dirac matrices)

$$
\begin{aligned}
\{Q^A_\alpha, \bar{Q}_{\dot\beta B}\} &= \frac{i}{\sqrt{2}} \delta^A_B (\gamma^\mu)_{\alpha\dot\beta} \, p_\mu , \\
\{Q^A_\alpha, Q^B_\beta\} &= C_{\alpha\beta} Z^{AB}
\end{aligned}
\qquad (14.40)
$$

translates immediately to the following operator product expansions:

$$
\begin{aligned}
\Sigma^A(z)\Sigma_B(w) &\sim (z-w)^{-\frac{3}{4}} \delta^A_B + (z-w)^{\frac{1}{4}} J_B{}^A(w) + \ldots \\
\Sigma^A(z)\Sigma^B(w) &\sim (z-w)^{-\frac{1}{4}} \psi^{AB} + O\big((z-w)^{\frac{3}{4}}\big).
\end{aligned}
\qquad (14.41)
$$

Here we have used that the momentum operator $i\partial X^\mu(z)$ is given by $-ie^{-\phi}\psi^\mu$ in the $q = -1$ picture. The dimension one fields $J^A{}_B$ can be shown to be currents of internal level one Kac-Moody algebras of rank k ($k = 1, 2, 3$ for $N = 1, 2, 4$). The dimension 1/2 fields $\psi^{AB} = -\psi^{BA}$ are related to the central charges in the $q = -1$ picture by $Z^{AB}_{-1} = \oint \frac{dz}{2\pi i} e^{-\phi}\psi^{AB}(z)$. Using the Frenkel-Kac construction, the currents $J^A{}_B$ can be expressed by k free internal bosons $H^{\text{int}}_i(z)$ ($i = 1, \ldots, k$). The internal vertex operators $V_{\text{int}}(z)$ can then always be written as (neglecting derivatives of H^{int}_i):

$$V_{\mathrm{int}}(z) = e^{i\boldsymbol{w}_{\mathrm{int}} \cdot \boldsymbol{H}_{\mathrm{int}}(z)} \tilde{V}_{\mathrm{int}}(z). \tag{14.42}$$

The vectors $\boldsymbol{w}_{\mathrm{int}}$ are the weights of the internal algebra g spanning the weight lattice Γ_k of g. The $\tilde{V}_{\mathrm{int}}(z)$ belong to the remaining conformal field theory with $\tilde{c} = 9 - k$ and commute with H_i^{int}.

Let us fill in some of the details for the phenomenologically most interesting case of $N = 1$ supersymmetry [12]. Eq.(14.41) then simplifies to

$$\Sigma(z)\Sigma^\dagger(w) \sim (z - w)^{-3/4} + \tfrac{1}{2}(z - w)^{1/4} J(w) ,$$
$$\Sigma(z)\Sigma(w) \sim O\big((z - w)^{3/4}\big). \tag{14.43}$$

Consider the four point correlation function

$$f(z_i) = \langle \Sigma(z_1)\Sigma^\dagger(z_2)\Sigma(z_3)\Sigma^\dagger(z_4)\rangle$$
$$= \Big(\frac{z_{13}z_{24}}{z_{12}z_{34}z_{14}z_{23}}\Big)^{3/4} \tilde{f}(x) \tag{14.44}$$

where $x = \frac{z_{12}z_{34}}{z_{13}z_{24}}$ (cf. eq.(4.72)). The leading orders of the operator product expansions eq.(14.43) determine the behavior of $f(z_i)$ as $z_{ij} \to 0$ for any pair i, j. One finds that $\tilde{f}(x)$ is an analytic function and constant at $x = 0, 1, \infty$, from which it follows that it is a constant. With $\langle 1 \rangle = 1$ we have $\tilde{f}(x) = 1$. In the limit $z_1 \to z_2$, eq.(14.44) becomes

$$f(z_i) \to \langle \Sigma(z_3)\Sigma^\dagger(z_4)\rangle\,(z_{12})^{-3/4} + \tfrac{1}{2}\langle J(z_2)\Sigma(z_3)\Sigma^\dagger(z_4)\rangle\,(z_{12})^{1/4}$$
$$= (z_{12}z_{34})^{-3/4} + \tfrac{3}{4}(z_{12}z_{34})^{1/4}(z_{23}z_{24})^{-1}. \tag{14.45}$$

From this we find, taking the limits $z_3 \to z_4$, $z_2 \to z_3$ and $z_2 \to z_4$:

$$J(z)J(w) \sim \frac{3}{(z - w)^2} + \text{finite} ,$$

$$J(z)\Sigma(w) \sim \frac{\tfrac{3}{2}\Sigma(w)}{z - w} + \text{finite} , \tag{14.46}$$

$$J(z)\Sigma^\dagger(w) \sim -\frac{\tfrac{3}{2}\Sigma^\dagger(w)}{z - w} + \text{finite} .$$

So $J(z)$ is a $U(1)$ Kac-Moody current which can be written in terms of a free boson as

$$J(z) = i\sqrt{3}\,\partial H(z)\,. \tag{14.47}$$

The operators Σ and $\Sigma^\dagger$ can be expressed as exponentials of H:

$$\Sigma(z) = e^{i\frac{\sqrt{3}}{2}H(z)} \quad , \quad \Sigma^\dagger = e^{-i\frac{\sqrt{3}}{2}H(z)}\,. \tag{14.48}$$

In fact, any operator with definite $U(1)$ charge Q, i.e.

$$J(z)V_Q(\bar{w},w) = \frac{Q\,V_Q(\bar{w},w)}{z-w} + \ldots \tag{14.49}$$

can be written as

$$V_Q(\bar{z},z) = e^{i\frac{Q}{\sqrt{3}}H(z)}P(J(z))\tilde{V}(\bar{z},z) \tag{14.50}$$

where $P(J)$ is a polynomial in J and its derivatives and $\tilde{V}$ is independent of the boson H. In particular, physical states are now characterized by vertex operators

$$V(\bar{z},z) = e^{i\lambda_R\cdot\phi(z)}e^{q\phi(z)}e^{i\frac{Q}{\sqrt{3}}H(z)}\tilde{V}_{\rm int}(\bar{z},z) \tag{14.51}$$

where $\tilde{V}_{\rm int}$ belongs to the remaining conformal field theory with $\bar{c} = 22$ and $\tilde{c} = 8$. (The boson H already contributes one unit to the central charge.) The numbers $\frac{Q}{\sqrt{3}}$ build the internal weight lattice Γ_1 of the $U(1)$ Kac-Moody algebra. BRST invariance of the gravitino vertex operator demands that the operator product between $T_F^{\rm int}(z)$ and $\Sigma(w)$, $\Sigma^\dagger(w)$ contains a branch cut of order $\frac{1}{2}$:

$$T_F^{\rm int}(z)\Sigma(w) = \frac{\hat{\Sigma}(w)}{(z-w)^{\frac{1}{2}}} + \ldots$$

$$T_F^{\rm int}(z)\Sigma^\dagger(w) = \frac{\hat{\Sigma}^\dagger(w)}{(z-w)^{\frac{1}{2}}} + \ldots \tag{14.52}$$

where $\hat{\Sigma}(w)$, $\hat{\Sigma}^\dagger(w)$ are operators of dimension 11/8. This equation, together with eq.(14.48) implies that the internal supercurrent does not have definite $U(1)$ charge but splits into two parts with $U(1)$ charges $Q = \pm 1$ respectively:

$$T_F^{\text{int}}(z) = e^{\frac{i}{\sqrt{3}}H(z)}\tilde{T}_F^+(z) + e^{-\frac{i}{\sqrt{3}}H(z)}\tilde{T}_F^-(z) = T_F^+(z) + T_F^-(z) \qquad (14.53)$$

with

$$J(z)T_F^\pm(w) \sim \pm\frac{T_F^\pm(w)}{z - w}. \qquad (14.54)$$

$\tilde{T}_F^\pm$ are fields of conformal dimension 4/3 of an internal conformal field theory with $\tilde{c} = 8$. Requiring T_F^{int} to satisfy eq.(14.23), we find

$$T_F^+(z)T_F^-(w) = \frac{\frac{3}{4}}{(z-w)^3} + \frac{\frac{1}{4}J(w)}{(z-w)^2} + \frac{\frac{1}{8}\partial J(w) + \frac{1}{4}T_{\text{int}}(w)}{z-w} + \dots$$

$$T_F^+(z)T_F^+(w) \sim T_F^-(z)T_F^-(w) \sim \text{ finite} \qquad (14.55)$$

where

$$T_{\text{int}}(z) = -\frac{1}{2}(\partial H(z))^2 + \tilde{T}_{\text{int}}(z). \qquad (14.56)$$

$\tilde{T}_{\text{int}}$ satisfies a Virasoro algebra with $\tilde{c} = 8$. Comparing now eqs.(14.46,53,54 and 55) with eq.(12.53) we find that T_{int}, $T_F^\pm$ and J generate a $n = 2$ superconformal algebra. Thus, $N = 1$ space-time supersymmetry requires an $n = 2$ extended superconformal algebra on the world-sheet. In addition, space-time supersymmetry demands that all states have quantized (integer or half-integer) $U(1)$ charge Q. This can be derived from the requirement that the operator product expansion of an arbitrary state eq.(14.51) with the gravitino vertex be local:

$$\left(\frac{1}{2}, \frac{1}{2}\right) \cdot \lambda_R + \frac{q}{2} + \frac{Q}{2} \in \mathbf{Z}. \qquad (14.57)$$

The following combinations of $D_{2,1}$ conjugacy classes and $U(1)$ charges are then allowed:

$$(0, \mathbf{Z}), \quad (V, 2\mathbf{Z} + 1), \quad (S, 2\mathbf{Z} + \tfrac{1}{2}), \quad (C, 2\mathbf{Z} - \tfrac{1}{2}). \qquad (14.58)$$

We are now ready to explain the connection between space-time supersymmetry and the appearance of the exceptional groups discussed in refs. [5, 13–18]. The key point is that the conjugacy classes eq.(14.58) build the weight lattice of the Lorentzian lattice $E_{3,1}$. To see this let us use instead of the Lorentzian lattice $D_{2,1}$ the Euclidean lattice D_5 as explained at the end of Chapter 13. Recall that D_5 can be decomposed to $D_2^{\text{Lorentz}} \otimes D_3^{\text{ghost}}$ where states in the canonical ghost picture correspond to fixed D_3^{ghost} lattice vectors as shown in eq.(13.98). Thus, all states eq.(14.51) are described by vectors w in the D_5 lattice. (In the canonical ghost picture the space-time and superconformal ghost contribution to the conformal weight of any state is $\frac{1}{2}w^2$.) In the Euclidean version we are now dealing with a level one $SO(10)$ Kac-Moody algebra. $N = 1$ space-time supersymmetry, i.e. the existence of the free internal boson $H(z)$, further enlarges $SO(10)$ to a $SO(10) \times U(1)$ Kac-Moody algebra with lattice vectors $(w, \frac{Q}{\sqrt{3}}) \in \Gamma_6 = D_5 \otimes U(1)$. The supercharges are represented by the Γ_6 lattice vectors $(\pm(\frac{1}{2}, \frac{1}{2}), \frac{1}{2}, \frac{1}{2}, -\frac{1}{2}, \frac{\sqrt{3}}{2})$ and $(\pm(\frac{1}{2}, -\frac{1}{2}), \frac{1}{2}, \frac{1}{2}, -\frac{1}{2}, -\frac{\sqrt{3}}{2})$. However these lattice vectors are among the "spinorial" roots of E_6 (cf. eq.(11.42). Their existence enlarges the $SO(10) \times U(1)$ Kac-Moody algebra to the level one E_6 Kac-Moody algebra. Locality of the operator algebra demands that all vectors $(w, \frac{\sqrt{3}}{3}Q) \in \Gamma_6$ have integer scalar product with these roots, i.e. Γ_6 is the weight lattice of E_6.

E_6 possesses three conjugacy classes denoted by 0, 1 and $\bar{1}$ with lowest representations $\underline{1}$, $\underline{27}$ and $\underline{\overline{27}}$ respectively (note that the 1 and $\bar{1}$ conjugacy classes are CPT conjugates of each other and therefore always appear together). Under $D_5 \otimes U(1)$ the conjugacy classes decompose as

$$0 = (0,0) \oplus (V, \sqrt{3}) \oplus (S, -\frac{\sqrt{3}}{2}) \oplus (C, \frac{\sqrt{3}}{2}),$$

$$1 = (0, \frac{2}{3}\sqrt{3}) \oplus (V, -\frac{\sqrt{3}}{3}) \oplus (S, \frac{\sqrt{3}}{6}) \oplus (C, -\frac{5}{6}\sqrt{3}), \qquad (14.59)$$

$$\bar{1} = (0, -\frac{2}{3}\sqrt{3}) \oplus (V, \frac{\sqrt{3}}{3}) \oplus (C, -\frac{\sqrt{3}}{6}) \oplus (S, \frac{5}{6}\sqrt{3}).$$

In this notation the 12 $U(1)$ conjugacy classes

$$q = -\sqrt{3}, \; -\frac{5}{6}\sqrt{3}, \ldots, \frac{5}{6}\sqrt{3} \qquad (14.60)$$

define the elements α_q of the one-dimensional $U(1)$ weight lattice by

$$\alpha_q = q + 2\sqrt{3}k \;\; (k \in \mathbf{Z}). \qquad (14.61)$$

The conjugacy classes in eq.(14.59) are just those allowed by locality (cf. eq.(14.58)) if we replace $D_{2,1}$ by D_5 and rescale the $U(1)$ charge by $\sqrt{3}$.

Space-time supersymmetry transformations act on a particular state (vertex operator) V_w ($w \in E_{3,1}$ resp. E_6) as

$$V'_{w'}(w) = \oint_{C_w} \frac{dz}{2\pi i} Q_\alpha(z) V_w(w). \qquad (14.62)$$

Since the supercharge corresponds to spinorial root vectors α of E_6, $V'_{w'}$ is characterized by the E_6 vector $w + \alpha$. Thus the supersymmetric partners correspond to vectors within the same E_6 conjugacy class. It follows that the supermultiplet structure is encoded in the representations of the exceptional group E_6.

Reversing arguments, it is the appearance of E_6 which is responsible for $N = 1$ space-time supersymmetry implying the $n = 2$ superconformal algebra together with the $U(1)$ quantization condition dictated by locality with the gravitino vertex. The quantization condition, the correlation between $U(1)$ charges and space-time transformation properties, are contained in the

$E_{3,1}$ or, equivalently E_6 weight lattice. The supercharge, being one of the roots, connects different states. It is also just the operator which generates the spectral flow between the NS and R sectors as described in Chapter 12.

Let us exemplify above by looking at the massless states of an arbitrary $N = 1$ supersymmetric (heterotic) string theory.[1] Here only the weights of the fundamental and adjoint representations of E_6 are relevant. Let us start with the latter. It decomposes under $SO(10) \times U(1)$ as

$$\underline{78} = (\underline{45}, 0) + (\underline{16}, -\frac{\sqrt{3}}{2}) + (\underline{\overline{16}}, \frac{\sqrt{3}}{2}) + (\underline{1}, 0). \qquad (14.63)$$

The $(\underline{\overline{16}}, \frac{\sqrt{3}}{2})$, $(\underline{16}, -\frac{\sqrt{3}}{2})$ representations correspond to a holomorphic spinor and antispinor respectively, namely just the supercharges Q_α, $\bar{Q}_{\dot\alpha}$. Group theoretically it is clear that the roots corresponding to the supercharges interpolate between the different $SO(10) \times U(1)$ conjugacy classes. But this can also be verified at the level of vertex operators if we make the truncation $D_5 \to D_{2,1}$. Indeed, acting with the supercharge on the holomorphic spinor we obtain the vector $\psi_\mu(z)e^{-\phi(z)}$. This state corresponds to the $(\underline{45}, 0)$ representation of $SO(10) \times U(1)$. But note that the spinor itself can be reached from a vertex operator $\sim i\sqrt{3}\partial H(z)$ which is just the $U(1)$ current. This state is not BRST invariant; it does not have a copy in the canonical ghost picture. It corresponds to an auxiliary field. The three states, the vector, the spinor and the auxiliary scalar are the off-shell degrees of freedom of a $N = 1$ vector supermultiplet. Multiplying these states with a left-moving Kac-Moody current $j^a(\bar{z})$ leads to a $N = 1$ gauge vector multiplet. On the other hand, multiplying with $\partial X_\mu(\bar{z})$ leads to the $N = 1$ supergravity multiplet where the auxiliary vector corresponds to the auxiliary gauge field of the $U(1)$ Kac-Moody algebra.

---

[1] It is helpful for this discussion to keep in mind the explicit example constructed in section 14.2.

The (massless) matter sector of the $N = 1$ supersymmetric heterotic string theory is obtained from the $\underline{27}$ ($\overline{\underline{27}}$) representation of E_6. The weights of this representation lead to conformal fields $e^{iw\cdot\phi(z)}$ ($w \in E_{3,1}$) of dimension $\frac{2}{3}$. (The weights in the $\underline{27}$ and $\overline{\underline{27}}$ of E_6 have (length)$^2 = \frac{4}{3}$.) Therefore, in order to obtain massless fields one has to multiply this operator by an internal conformal field $\tilde{G}^i(\bar{z}, z)$ ($i = 1, \ldots, N$, N being the number of "families") of conformal dimension $\bar{h} = 1$, $\tilde{h} = \frac{1}{3}$, where the $\tilde{G}^i(\bar{z}, z)$ are fields of the internal conformal field theory with $\bar{c} = 22$ and $\tilde{c} = 8$. The fundamental representation of E_6 decomposes under $SO(10) \times U(1)$ as

$$\underline{27} = (\underline{16}, \frac{\sqrt{3}}{6}) + (\underline{10}, -\frac{\sqrt{3}}{3}) + (\underline{1}, \frac{2\sqrt{3}}{3}). \tag{14.64}$$

The first term corresponds to a space-time spinor. Acting with the supercharge on this state gives the second term in eq.(14.64) which corresponds to a physical massless scalar in the -1 ghost picture. We should stress that when determining the space-time properties of a state given by a particular $SO(10) \times U(1)$ representation, one always has to make the truncation $D_5 \to D_2$. The scalar vertex operators $V_i(\bar{z}, z) \sim e^{-i\frac{\sqrt{3}}{3}H_{\text{int}}(z)}\tilde{G}^i(\bar{z}, z)$ of conformal dimension $(\bar{h}, h) = (1, \frac{1}{2})$ and $U(1)$ charge $Q = -1$ are just the chiral primary fields of the internal $n = 2$ superconformal algebra, as defined in Chapter 12. Finally, the last term in eq.(14.64) corresponds to an unphysical, i.e. not BRST invariant, scalar in the 0 ghost picture. Together, these three fields build the off-shell degrees of freedom of a $N = 1$ chiral multiplet (if we add also the CPT conjugate states in the $\overline{\underline{27}}$ representation). The $U(1)$ charge is just the R-charge of supersymmetric field theory.

Let us now describe the appearance of the exceptional group E_6 in four-dimensional $N = 1$ supersymmetric heterotic string from a somewhat different point of view. Consider the four-dimensional Poincaré algebra with $L_{\mu\nu}$ the generators of the Wick rotated Lorentz group $SO(4)$ and p_μ the space-time momenta:

$$[L_{\mu\nu}, L_{\rho\sigma}] \sim \delta_{\mu\rho} L_{\nu\sigma} - \delta_{\nu\rho} L_{\mu\sigma} + \delta_{\nu\sigma} L_{\mu\rho} - \delta_{\mu\sigma} L_{\nu\rho} \,,$$
$$[L_{\mu\nu}, p_\rho] \sim \delta_{\mu\rho} p_\nu - \delta_{\nu\rho} p_\mu \,. \tag{14.65}$$

Adding the space-time supercharge one obtains the following graded Lie algebra

$$[Q, L_{\mu\nu}] \sim \gamma_{\mu\nu} Q \,,$$
$$[Q, p_\mu] \sim 0 \,, \tag{14.66}$$
$$\{Q, \bar{Q}\} \sim \gamma^\mu p_\mu \,.$$

The generators $L_{\mu\nu}$, p_μ and Q can be represented as the contour integrals of the following operators:

$$L^{\mu\nu} = \psi^\mu \psi^\nu \,,$$
$$p^\mu = \psi^\mu e^{-\phi} \,, \tag{14.67}$$
$$Q_\alpha = S_\alpha e^{-\phi/2} e^{i\frac{\sqrt{3}}{2} H} \,.$$

In the language of the covariant lattice $D_{2,1} \otimes U(1)^{\text{int}} = D_2^{\text{Lorentz}} \otimes D_1^{\text{ghost}} \otimes U(1)^{\text{int}}$ they are given by the following lattice vectors.

$$(\lambda, q, \frac{Q}{\sqrt{3}})_{L_{\mu\nu}} = (\pm 1, \pm 1, 0, 0) \,,$$
$$(\lambda, q, \frac{Q}{\sqrt{3}})_{p_\rho} = (\pm 1, 0, -1, 0), \ (0, \pm 1, -1, 0) \,, \tag{14.68}$$
$$(\lambda, q, \frac{Q}{\sqrt{3}})_{Q} = (\pm(\frac{1}{2}, \frac{1}{2}), -\frac{1}{2}, \frac{\sqrt{3}}{2}) \,.$$

(For the Lorentz generators $L_{\mu\nu}$ we have also to include the Cartan subalgebra generators $\partial\phi_1(z)$, $\partial\phi_2(z)$.) One can show that up to unphysical terms these operators generate the graded Lie algebra eqs.(14.65,66). For the supercharges we get an anticommutator since $(\lambda, q, \frac{Q}{\sqrt{3}})_Q^2 = 1$. It is then evident that the supercharge vertex operator corresponds to those "root" vectors which extend the algebra $D_{2,1} \times U(1)_{\text{int}}$ to the exceptional algebra $E_{3,1}$. Finally, replacing D_1^{ghost} by D_3^{ghost} and therefore $D_{2,1}$ by D_5 the exceptional algebra E_6 becomes, in the sense discussed above, equivalent to the graded Lie algebra eqs.(14.65,66). The anticommutator in eq.(14.66) is converted into a commutator since the supercharges now correspond to $(\text{length})^2 = 2$ roots of E_6.

The appearance of the exceptional group can also be used to show that the one-loop partition function of any four-dimensional $N = 1$ supersymmetric heterotic string theory vanishes. Since every state is characterized by an E_6 weight vector it is clear that the partition function contains the sum over the E_6 weight lattice, i.e. over the three level one characters of E_6 which are in one-to-one correspondence to the three different conjugacy classes (the level one Kac-Moody characters were introduced in Chapter 11). The only subtlety which arises is the fact that one must sum only over physical transverse states, namely those states which have fixed D_4 lattice vectors x_0 (see eq.(13.99)) when decomposing E_6 to $D_1 \otimes D_4 \otimes U(1)$. We will call these restricted E_6 characters $Ch_i(\tau)$ where i denotes the three conjugacy classes 0, 1 and $\bar{1}$. Then the partition function has the following general form:

$$\chi(\bar{\tau}, \tau) \sim \frac{1}{\mathrm{Im}\tau} \frac{1}{|\eta(\tau)|^4} Ch_i(\tau) Ch_j^{\tilde{c}=8, \bar{c}=22}(\tau, \bar{\tau}) a_{ij} \qquad (14.69)$$

$(i = 0, 1, \bar{1})$, where $Ch_j^{\tilde{c}=8, \bar{c}=22}$ are the characters of the internal conformal field theory without the free boson H. The choice of the coefficients a_{ij} depends on the particular model and has to satisfy the constraints of modular invariance and spin statistics. The "true" (unrestricted) E_6 characters were given in eq. (11.103). From them it is easy to derive the restricted, physical E_6 characters if we make the truncation as described at the end of Chapter 13. We find

$$Ch_0(\tau) = \frac{1}{\eta^2(\tau)} \Big\{ \theta_3(0|3\tau)\theta_3(0|\tau) - \theta_4(0|3\tau)\theta_4(0|\tau) - \theta_2(0|3\tau)\theta_2(0|\tau) \Big\},$$

$$Ch_1(\tau) = \frac{1}{\eta^2(\tau)}\left\{-\theta\!\left[{}^{1/6}_{\ 0}\right](0|3\tau)\theta_2(0|\tau) + \theta\!\left[{}^{2/3}_{\ 0}\right](0|3\tau)\theta_3(0|\tau)\right. \tag{14.70}$$

$$\left. - e^{-2\pi i/3}\theta\!\left[{}^{2/3}_{1/2}\right](0|3\tau)\theta_4(0|\tau)\right\},$$

$$Ch_{\bar{1}}(\tau) = \frac{1}{\eta^2(\tau)}\left\{-\theta\!\left[{}^{5/6}_{\ 0}\right](0|3\tau)\theta_2(0|\tau) + \theta\!\left[{}^{1/3}_{\ 0}\right](0|3\tau)\theta_3(0|\tau)\right.$$

$$\left. + e^{-\pi i/3}\theta\!\left[{}^{1/3}_{1/2}\right](0|3\tau)\theta_4(0|\tau)\right\}.$$

Now, because of space-time supersymmetry, these characters (not however the original characters eq.(11.103)) are supposed to vanish identically. This can indeed be proven using the theory of modular forms [16].

By similar arguments as for the case of $N = 1$ supersymmetry one can also show that the presence of two holomorphic supercharges ($N = 2$ space-time supersymmetry) in four dimensions implies the existence of an internal right-moving $SU(2) \times U(1)$ Kac-Moody algebra which now extends $D_{2,1}$ to $E_{4,1}$ or D_5 to E_7. The supersymmetry on the world-sheet is now extended to a $n = 4$, $c = 6$ superconformal algebra plus a superconformal system with $n = 2$, $c = 3$. Finally $N = 4$ space-time supersymmetry implies an internal $SO(6)$ Kac-Moody algebra which extends $D_{2,1}$ to $E_{5,1}$ respectively D_5 to E_8. We should mention that the extended world-sheet algebras can only correspond to global world-sheet symmetries. The reason for this was given in Chapter 12.

REFERENCES

[1] L. Alvarez-Gaumé, P. Ginsparg, G. Moore and C. Vafa, *"An $O(16) \times O(16)$ heterotic string"*, *Phys. Lett.* **171 B** (1986) 155.

[2] L.J. Dixon and J.A. Harvey, *"String theories in ten dimensions without space-time supersymmetry"*, *Nucl. Phys.* **B274** (1986) 93.

[3] W. Lerche and D. Lüst, *"Covariant heterotic strings and odd self-dual lattices"*, *Phys. Lett.* **187B** (1987) 45.

[4] W. Lerche, D. Lüst and A.N. Schellekens, *"Ten-dimensional heterotic strings from Niemeier lattices"*, *Phys. Lett.* **B181** (1986) 71; Erratum *Phys. Lett.* **B184** (1987) 419.

[5] W. Lerche, D. Lüst and A.N. Schellekens, *"Chiral four-dimensional heterotic strings from self-dual lattices"*, *Nucl. Phys.* **B287** (1987) 477.

[6] W. Lerche, A.N. Schellekens and N. Warner, *"Lattices and Strings"*, *Phys. Rep.* **177** (1989) 1.

[7] P. Candelas, G.T. Horowitz, A. Strominger and E. Witten, *"Vacuum configurations for superstrings"*, *Nucl. Phys.* **B258** (1985) 46.

[8] L. Dixon, J. Harvey, C. Vafa and E. Witten, *"Strings on orbifolds"*, *Nucl. Phys.* **B261** (1985) 651; *"Strings on orbifolds 2"*, *Nucl. Phys.* **B274** (1986) 285.

[9] H. Kawai, D. Lewellen and S. Tye, *"Construction of four-dimensional string models"*, Phys. Rev. Lett. **57** (1986) 1832; *"Construction of fermionic string models in four dimensions"*, Nucl. Phys. **B288** (1987) 1.

[10] I. Antoniadis, C. Bachas and C. Kounnas, *"Four-dimensional superstrings"*, Nucl. Phys. **B289** (1987) 87.

[11] J.H. Schwarz, *"Superconformal symmetry and superstring compactification"*, preprint CALT-68-1531 (1989).

[12] T. Banks, L. Dixon, D. Friedan and E. Martinec, *"Phenomenology and conformal field theory or can string theory predict the weak mixing angle?"*, Nucl. Phys. **B299** (1988) 613.

[13] J. Lauer, D. Lüst and S. Theisen, *"Four-dimensional supergravity from four-dimensional strings"*, Nucl. Phys. **B304** (1988) 236.

[14] J. Lauer, D. Lüst and S. Theisen, *"Supersymmetric string theories, superconformal algebras and exceptional groups"*, Nucl. Phys. **B309** (1988) 771.

[15] W. Lerche, *"Elliptic index and superstring effective action"*, Nucl. Phys. **B308** (1988) 397.

[16] D. Lüst and S. Theisen, *"Superstring partition functions and the characters of exceptional groups"*, preprint MPI-PAE/PTh 41/88, to appear in *Phys. Lett* **B**.

[17] W. Lerche, A.N. Schellekens and N. Warner, *"Ghost triality and superstring partition functions"*, Phys. Lett. **B214** (1988) 41.

[18] D. Lüst and S. Theisen, *"Exceptional groups in string theory"*, preprint CERN-TH.5256/88, to appear in *Int. J. Mod. Phys.*

Chapter 15

Low Energy Field Theory

String theory is claimed to be a unifying framework for the description of all particles and their interactions, including gravity. However, up to now our exposition of the subject was rather formal and it is not at all transparent how it can be relevant for low energy phenomenology. The only hint we got so far was from looking at the spectrum. There especially the occurrence of a spin two tensor particle indicated that gravity might be contained in string theory. We have learned to formulate string theory directly in four dimensions and we know how to get massless vectors that transform in the adjoint representation of a simply-laced gauge group. This gives us hope that string theory might provide a description of gauge theories. Also, the four-dimensional string theories contain plenty of massless scalars and fermions in non-trivial representations of the gauge group.

In this final chapter we will show how contact with the real world is made. This will be done by deriving a point particle field theory Lagrangian which represents the low energy description of string theory in the sense that is reproduces string scattering amplitudes. Here low energy means that the string scale $\alpha' \to 0$; all massive modes have been integrated out and we obtain a description of the massless modes only. Since, as we will see below, the string scale turns out to be the Planck scale, all known particles have to be contained among the massless string states. Also, the couplings of the states will be related to the string coupling constant (cf. below) and the string scale which is set by the string tension. Of course we are still far away from extracting the standard model Lagrangian with all its particular masses and coupling constants but at least we can see whether we find something semi-realistic, i.e. a field theory with generically the correct features. This is in fact all we can hope for at the moment.

The first indication that string theory is a low energy expansion about a point particle theory arose in a paper by Scherk [1]. Neveu and Scherk [2] showed that in the limit where the Regge slope $\alpha' \to 0$, the string scattering amplitudes of the massless spin one particle of the open bosonic string can be reproduced by a Yang-Mills field theory. Subsequently it was shown by Yoneya [3] and Scherk and Schwarz [4] that the effective action of the massless spin two state is, in the zero slope limit, given by the Einstein-Hilbert action of gravitation. Later the effective action for the superstring [5] and the bosonic part of the heterotic string [6, 7, 8] in ten dimensions was derived. The formulation of four-dimensional heterotic strings allows to calculate directly from string scattering amplitudes the effective action of the massless modes in four dimensions. This was done in [9, 10] for supersymmetric as well as non-supersymmetric four-dimensional string theories, focussing mainly on the covariant lattice formulation. The calculation of string amplitudes and the effective action for orbifold theories can be found in [11, 12]. We will try to present our results as model independent as possible. However, whenever we need to be more specific, we will use as examples models based on the covariant lattice construction. In these models all vertex operators can be expressed, via bosonization, in terms of free fields and an explicit evaluation of all correlation functions is straightforward, if tedious at times[1]. This is done using Wick's theorem and the basic free field

---

[1] This is mainly due to the loss of manifest covariance through bosonization. The covariant structure has to be reassembled from the cocycle factors. However, it is often easy to guess the structure and fix coefficients by looking at particular

two-point functions which we collect below for convenience:

$$\langle X^i(z)X^j(w)\rangle = -\delta^{ij}\ln(z-w)$$

$$\rightarrow \langle e^{i\lambda\cdot X(z)}e^{i\mu\cdot X(w)}\rangle = (z-w)^{\lambda\cdot\mu},$$

$$\langle \phi(z)\phi(w)\rangle = -\ln(z-w) \tag{15.1}$$

$$\rightarrow \langle e^{q\phi(z)}e^{q'\phi(w)}\rangle = (z-w)^{-qq'},$$

$$\langle \psi^\mu(z)\psi^\nu(w)\rangle = -g^{\mu\nu}(z-w)^{-1}.$$

(ϕ arises from the bosonization of the superconformal ghosts.) The lattice vectors of the various vertex operators in a correlation function have to add up to zero. This lattice momentum conservation is nothing but invariance under the symmetries (Lorentz invariance, gauge invariance).

The general procedure to extract the low energy field theory from string theory is the following: one first calculates various string scattering amplitudes of massless string states, represented by their vertex operators. Then one writes down a field theory Lagrangian which reproduces these amplitudes. This is done in a perturbative fashion. One starts by writing down the effective Lagrangian that describes the massless free particles, $\mathcal{L}_{2pt}$. Then one adds $\mathcal{L}_{3pt}$ to reproduce the three point string amplitudes. $\mathcal{L}_{3pt}$ already allows to relate various coupling constants of the effective action to the string coupling constant and the string tension α' if we reintroduce powers of $\sqrt{\frac{\alpha'}{2}}$ to give dimensionally correct expressions. Especially higher powers of the external momenta will be accompanied by powers of $\sqrt{\alpha'}$ via the substitution $k \rightarrow \sqrt{\frac{\alpha'}{2}}k$. In this way the expansion of the effective action in numbers of derivatives is an expansion in powers of $\sqrt{\alpha'}$. At the next step one considers the four-point amplitudes. Unitarity guarantees that the

---

components.

massless poles will be those generated by the tree graphs of $\mathcal{L}_{3pt}$. This allows us to check again the relation between the relevant coupling constants. The remainder is in general due to massive particle exchanges and will be reproduced by $\mathcal{L}_{4pt}$. The contribution to the string four point amplitudes which are due to massive particle exchange can be expanded in powers of the external momenta. Each term in this expansion generates a local four point vertex in $\mathcal{L}_{4pt}$. This procedure can now be carried on to arbitrary order. At each order we will write $\mathcal{L}$ in a form invariant with respect to all the global and local symmetries so that for instance the kinetic energy term for a charged scalar, which is part of $\mathcal{L}_{2pt}$, does already contain the three and four point couplings of the scalars to gauge fields (via gauge invariance) and couplings to arbitrary order of the scalar to gravitons (general coordinate invariance). These couplings do of course have to reproduce the corresponding string amplitudes.[2]

It is however only possible to calculate string amplitudes with external on-shell states; we recall that this follows from the condition of BRST invariance of the vertex operators. This means that the amplitudes do not fix the field theory Lagrangian completely. However, appealing to symmetries such as general coordinate invariance, gauge symmetries and supersymmetry (if present) allows us to derive a unique (up to field redefinitions) Lagrangian which furnishes a low energy description of a given string model. Here we will only compute string tree level amplitudes. This restriction is not one of principle but rather one of simplicity: one and higher loop amplitudes are much harder to evaluate.

---

[2] An alternative way of arriving at an effective action is the so-called sigma-model approach [13, 14, 15]. In this language what we are doing is a calculation to all orders in sigma-model perturbation theory and to lowest order in string perturbation theory.

We know from previous chapters that external physical on-shell states can be represented by vertex operators $V(z, \bar{z}, k)$ where k is the momentum of the state. Then, in the language of conformal field theory, a N-particle amplitude is simply given by the correlation function

$$A \sim g^{N-2} \langle V_1 \ldots V_N \rangle \tag{15.2}$$

where g is the string coupling constant. There is one power of the string coupling constant for each splitting of one string into two or two strings merging into one.

As in ordinary field theory, where in the calculation of matrix elements we have to integrate over the positions of the field operators, we have to integrate over the insertion points of the vertex operators. The integrated vertex operators are the momentum space representatives of vertex operators in position space. For instance, a general vertex operator has the form $V(k) = \int d^2 z \mathcal{V}(\bar{z}, z) e^{ik_\mu X^\mu(\bar{z}, z)}$ where $\mathcal{V}(\bar{z}, z)$ carries all the quantum numbers of the state, such as the behavior under Lorentz transformations and the charges with respect to internal symmetries (gauge symmetries). Fourier transforming it we get $V(x) = \int d^d k \int d^2 z \mathcal{V}(\bar{z}, z) e^{ik_\mu X^\mu(\bar{z}, z)} e^{ik_\nu x^\nu} = \int d^2 z \mathcal{V}(\bar{z}, z) \delta^{(d)}(X(\bar{z}, z) - x)$. In ordinary field theory translation invariance results in an infinite volume factor. Here we have to deal with the invariance under SL_2. We have learnt in Chapter 6 how to factor out the $SL(2, \mathbb{C})$ volume. That led to the following prescription. We formally insert a factor

$$\frac{c(z_i, \bar{z}_i) c(z_j, \bar{z}_j) c(z_k, \bar{z}_k)}{\int dz_i^2 dz_j^2 dz_k^2}. \tag{15.3}$$

This means that we arbitrarily fix the positions of three of the vertex operators, drop the corresponding integrations and insert ghost fields at these positions. One usually chooses for the fixed positions $z = 0, 1$ and ∞. For three point amplitudes it is however advantageous to keep them at arbitrary (but fixed) values and the fact that the final amplitude has to be independent of them serves as a check. The ghost insertions do not change the conformal weights of the vertex operators since both dz and $c(z)$ have weight -1. Also, we have shown in Chapter 5 that if $\int dz V(z)$ is BRST invariant then so is $c(z)V(z)$. Thus, for three point functions no integral is left to do.

We also have to worry about the zero modes of the superconformal ghosts. At tree level there are two γ zero modes. As discussed in Chapter 13 they are taken care of by choosing the ghost pictures for the vertex operators such that their superconformal ghost charges satisfy $\sum_i q_i = -2$. As explained there, the amplitudes are independent of the way how the superconformal ghost charges are distributed over the various vertex operators. For example, consider the four-point amplitude with two fermions Ψ and two scalars Φ. Then it follows that $\langle V^{\Psi}_{+\frac{1}{2}}(1)V^{\Psi}_{-\frac{1}{2}}(2)V^{\Phi}_{-1}(3)V^{\Phi}_{-1}(4)\rangle = \langle V^{\Psi}_{-\frac{1}{2}}(1)V^{\Psi}_{-\frac{1}{2}}(2)V^{\Phi}_{0}(3)V^{\Phi}_{-1}(4)\rangle$, where the subscripts denote the ghost charges. The particular choice is a matter of convenience.

We are now ready to present a few examples. At tree level correlation functions always factorize into an holomorphic and an anti-holomorphic part which can be evaluated separately. We will start with the three gauge boson amplitude. To satisfy the superconformal ghost charge condition we need two of the graviton vertex operators in the $q = -1$ and one in the $q = 0$ ghost picture. They are given by

$$V_{-1}(\bar{z}, z, k) = \epsilon_\mu(k)\, J^i(\bar{z})\psi^\mu e^{-\phi}(z)\, e^{ik_\rho X^\rho(\bar{z},z)}\,,$$

$$V_0(\bar{z}, z, k) = \epsilon_\mu(k)\, J^i(\bar{z})\, [\partial X^\mu + i(k\cdot\psi)\psi^\mu](z)\, e^{ik_\rho X^\rho(\bar{z},z)}\,,$$

$$(15.4)$$

where $\epsilon_\mu(k)$ are polarization vectors which satisfy the on-shell condition $k^\mu\epsilon_\mu(k) = 0$. Let us look at a specific model with gauge group $SO(2n)$ at level one. We can then write the currents as $J^i = \frac{1}{2} :\psi^M\psi^N: (T^i)^{MN}$ where the representation matrices are normalized as $\mathrm{Tr}(T^iT^j) = 2\delta^{ij}$. (This corresponds to normalizing the root vectors to have $(\text{length})^2 = 2$.) The evaluation of the three gauge boson amplitude

$$A = g\langle c\bar{c}V_{-1}(\bar{z}_1, z_1, k_1)\, c\bar{c}V_{-1}(\bar{z}_2, z_2, k_2)\, c\bar{c}V_0(\bar{z}_3, z_3, k_3)\rangle \tag{15.5}$$

is now straightforward. It factors into several independent pieces. Most of them a trivial to evaluate. (We will give some details in our next example.) We find[3]

$$A^{ijk} = \frac{g}{2}\,\mathrm{Tr}\{[T^i, T^j]T^k\}\, t^{\mu_1\mu_2\mu_3}\, \epsilon^{(1)}_{\mu_1}\epsilon^{(2)}_{\mu_2}\epsilon^{(3)}_{\mu_3} \tag{15.6}$$

where we have abbreviated

$$t^{\mu_1\mu_2\mu_3} = \delta^{\mu_1\mu_2}\, k_2^{\mu_3} + \delta^{\mu_3\mu_2}\, k_3^{\mu_1} + \delta^{\mu_3\mu_1}\, k_1^{\mu_2}. \tag{15.7}$$

In the derivation of this result we have used momentum conservation $k_1 + k_2 + k_3 = 0$, from which, together with $k_i^2 = 0$ for massless particles, it follows that $k_i \cdot k_j = 0$. We have also used the fact that the momentum vectors will be contracted with transverse polarization vectors (i.e. we can replace e.g. $k_1^{\mu_2} \to -k_3^{\mu_2}$ since $k_2^{\mu_2}\epsilon_{\mu_2} = 0$).

Our next example is the scattering of three gravitons. The vertex operators differ from those of the gauge bosons only in their left-moving part. They are

$$V_{-1}(\bar{z}, z, k) = \epsilon_{\mu\nu}(k)\,\bar{\partial}X^\mu(\bar{z})\psi^\nu e^{-\phi}(z)\, e^{ik_\rho X^\rho(\bar{z},z)}\,,$$

$$V_0(\bar{z}, z, k) = \epsilon_{\mu\nu}(k)\,\bar{\partial}X^\mu(\bar{z})\,[\partial X^\nu + i(k\cdot\psi)\psi^\nu](z)\, e^{ik_\rho X^\rho(\bar{z},z)}\,, \tag{15.8}$$

where $\epsilon_{\mu\nu}(k)$ is a polarization tensor which satisfies the on-shell condition $k^\mu\epsilon_{\mu\nu}(k) = k^\nu\epsilon_{\mu\nu}(k) = 0$. In fact, above vertex operators represent either a graviton $(h_{\mu\nu})$, an anti-symmetric tensor $(B_{\mu\nu})$ or a dilaton (D), depending on whether the polarization tensor is symmetric traceless, anti-symmetric or transverse diagonal:

---

[3] We have to comment on the normalization of the amplitudes. They must be chosen such that massless pole terms in four point amplitudes factorize into products of three point amplitudes. Alternatively, one can fix the normalizations of the vertex operators by looking at a few amplitudes, for instance the coupling of various fields to the graviton. This should then give consistent results for all other amplitudes. For details we refer to the literature.

$$
h_{\mu\nu} = \frac{1}{2}(\epsilon_{\mu\nu} + \epsilon_{\nu\mu}) - \frac{1}{d-2}\epsilon^{\rho}{}_{\rho}(\eta_{\mu\nu} - k_{\mu}\bar{k}_{\nu} - \bar{k}_{\mu}k_{\nu}),
$$

$$
B_{\mu\nu} = \frac{1}{2}(\epsilon_{\mu\nu} - \epsilon_{\nu\mu}),
$$

$$
D_{\mu\nu} = \frac{1}{\sqrt{d-2}}(\eta_{\mu\nu} - k_{\mu}\bar{k}_{\nu} - \bar{k}_{\mu}k_{\nu}),
$$

(15.9)

where we have defined a vector $\bar{k}$ such that $\bar{k}^2 = 0$ and $k \cdot \bar{k} = 1$. We sometimes refer to these states collectively as $G_{\mu\nu}$. The vertex operators as given in eq.(15.8) are valid for all heterotic string theories in any dimension since they only involve the "space-time" degrees of freedom X^{μ}, ψ^{μ} which are free conformal fields. The normalization of the dilaton is such as to lead to a canonically normalized kinetic energy (c.f. below). We will restrict ourselves in the following to the phenomenologically most interesting case of $d = 4$. The evaluation is again straightforward. Let us give some details for the correlator $\langle \prod_{i=1}^{3} \bar{\partial}X^{\nu_i}(\bar{z}_i)e^{ik_i \cdot X(\bar{z}_i)}\rangle$. An easy way to evaluate it is to use the representation $\rho^i_{\mu}\bar{\partial}X^{\mu}e^{ik_i \cdot X(\bar{z}_i)} = \exp i[k_{i\mu}X^{\mu}(\bar{z}_i) - i\rho^i_{\mu}\bar{\partial}X^{\mu}(\bar{z}_i)]|_{\text{linear in }\rho_i}$. We then use

$$
\langle \prod_i \exp i(k_{i\mu}X^{\mu}(\bar{z}_i) - i\rho^i_{\mu}\bar{\partial}X^{\mu}(\bar{z}_i)))\rangle
$$

$$
= \prod_{j>i}(\bar{z}_i - \bar{z}_j)^{k_i \cdot k_j} \exp\left\{-\sum_{j>i}\frac{\rho^i \cdot \rho^j}{(\bar{z}_i - \bar{z}_j)^2} + i\sum_{i \neq j}\frac{k_i \cdot \rho^j}{\bar{z}_i - \bar{z}_j}\right\}
$$

(15.10)

from which we easily get by expansion

$$
\langle \prod_{i=1}^{3} \bar{\partial}X^{\nu_i}(\bar{z}_i)e^{ik_i \cdot X(\bar{z}_i)}\rangle
$$

$$
= -i(t^{\nu_1\nu_2\nu_3} + k_1^{\nu_2}k_2^{\nu_3}k_3^{\nu_1})(\bar{z}_{12}\bar{z}_{13}\bar{z}_{23})^{-1}
$$

(15.11)

where we have defined $z_{ij} = z_i - z_j$. The final expression for the amplitude is then

$$
A^{GGG} = g\, t^{\mu_1\mu_2\mu_3}\, t^{\nu_1\nu_2\nu_3}\, \epsilon^{(1)}_{\mu_1\nu_1}(k_1)\,\epsilon^{(2)}_{\mu_2\nu_2}(k_2)\,\epsilon^{(3)}_{\mu_3\nu_3}(k_3) + O(k^4). \quad (15.12)
$$

This result is valid for heterotic strings in any number of dimensions. If we drop the $O(k^4)$ terms it is also valid for type II strings.

Our next example is the scattering of two space-time fermions and one scalar (Yukawa coupling). For this amplitude all three vertex operators can be chosen in the canonical ghost picture. In any lattice model the vertex operator for the fermions in the canonical ghost picture is given by

$$V_{-\frac{1}{2}}(\bar{z}, z, k) = u_\alpha(k)\, e^{i w_L \cdot X_L(\bar{z})}\, e^{i w_R \cdot X_R} S^\alpha e^{-\frac{1}{2}\phi}(z)\, e^{i k_\rho X^\rho(\bar{z}, z)}. \qquad (15.13)$$

Here, $u_\alpha(k)$ is a polarization spinor satisfying the on-shell condition (Dirac equation) $\slashed{k} u(k) = 0$. w_L and w_R specify the transformation properties of the fields under the internal conformal field theory: w_L describes the charges of the states under the gauge group of the particular model. We have not written explicitly the cocycle factors. Masslessness of the fermions requires $w_L^2 = 2$ and $w_R^2 = 3/4$. The scalar vertex operator takes the following generic form

$$V_{-1}(\bar{z}, z, k) = e^{i w_L \cdot X_L(\bar{z})}\, e^{i w_R \cdot X_R} e^{-\phi}(z)\, e^{i k_\mu X^\mu(\bar{z}, z)}. \qquad (15.14)$$

Masslessness of the scalar requires that $w_L^2 = 2$ and $w_R^2 = 1$. For the amplitude to be non-zero, we need $\sum_{i=1}^3 w_{iL} = \sum_{i=1}^3 w_{iR} = 0$ from which it follows that $w_{iL} \cdot w_{jL} = -1$, $w_{1R} \cdot w_{3R} = w_{2R} \cdot w_{3R} = -1/2$, and $w_{1R} \cdot w_{2R} = -1/4$. Then

$$\langle \prod_{i=1}^3 e^{i w_{iL} \cdot X_L(\bar{z}_i)} \rangle = (\bar{z}_{12} \bar{z}_{13} \bar{z}_{23})^{-1},$$

$$\langle \prod_{i=1}^3 e^{i w_{iR} \cdot X_R(z_i)} \rangle = z_{12}^{-1/4} z_{13}^{-1/2} z_{23}^{-1/2}. \qquad (15.15)$$

Also, the two fermions have to be of the same chirality to get a non-vanishing amplitude:

$$\langle S^\alpha(z_1) S^\beta(z_2) \rangle = C^{\alpha\beta} z_{12}^{-1/2}. \qquad (15.16)$$

The conformal ghost part was given in eq.(6.50) and the superconformal ghost contribution is:

$$\langle e^{-\phi/2}(z_1) e^{-\phi/2}(z_2) e^{-\phi}(z_3) \rangle = z_{12}^{-1/4} z_{13}^{-1/2} z_{23}^{-1/2}. \qquad (15.17)$$

The final result for the Yukawa amplitude is then

$$A = g\left(u_\alpha^{(1)} C^{\alpha\beta} u_\beta^{(2)}\right) C^{MNP} \qquad (15.18)$$

where C^{MNP} is a Clebsch-Gordan coefficient which couples the three states to a singlet under the internal symmetries. It arises from the cocycle factors and the lattice momentum conservation constraint $\delta(\sum w_{iL})\delta(\sum w_{iR})$. Its appearance is obvious.

The calculation of four-point amplitudes is rather involved, even though straightforward. We will not present an example here and only make a few comments. After setting the insertion points of three of the vertex operators to $0, 1$ and ∞, there is one integration left which can be done explicitly. Most of the four-point amplitudes are model dependent. The only exceptions are the ones with only external gravitons, anti-symmetric tensors or dilatons, and, modulo the differences in the gauge group and the level of the corresponding Kac-Moody algebras, with external gauge particles. (In supersymmetric theories their supersymmetric partners are also allowed). This is easy to understand from a low-energy field theory point of view. Since no single matter field couples to two fields in the gravitational or gauge sector, they cannot be exchanged in tree diagrams with no external matter fields.

Let us now turn to the problem of finding the field-theory Lagrangian that reproduces above string amplitudes. The gravitational sector contains a spin two particle which we identify with the graviton, an anti-symmetric tensor and a dilaton. Their action up to second order in derivatives is

$$\mathcal{L} = \sqrt{-g}\left\{\frac{1}{2\kappa_4^2}R - \frac{1}{6}e^{-2cD}H_{\mu\nu\rho}H^{\mu\nu\rho} - \frac{1}{2}g^{\mu\nu}\partial_\mu D\partial_\nu D\right\} \qquad (15.19)$$

where $\kappa_4 = \sqrt{8\pi G}$ is the four-dimensional gravitational coupling constant (G is Newton's constant), R the curvature scalar and $H_{\mu\nu\rho} = \partial_\mu B_{\nu\rho} + \partial_\rho B_{\mu\nu} + \partial_\nu B_{\rho\mu}$ the totally antisymmetric field strength associated with the anti-symmetric tensor. The graviton field is defined by $g_{\mu\nu} = \eta_{\mu\nu} + 2\kappa_4 h_{\mu\nu}$.

A comment on the dilaton coupling is in order. It expresses the fact that the classical action is scale invariant and that a constant shift of the dilaton field amounts to a change of the overall constant in front of the action. This means that each term in the Lagrangian must be multiplied by a factor $\exp[c(w-1)D]$ where w is its conformal weight (in a four-dimensional sense) and c a constant to be determined later. The conformal weights are determined as follows. $g_{\mu\nu}(\gamma_\mu)$ and $g^{\mu\nu}(\gamma^\mu)$ have weights $+1(+1/2)$ and $-1(-1/2)$ respectively; matter fermions have $w = -1/4$. For space-time supersymmetric theories we also have a gravitino and a dilatino with weights $+1/4$ and $-1/4$ respectively.

What we have to check now is whether this action reproduces the corresponding string theory amplitudes. Only if it does can we identify the massless spin two string mode with the graviton and have justified the claim that string theory automatically incorporates gravity, one of its major achievements. For three external gravitons the amplitude eq.(15.12) is to $O(k^2)$

$$A^{hhh} = g((k_2 h_1 k_2)(h_2 h_3) + 2(k_2 h_1 h_2 h_3 k_1) + \text{cyclic perms.}) \qquad (15.20)$$

where the notation implies index contractions in an obvious way. If we now expand the Einstein-Hilbert action to third order in the graviton field we get

$$\frac{1}{2\kappa_4^2}\sqrt{-g}R\Big|_{3\text{pt}} = -\kappa_4(h^{\mu\nu}h^{\rho\sigma}\partial_\mu\partial_\nu h_{\rho\sigma} + 2\partial^\sigma h_{\mu\nu}\partial^\mu h^{\nu\rho}h_{\rho\sigma}) \qquad (15.21)$$

where we have used the on-shell conditions for the gravitons, i.e. $k^\mu h_{\mu\nu} = 0$, $h^\mu{}_\mu = 0$. This does indeed reproduce A^{hhh} provided we set

$$\kappa_4 = \frac{1}{2}g\sqrt{\frac{\alpha'}{2}} \qquad (15.22)$$

where we have reintroduced the slope parameter to get a dimensionally correct expression. We have thus related the four-dimensional gravitational coupling constant to the string coupling constant and the string tension. We now expand the $H_{\mu\nu\rho}$ term in $\mathcal{L}$ to third order:

$$-\frac{1}{6}\sqrt{g}\,e^{-2cD}H_{\mu\nu\rho}H^{\mu\nu\rho}\Big|_{3pt} = cD\left(\partial_\mu B_{\nu\rho}\partial^\mu B^{\nu\rho} + 2\partial_\mu B_{\nu\rho}\partial^\nu B^{\rho\mu}\right)$$

$$+ \kappa_4 h^{\mu\nu}\left(\partial_\mu B_{\sigma\rho}\partial_\nu B^{\sigma\rho} + 4\partial_\rho B_{\sigma\mu}\partial_\nu B^{\rho\sigma} + 2\partial_\rho B_{\sigma\mu}\partial^\rho B^\sigma{}_\nu\right). \tag{15.23}$$

The first part reproduces (on-shell)

$$A^{BBD} = -\frac{4}{\sqrt{2}}g(k_2 B_1 B_2 k_1) \tag{15.24}$$

which follows from eq.(15.12) with appropriate choice for the polarization tensors. Comparison determines the constant c to

$$c = \frac{1}{2}\sqrt{\alpha'}g = \sqrt{2}\kappa_4. \tag{15.25}$$

The second part of eq.(15.23) is to be compared with

$$A^{hBB} = g\Big(-(k_3 h_1 k_3)(B_3 B_2) + 2(k_1 B_2 h_1 B_3 k_2)$$

$$- 2(k_2 B_3 B_2 h_1 k_3) - 2(k_3 h_1 B_3 B_2 k_1)\Big). \tag{15.26}$$

We again find the relation eq.(15.22). Finally,

$$-\frac{1}{2}\sqrt{-g}\,g^{\mu\nu}\partial_\mu D\partial_\nu D\Big|_{3pt} = \kappa_4 h^{\mu\nu}\partial_\mu D\partial_\nu D \tag{15.27}$$

reproduces

$$A^{hDD} = -g(k_3 \epsilon_1 k_2) \tag{15.28}$$

yielding eq.(15.22) once more. For all other choices of the external fields the string amplitude vanishes to $O(k^2)$ and it is easy to see that $\mathcal{L}$ does not lead to any other three-point on-shell amplitudes either. The terms $O((\alpha')^{3/2})$ or, alternatively, $O(k^4)$ lead to higher derivative terms in the action; we will not discuss them here.

One can next check that the three gauge boson amplitude is reproduced by the term in

$$\mathcal{L} = -\frac{1}{4}\sqrt{-g}\,e^{-\sqrt{2}\kappa_4 D}\mathrm{Tr}\,F_{\mu\nu}F^{\mu\nu} \qquad (15.29)$$

which is cubic in the gauge field. We find the relation

$$g_4 = \frac{1}{2}g = \sqrt{\frac{2}{\alpha'}}\kappa_4. \qquad (15.30)$$

g_4 is the gauge coupling constant in four dimensions which we have thus related to the gravitational constant and the string tension. With a gauge coupling constant of $O(1)$ we find that the string scale (set by the string tension $T \sim 1/\alpha'$) has to be identified with the Planck scale.

Let us now turn to the Yukawa coupling. The amplitude eq.(15.18) follows from the direct coupling term

$$\mathcal{L} = y\sqrt{-g}\,e^{\frac{1}{\sqrt{2}}\kappa_4 D}\bar{\Psi}^M\Psi^N\Phi^P C^{MNP} + \mathrm{h.c.} \qquad (15.31)$$

Comparison with the string amplitude shows that the Yukawa coupling constant y obeys

$$y \sim g_4. \qquad (15.32)$$

We see that the non-vanishing Yukawa couplings are of the same order as the gauge couplings since in the covariant lattice construction the Clebsch-Gordan coefficients are either zero or $O(1)$.

Proceeding in this way one can relate all field theory coupling constants to the string coupling constant and the string tension; especially, all field theory couplings will be related. There are of course many more terms in $\mathcal{L}_{3pt}$ which we would recover if we calculated other string amplitudes. But, for instance, all amplitudes with only gravitational external fields will, to $O(\sqrt{\alpha'})$, be reproduced by $\mathcal{L}$ as given in eq.(15.19), partially through direct interactions and partially through graviton exchange.

The comparison of string four-point amplitudes with field theory is quite cumbersome; it turns out that the string theory calculation is much easier than the field theory calculation, since in the former case there is only one correlation function to compute as compared to several field theory diagrams. In the low energy field theory the four-point amplitudes are generated by the exchange of massless particles and by new local four point vertices. Therefore, to extract these new contact interactions from a given string four-point amplitude one has to subtract first all poles coming from the exchange of massless states. The exchange diagrams must be computed from $\mathcal{L}_{3pt}$ which was derived from the three point string amplitudes.

At the end of this chapter let us discuss some features of string scattering amplitudes and the resulting low energy effective action which are valid for any four-dimensional heterotic string construction. We will only use general information obtained from conformal field theory. As we have already mentioned, the vertex operators of the graviton, antisymmetric tensor field and dilaton are model independent since they involve only the free two-dimensional fields X^μ and ψ^μ. Therefore, all N-point tree amplitudes involving only these fields are the same in any four-dimensional heterotic string theory. It trivially follows that the effective action eq.(15.19) in the gravitational sector is unique (up to field redefinitions), and the relation between the gravitational coupling constant and the string coupling constant is universal. Indeed, this is true in any number of dimensions up to powers of $\sqrt{\alpha'}$ which are necessary to make this relation dimensionally correct.

Let us now reconsider amplitudes involving gauge boson vertex operators. We assume that the gauge group G arises from the left-moving sector of the heterotic string and is due to a level k Kac-Moody algebra with currents $J^a(\bar{z})$:

$$J^a(\bar{z}) J^b(\bar{w}) = \frac{k\delta^{ab}}{(\bar{z} - \bar{w})^2} + \frac{if^{abc} J^c(\bar{w})}{(\bar{z} - \bar{w})} + \dots \tag{15.33}$$

This equation shows that the normalization of the Kac-Moody currents depends on the level of the affine algebra:

$$\langle J^a(\bar{z}) J^b(\bar{w}) \rangle = \frac{k\delta^{ab}}{(\bar{z} - \bar{w})^2} . \tag{15.34}$$

Therefore, to relate the three gauge boson amplitude for level k to the one computed before, which was at level one, we need to rescale the currents by $\tilde{J}^a(\bar{z}) = J^a(\bar{z})/\sqrt{k}$ and the gauge boson vertex operator in the canonical ghost picture has the general form:

$$V_{-1}^a(\bar{z}, z, k) = \frac{1}{\sqrt{k}} \epsilon_\mu(k)\, J^a(\bar{z})\, \psi^\mu e^{-\phi}(z)\, e^{ik_\rho X^\rho(\bar{z}, z)}. \qquad (15.35)$$

In the zero ghost picture it is given by

$$V_0^a(\bar{z}, z, k) = \frac{1}{\sqrt{k}} \epsilon_\mu(k)\, J^a(\bar{z})\, [\partial X^\mu + i(k \cdot \psi)\psi^\mu](z) e^{ik_\rho X^\rho(\bar{z}, z)}. \qquad (15.36)$$

We can now easily compute the three gauge boson amplitude. The three-point correlation function of the left-moving internal currents $\tilde{J}^a(\bar{z})$ follows from the Kac-Moody algebra eq.(15.33):

$$\langle \tilde{J}^a(\bar{z}_1)\tilde{J}^b(\bar{z}_2)\tilde{J}^c(\bar{z}_3)\rangle = \frac{if^{abc}/\sqrt{k}}{\bar{z}_{12}\bar{z}_{13}\bar{z}_{23}}. \qquad (15.37)$$

The contribution from the right-moving sector is as before and the three gauge boson amplitude becomes

$$A^{ijk} = \frac{g}{2\sqrt{k}} \operatorname{Tr}\{[T^i, T^j]T^k\}\, t^{\mu_1\mu_2\mu_3}\, \epsilon_{\mu_1}^{(1)}\epsilon_{\mu_2}^{(2)}\epsilon_{\mu_3}^{(3)}. \qquad (15.38)$$

The gauge coupling constant is now related to the gravitational coupling constant and the string tension via

$$g_4 = \frac{1}{2\sqrt{k}}g = \sqrt{\frac{2}{k\alpha'}}\kappa_4. \qquad (15.39)$$

We see that the level of the Kac-Moody algebra determines the ratio between the gauge and the gravitational coupling constant.

As a final example, we want to give the general expression for the Yukawa couplings in any string model. The vertex operators of the massless fermions Ψ^i $(i = 1, \ldots, N_F)$ involve internal conformal fields $G^{\Psi_i}(\bar{z}, z)$ (see eq.(14.32)) of conformal dimension $(\bar{h}, h) = (1, \frac{3}{8})$ which depend on the specific model. Analogously, the vertex operators eq.(14.34) of the massless scalars Φ_i $(i = 1, \ldots, N_S)$ contain internal conformal fields $G^{\Phi_i}(\bar{z}, z)$ of conformal dimension $(\bar{h}, h) = (1, \frac{1}{2})$. (For supersymmetric models $N_S = N_F$.) Using the information about the conformal dimensions, the three-point function of these internal operators has the following form:

$$\langle G^{\Psi_i}(\bar{z}_1, z_1) G^{\Psi_j}(\bar{z}_2, z_2) G^{\Phi_k}(\bar{z}_3, z_3)\rangle = \frac{C_{ijk}}{\bar{z}_{12}\bar{z}_{13}\bar{z}_{23} z_{12}^{1/4} z_{13}^{1/2} z_{23}^{1/2}} \cdot \tag{15.40}$$

The coefficients C_{ijk} are the operator product coefficients and contain all information about the non-vanishing couplings between these fields. The contribution of the space-time spin fields and of the ghost fields is as before; they cancel the z-dependence in eq.(15.40). Thus, the final result for this amplitude is

$$A_{ijk} = g C_{ijk} u_\alpha^{(1)} C^{\alpha\beta} u_\beta^{(2)}. \tag{15.41}$$

We see that the Yukawa coupling constants $y_{ijk} \sim g C_{ijk}$ are determined by the operator product coefficients of the internal conformal field theory. This is in fact the case for all three point amplitudes as we know from Chapter 4. Which scalars and fermions are present and their operator product coefficients does, of course, depend on the specific model under considerations. In the covariant lattice construction the C_{ijk} were always Clebsch-Gordan coefficients of some simply laced Lie algebra which are all of $O(1)$. We should note, however, that in general the C_{ijk} are not necessarily all of the same order of magnitude. Therefore there might be a possible hierarchy between the various Yukawa couplings. For example, this is actually the case in orbifold compactifications where the Yukawa coupling constants may depend exponentially on the size of the orbifold the string is compactified on.

This concludes our contact with low energy physics. Of course, the relation among the various coupling constants should be understood to hold at some scale close to the Planck scale. From there to low energies, they will evolve according to the renormalization group equations. Also, string loop effects can give masses to some of the matter fields. With these remarks,

our exposition in this chapter can only be understood as an illustration of
how to get familiar physics (point particle field theory) from string physics.

REFERENCES

[1] J. Scherk, *"Zero slope limit of the dual resonance model"*, *Nucl. Phys.* **B31** (1971) 222.

[2] A. Neveu and J. Scherk, *"Connection between Yang-Mills fields and dual models"*, *Nucl. Phys.* **B36** (1972) 155.

[3] T. Yoneya, *"Connection of dual models to electrodynamics and gravidynamics"*, *Prog. Theor. Phys.* **51** (1974) 1907.

[4] J. Scherk and J.H. Schwarz, *"Dual models for non-hadrons"*, *Nucl. Phys.* **B81** (1974) 118.

[5] M.B. Green, J.H. Schwarz and L. Brink, *"$N = 4$ Yang-Mills and $N = 8$ supergravity as the limit of string theories"*, *Nucl. Phys.* **B198** (1982) 474.

[6] D.J. Gross, J. Harvey, E. Martinec and R. Rohm, *"Heterotic string"*, *Phys. Rev. Lett.* **54** (1985) 502; *"Heterotic string theory (I). The free heterotic string"*, *Nucl. Phys.* **B256** (1985) 253; *"Heterotic string theory (II). The interacting heterotic string"*, *Nucl. Phys.* **B267** (1986) 75.

[7] N. Cai and C.A. Nuñez, *"Heterotic string covariant amplitudes and low energy effective action"*, *Nucl. Phys.* **B287** (1987) 279.

[8] D. Gross and J. Sloan, *"The quartic effective action for the heterotic string"*, *Nucl. Phys.* **B291** (1987) 41.

[9] D. Lüst, S. Theisen and G. Zoupanos, *"Four-dimensional heterotic strings and conformal field theory"*, *Nucl. Phys.* **B296** (1988) 800.

[10] J. Lauer, D. Lüst and S. Theisen, *"Four-dimensional supergravity from four-dimensional strings"*, Nucl. Phys. **B304** (1988) 236.

[11] S. Hamidi and C. Vafa, *"Interactions on orbifolds"*, Nucl. Phys. **B279** (1987) 465.

[12] L. Dixon, D. Friedan, E. Martinec and S. Shenker, *"The conformal field theory of orbifolds"*, Nucl. Phys. **B282** (1987) 13.

[13] E.S. Fradkin and A.A. Tseytlin, *"Effective field theory from quantized strings"*, Phys. Lett. **158B** (1985) 316.

[14] C.G. Callan, D. Friedan, E. Martinec and M. Perry, *"Strings in background fields"*, Nucl. Phys. **B262** (1985) 593.

[15] A. Sen, *"Heterotic string in arbitrary background fields"*, Phys. Rev. **D32** (1985) 2102; *"Equations of motion for the heterotic string theory from the conformal invariance of the sigma model"*, Phys. Rev. Lett. **55** (1985) 1846.

Lecture Notes in Mathematics

Lecture Notes in Physics

GRADUATE TEXTS IN CONTEMPORARY Physics

This is a series of high-level texts based mostly on lectures and courses given at the University of Maryland, College Park, USA. Each text gives a coherent introduction to a leading-edge topic in contemporary physics.

Series Editors: J. L. Birman, H. Faissner, J. W. Lynn

R. N. Mohapatra

Unification and Supersymmetry

The Frontiers of Quark-Lepton Physics

1986. XIII, 309 pp. 49 figs.
ISBN 3-540-96285-9

Contents: Important Basic Concepts in Particle Physics. – Spontaneous Symmetry Breaking, Nambu-Goldstone Bosons, and the Higgs Mechanism. – The $SU(2)_L$ x U(1) Model. – CP-Violation: Weak and Strong. – Grand Unification and the SU(5) Model. – Left-Right Symmetric Models of Weak Interactions. – SO(10) Grand Unification. – Technicolor and Compositeness. – Global Supersymmetry. – Field Theories with Global Supersymmetry. – Broken Supersymmetry and Application to Particle Physics. – Phenomenology of Supersymmetric Models. – Supersymmetric Grand Unification. – Local Supersymmetry (N = 1). – Application of Supergravity (N = 1) to Particle Physics. – Beyond N = 1 Supergravity.

M. Kaku

Introduction to Superstrings

1988. XVI, 568 pp. 48 figs.
ISBN 3-540-96700-1

Contents: I. FIRST QUANTIZATION AND PATH INTEGRALS: Path Integrals and Point Particles. Nambu-Goto Strings. Superstrings. Conformal Field Theory and Kac-Moody Algebras. Multi-Loops and Teichmuller Space. II. SECOND QUANTIZATION AND THE SEARCH FOR GEOMETRY: Light Cone Field Theory. BRST Field Theory. Geometric String Field Theory. III. MODEL BUILDING AND PHENOMENOLOGY: Anomalies and the Atiyah-Singer Theorem. Heterotic Strings and Compactification. Calabi-Yau Spaces and Orbifolds. References. Appendix.

H.-V. Klapdor (Ed.)

Neutrinos

With contributions by numerous experts
1988. VIII, 339 pp. 164 figs.
ISBN 3-540-50166-5

Contents: Neutrino Properties. – Neutrino Reactions and the Structure of the Neutral Weak Current. – Massive Neutrinos in Gauge Theories. – Neutrinos in Left-Right Symmetric, SO(10) and Superstring Inspired Models. – Double Beta Decay Experiments and Searches for Dark Matter Candidates and Solar Axions. – Double Beta Decay, Neutrino Mass and Nuclear Structure. – Neutrino Oscillations in Vacuum and Matter. – Searches for Lepton-Flavour Violation. – Neutrino Physics and Supernovae: What have we learned from SN 1987A? – Neutrinos in Cosmology. – Index of Contributors.

H. Mitter, F. Widder (Eds.)

Particle Physics and Astrophysics

Current Viewpoints

Proceedings of the XXVII Int. Universitätswochen für Kernphysik, Schladming, Austria, February 1988
1989. X, 309 pp. 100 figs. ISBN 3-540-50699-3

Contents: Introduction and Survey I: Concepts from Particle Physics. – Introduction and Survey II: Concepts from Astrophysics and Cosmology. – Physics of Type II Supernova Explosions. – Evidence for Black Holes. – Principles of Black Hole Energetics. – Quantum Gravity Motivated Computer Simulations. – Neutrino Oscillations and Their Damping. – Phase Transitions as the Origin of Large Scale Structure in the Universe.

Springer-Verlag
Berlin Heidelberg New York
London Paris Tokyo Hong Kong